Springer Series in
Experimental Entomology

Thomas A. Miller, Editor

Springer Series in Experimental Entomology
Editor: T.A. Miller

Insect–Plant Interactions

Edited by
James R. Miller
Thomas A. Miller

With Contributions by

M. Berenbaum   S. Finch   J.L. Frazier   F.J. Hanke
F.E. Hanson   I. Ishaaya   M. Kogan   I. Kubo
A.C. Lewis   S.B. Opp   R.J. Prokopy   M.C. Singer
W.M. Tingey   H.F. van Emden

With 65 Figures

Springer-Verlag
New York Berlin Heidelberg
London Paris Tokyo

James R. Miller
Michigan State University
Pesticide Research Center
East Lansing, Michigan 48824
U.S.A.

Thomas A. Miller
Department of Entomology
University of California
Riverside, California 92521
U.S.A.

Library of Congress Cataloging in Publication Data
Main entry under title:
Insect–plant interactions.
   (Springer series in experimental entomology)
   Bibliography: p.
   Includes index.
   1. Insect–plant relationships.   I. Miller, James R.
(James Ray), 1948–   . II. Miller, Thomas A.
III. Berenbaum, M. (May)   IV. Series.
QL496.I388   1986   595.7′0524   85-30308

Typeset by David E. Seham Associates Inc., Metuchen, New Jersey.
Printed and bound by R.R. Donnelley and Sons, Harrisonburg, Virginia.
Printed in the United States of America.

9 8 7 6 5 4 3 2 1

ISBN 0-387-96260-3 Springer-Verlag New York Berlin Heidelberg
ISBN 3-540-96260-3 Springer-Verlag Berlin Heidelberg New York

Series Preface

Insects as a group occupy a middle ground in the biosphere between bacteria and viruses at one extreme, amphibians and mammals at the other. The size and general nature of insects present special problems to the study of entomology. For example, many commercially available instruments are geared to measure in grams, while the forces commonly encountered in studying insects are in the milligram range. Therefore, techniques developed in the study of insects or in those fields concerned with the control of insect pests are often unique.

Methods for measuring things are common to all sciences. Advances sometimes depend more on how something was done than on what was measured; indeed a given field often progresses from one technique to another as new methods are discovered, developed, and modified. Just as often, some of these techniques find their way into the classroom when the problems involved have been sufficiently ironed out to permit students to master the manipulations in a few laboratory periods.

Many specialized techniques are confined to one specific research laboratory. Although methods may be considered commonplace where they are used, in another context even the simplest procedures may save considerable time. It is the purpose of this series (1) to report new developments in methodology, (2) to reveal sources of groups who have dealt with and solved particular entomological problems, and (3) to describe experiments which may be applicable for use in biology laboratory courses.

THOMAS A. MILLER
Series Editor

Preface

The topic of insect–plant relationships has captured human attention since antiquity. Throughout recorded history, certain insect herbivores (e.g., locusts) have demanded attention by periodically devastating the plants upon which human populations depend. In some instances, such plagues may have actually shaped human history (e.g., Exodus 10:12–19).

Approximately 8,000 years of agricultural practice have not changed the overall picture; insects remain and may have even gained ground as our chief competitors for food and fiber despite the energy-driven escalation in pest control technologies. The prospect that human populations worldwide continue to increase in the face of limited energy resources, combined with the sheer numbers, diversity, and adaptability of insect herbivores, makes knowledge of insect–plant relationships a global practical concern.

Insect–plant relationships are also tremendously intriguing from the perspective of fundamental knowledge of Earth's biosphere. Scholars of evolution continue to be drawn to profound questions such as: 1. To what extent has the appearance of insect fauna influenced flora via pollenation, seed dispersal, and herbivory? 2. What types of insect–plant interactions promote generalization, specialization, diversification, and speciation?

Steadily, increasing numbers of researchers are being drawn to the more proximate behavioral/physiological puzzles like: 1. Precisely how do insects find host plants in space and time? 2. How do they sense and recognize hosts? 3. What are the behavioral sequences leading to sustained feeding and oviposition? 4. How is herbivore fitness influenced by the plant's structure, as well as its composition of nutrients and defensive chemicals? The knowledge that insects are earth's most abundant animal group and that half of all insects are herbivores reinforces the conviction that such questions are central to biology.

As part of the Springer Series in Experimental Entomology, the theme of this book is approaches and methods for entomological research. The focus in this case is upon the more mechanistic aspects of insect–plant relationships, or *insect–plant interactions* as this specialty has become known. Clearly, this topic is too extensive to be covered comprehensively in one volume. We do believe, however, that this book represents a unique transect across the spectrum of methodologies appropriate for current research in insect–plant interactions. Many of the contributing experts have managed to offer an authoritative overview of their respective area as well as address their primary charge of covering approaches and methods at a level useful to nonexpert and expert researchers. Where treatments are not comprehensive, citations provided are intended to aid pursuit of additional relevant literature.

Because of its importance to biology, the field of insect–plant interactions must be dynamic. Standards for approaches and methods of research expounded today need to yield to tomorrow's new insights. If it helps stimulate research that dates it, this volume will have fulfilled the hopes of its planners.

JAMES R. MILLER
THOMAS A. MILLER

Contents

Contributors

M. BERENBAUM
Department of Entomology, University of Illinois, Urbana, Illinois 61801, U.S.A.

S. FINCH
Entomology Section, National Vegetable Research Station, Wellesbourne, Warwick CV35 9EF, United Kingdom

J.L. FRAZIER
Biochemicals Department, E.I. DuPont de Nemours Co., Experimental Station, Wilmington, Delaware 19898, U.S.A.

F.J. HANKE
Division of Entomology, University of California, Berkeley, California 94720, U.S.A.

F.E. HANSON
Department of Biology, University of Maryland, Baltimore, Maryland 21228, U.S.A.

I. ISHAAYA
Division of Entomology, Volcani Center, Institute of Plant Protection, P.O.B. 6, Bet Dagan 50-250, Israel

M. KOGAN
Department of Agricultural Entomology, University of Illinois, Urbana, Illinois 61801, U.S.A.

I. KUBO
Division of Entomology, University of California, Berkeley, California
94720, U.S.A.

A.C. LEWIS
EPO Biology, University of Colorado, Boulder, Colorado 80309, U.S.A.

S.B. OPP
Department of Entomology, University of Massachusetts, Amherst, Massachusetts 01003, U.S.A.

R.J. PROKOPY
Department of Entomology, University of Massachusetts, Amherst, Massachusetts 01003, U.S.A.

M.C. SINGER
Department of Zoology, University of Texas, Austin, Texas 78712, U.S.A.

W.M. TINGEY
Department of Entomology, Cornell University, Ithaca, New York 14853,
U.S.A.

H.F. VAN EMDEN
Departments of Agriculture and Horticulture, Earley Gate, Reading RG6
2AT, United Kingdom

Chapter 1

Approaches and Methods for Direct Behavioral Observation and Analysis of Plant–Insect Interactions

Susan B. Opp and Ronald J. Prokopy[1]

Description is never, can never be, random; it is in fact highly selective, and selection is made with reference to the problems, hypotheses and methods the investigator has in mind.

Tinbergen, 1963, p. 412

I. Introduction

Perhaps the greatest contribution of Konrad Lorenz to the discipline of ethology was to advocate use of biological investigatory methods for behavioral studies of animals (Tinbergen, 1963). Observation conducted as a systematic biological investigation differs from merely watching the behavior of an animal in that observation utilizes the uniquely human capacity for thought. Thought gives rise to a framework of important questions to consider during observation. What, when, how, why, and where are all highly pertinent aspects of studying the activity of an animal (Lehner, 1979). Other behaviorists have suggested consideration of proximate and ultimate factors affecting behavior as an advisable approach to behavioral investigation (e.g., Drickamer and Vessey, 1982). Yet another way of visualizing the study of animal behavior focuses on the function, causation, ontogeny, and evolution of observed behaviors (Tinbergen, 1963). But no matter how the study of behavior is conceptualized, the necessity, when

[1]Department of Entomology, University of Massachusetts, Amherst, Massachusetts 01003, U.S.A.

conducting observations, for continual thought and reflection, particularly within the context of behavioral-ecological theory, is an indispensible consideration.

Here, we are concerned primarily with direct behavioral observations that pertain to plant–insect interactions. As in any behavioral research, the key to a sound and rewarding study lies in striking a balance between observation and experimentation. Irrespective of the species of insect, both observation and experimentation are necessary to develop a thorough understanding of behavior.

During design of an observational or experimental study, the ultimate goals of the research are of central importance. Still, these goals should never be allowed to overshadow the true reality of what is being observed. Flexibility of observation and interpretation is of utmost importance because often preliminary results will point to a more fruitful approach to addressing a particular question. At the same time, however, allowing a study to follow too many tangential questions can dilute and confuse the purpose of the original work. Obviously, one should strive for a balance between flexibility and steadfastness of goals.

Since the object of behavioral observation is to study the activities of an animal, the suitability of the study animal should be given close consideration. Optimally, the animal should be reasonably large and conspicuous to facilitate visual identification. If insect flight activity is of interest, flights of the study species should be neither too rapid nor too long if they are to be observed from start to finish. Because the proximity of the observer may disrupt natural behavior, an animal that can be observed readily from some distance may prove the more appropriate type to study. Depending on the observer's goals, such factors as ease of collecting and rearing, daily activity patterns, and population sizes of animals may be important. Even the most well-formulated investigation, if conducted on an animal not amenable to observation, will encounter difficulties.

A significant part of an investigation is attitude and interest. Systematic behavioral observations, whether conducted in the field, under some sort of seminatural conditions, or in the laboratory, require patience, imagination, and careful planning. If observations are conducted in conjunction with colleagues, constant communication and comparability of approaches must be maintained among researchers. When carried out thoroughly, behavioral observation can be an exciting and enlightening undertaking, providing the basis for long-term, in-depth understanding of interactions between an animal and its surroundings.

II. Categorization of Insect Behavior

As a first step in behavioral studies, many ethologists (e.g., Slater, 1978; Lehner, 1979; Drickamer and Vessey, 1982) suggest developing a set of descriptions of characteristic behavior patterns of a species, called an eth-ogram. Simple cursory description of an animal's behavior may not provide enough information to delimit among its different patterns of behavior. Rather, to develop an ethogram, some method of behavioral quantification may be necessary (Slater, 1978). A list of all known behaviors of an animal, called a catalog, is a sample from the animal's behavioral repertoire, which consists of all behaviors the animal is potentially capable of performing (Fagen, 1978; Lehner, 1979). The terminology associated with lists and descriptions of behavior is not nearly so important, however, as an ap-preciation of the complex nature of behavior, which necessitates an or-ganized means of describing and classifying behavioral acts.

Although ethologists' opinions differ regarding the importance of quan-tification for behavioral category delimitation, certain requisites are nec-essary for designing any behavioral study, regardless of viewpoint or the animal being considered. One must focus on the information being sought, which hinges on development of reasonable hypotheses and goals. As mentioned previously, the complexity of behavior renders an organized scheme of behavioral classification a must if a researcher hopes to develop an understanding of an animal's activities. Some behaviors of an animal must, therefore, be selected and others neglected in order to segregate and identify the components of a complex series of events (Hinde, 1970).

Slater (1978) suggests several characteristics of behavior patterns that should help facilitate categorization. The patterns should be: (1) species typical, (2) repeatedly recognizable, and (3) made up of movements that occur predictably as a unit. For example, Slater and Ollason (1973) denoted a number of behavioral categories of male zebra finches, such as feeding, drinking, sand taking, cuttlefish bone taking, locomotion, bill-wiping, ruf-fling, wing-shaking, preening, scratching, stretching, singing, and gaps in behavior. This categorization facilitated analysis of transitions that oc-curred between types of behavior. Still, not all zebra finch behaviors could be separated readily by these workers into discrete categories, adding some degree of uncertainty to the analysis.

Additional helpful guidelines, suggested by Slater (1978), to follow when developing behavioral categories include: a category should be discrete and internally homogeneous; the title of a category should not contain causal or functional implications; and it is more useful to split than to lump categories, although the number of categories must be manageable —categories may always be coalesced later. To illustrate, Jackson (1983) recognized and described the complex mating display repertoire, consisting of 26 different behavior categories, of the male salticid spider *Mopsus mormon*. These categories were retract palps, arch palps, lower body, elevate abdomen, twitch abdomen, elevated legs display, posture, gesti-culate, spurt, elevate cephalothorax, bend abdomen, hunched legs display, pose and wag, sweep, mount, watch and follow, decamp, chew, probe, palpate, pull, heave and bump, tap with legs, stroke, scrape with palp, and apply palp. To clarify this array of behaviors, Jackson chose to organize the categories into three mating tactics that depended on the location of the female spider when an encounter with a male occurred.

It is evident, then, that one must develop a systematic approach to classifying and recording the behaviors being observed. The necessity for good organization cannot be stressed too heavily in the pursuit of behavioral understanding.

III. Nature of Plant–Insect Interactions

Within the context of plant–insect interactions, the process of resource acquisition by the insect holds a central place. The habitat of an insect may be divided into resource and nonresource areas. The location, density, and quality of resources as well as the structure of the habitat all influence insect activity (Thorsteinson, 1960; Thompson and Price, 1977; Finch, 1980; Kareiva, 1982; Scriber, 1983; Stanton, 1983) and thus are critical to studies of insect foraging behavior. When designing behavioral studies in seminatural or laboratory situations, therefore, one should attempt to duplicate as closely as possible the resource composition and structure of the natural insect habitat to obtain results with applicability to natural field situations.

The plant may be thought of as providing one or more of four essential resources for the insect: feeding sites, mating sites, egg-laying sites, and/ or refugia (Prokopy et al., 1984). Using this scheme as an organizational basis of plant–insect interaction studies, observed behaviors are categorized according to which essential resource is being exploited and on what sorts of plants or plant structures such exploitation is occurring. Owens et al. (1982), for example, classified the behavior on host trees of adult plum curculios, *Conotrachelus nenuphar,* according to which resource-acquisition activities were taking place (feeding, mating, egg-laying) and which sorts of plant structures (fruits, leaves, twigs, clusters) were involved in providing each resource. They observed plum curculio individuals on randomly selected tree branches for a median time of 30 min each and recorded, in addition to the above, time of day of activity occurrence, temperature conditions, and whether insects arrived at a resource site by crawling or flying. The type of information gained from such observations is fundamental for developing an understanding of the nature of plant–insect interactions.

IV. Techniques in Observational Studies

Many techniques have been developed for recording behavioral observations ranging from diligent note-taking by hand to computer-assisted analysis of activities recorded on a videotape deck. Not all of the techniques discussed here have been applied to direct observation of plant–insect interactions, but technologies developed for other behavioral disciplines can serve as examples of what could be used to facilitate plant-insect observations. We caution that increased technology does not necessarily increase understanding; often the simplest possible method yields a great deal of information. Appropriate technology for behavioral observation depends on the type of behavior being studied and the conditions under which the study will take place.

Perhaps the simplest technique for recording observations is note-taking by hand (e.g., Prokopy et al., 1972). An advancement over this technique is use of a portable tape recorder to record observations verbally for later transcription (e.g., Prokopy, 1976; Woodell, 1978; Smith and Prokopy, 1981; Fitzpatrick and Wellington, 1983; Stanton, 1983), thus enabling the observer to watch the insect continuously without looking away to take notes. As an example, to assess effects of environmental factors on foraging bumblebees in the field, Woodell (1978) used a compass in conjunction with a portable cassette recorder to determine and record flight direction, while at the same time gaining a record of wind direction and speed and the time of day. Fitzpatrick and Wellington (1983), on the other hand, utilized a camera and stop watch in addition to tape recorder to study territorial behaviors of syrphid (hover) flies in the field. They recorded the locations and durations of particular kinds of behaviors, and then later were able to create detailed maps of individual insect home ranges and territories. *Colias p. eriphyle* food-plant selection behavior in the field was studied by Stanton (1983), who followed the movements of individual ovipositing females and verbally recorded on tape the origin, length, and destination of each flight, the time of day, and the prevailing weather conditions. In addition, Stanton marked all butterfly landing spots with sequentially numbered flags so that distances between landing spots could be compared with food-plant abundance. Together, these examples indicate the usefulness of tape recorders for field observation, especially when insect activity is being monitored continuously or a number of factors are being considered jointly. One drawback of verbal tapes, the necessity for transcription at a later time, may contribute to the comparatively infrequent use of tape recorders in laboratory situations, where a written form of recording often is just as easily accomplished.

Visual recordings are commonly used in behavioral investigations. Super 8 movies and 35-mm photographs have been employed to establish and study mating processes of adult western corn rootworms, *Diabrotica virgifera,* both in the laboratory and in the field (Lew and Ball, 1979). Video recordings have been used primarily for detailed analyses of behavioral sequences that may occur very rapidly, as in lepidopteran mating behavior (e.g., Baker and Cardé, 1979). On video playback at slower speeds, behavioral sequence details become more apparent. As an elaboration on

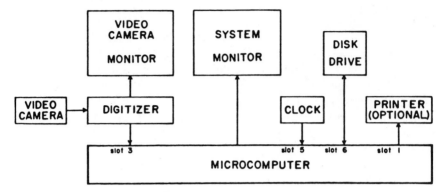

Figure 1. Schematic diagram of the components of an electronic arthropod tracking system indicating position of digitizer. (From Hoy, et al., 1983.)

the use of video recorders to track the path of small animals, Hoy et al. (1983) advocate use of a computer digitizer that can record the two-dimensional locomotory path of an insect (Fig. 1), thereby eliminating the tedium, bias, and potential errors associated with hand recording and intuitive analysis of movements. The information is stored in the computer as a time series of x and y coordinates. One further technological step in recording animal movements involves use of two recording devices that will generate $x, y,$ and z coordinates when used in conjunction with a computer. Zalucki and Kitching (1982) developed this type of three-dimensional approach to study the movements of *Danaus plexippus* butterflies among food plants under seminatural conditions in the field.

Once the behavioral categories and precise activities of interest have been determined, the researcher may concentrate on location and frequency of certain behaviors. In such cases, an event recorder to tally the location, occurrence, and duration of behaviors may be useful. Zalucki and Kitching (1982), for example, recognized four different behavioral states of *D. plexippus* during host plant visits in the field, and summed information on each state using an event recorder. In the laboratory, Waage (1979) observed the parasitoid, *Nemeritis canescens,* foraging for hosts and recorded the time spent in various behaviors with an 8-channel event recorder.

Insects have different visual sensitivities than humans; techniques used in conducting a behavioral study must sometimes be altered accordingly. For some crepuscular or nocturnal species, use of a photomultiplication device for field observations may be necessary (Faegri, 1978; Lingren et al., 1978). For example, many nocturnally active lepidopterous cotton pests have been observed orienting to host plants and pheromone sources in the field with the use of night vision goggles (Lingren et al., 1978). Such goggles can be used successfully under passive moonlight or starlight conditions, or, on very dark nights, be supplemented with artificial light from a flashlight or headlamp equipped with filters to eliminate the portion of the light spectrum to which the insect reacts. In certain cases, however, photomultiplication devices may be unusable because the insect is highly cryptic and thus cannot be detected easily (e.g., Owens et al., 1982). Some insects are disturbed by the use of a flashlight or red light. Also, it is conceivable that certain insects may react to virtually all wavelengths of light visible to humans, thereby ruling out use of filtered artificial light in conjunction with night vision goggles.

If only the laboratory activity cycle but not the precise behavioral repertoire exhibited by a crepuscularly or nocturnally active insect is of interest, a device developed by Chabora and Shukis (1979) may be useful. These workers employed a capacitance discharge device in conjunction with a strip chart recorder to record each time an insect bridged a pair of conductive plates. Using this device, Chabora and Shukis (1979) found in the house fly, *Musca domestica,* a previously unrecorded increased rate of activity toward the end of the dark period, when direct visual observation was difficult.

Because the visual sensitivity spectrum of most insects (ca. 350–650 nm) differs from that of humans (ca. 400–700 nm), the observer cannot always be sure that the insect views its environment in the same manner as a human. If, for instance, an insect is suspected of responding to ultraviolet light, then use of a portable video camera equipped with an ultraviolet-transmitting lens can provide valuable information from a different sensory perspective (Eisner et al., 1969). Always, one should attempt to remember that the insect may be reacting to types of stimuli (visual as well as olfactory or auditory) that are virtually undetectable by human senses. The fact that humans do not detect such stimuli in no way discounts the possibility of their impact on an insect.

An approach employed to aid in tracking the activities of individual insects temporarily lost from direct view in field observation studies has been to mark individuals conspicuously so they can be identified easily later. Fluorescent powders to mark insects have been utilized in projects ranging from mark-recapture studies of mosquitoes (Bennett et al., 1981) to studies of intertree foraging behaviors of solitary bees (Frankie et al., 1976). Another technique uses a Geiger counter to track intra- or interplant movements of foraging individuals tagged with a low dosage of ^{32}P or some other radioactive material (Averill and Prokopy, unpublished observations). Still another technique involves marking an individual with paint in a distinct pattern or attaching a numbered and/or colored tag to the body to facilitate individual identification (Figs. 2 and 3). Heinrich (1976) attached color-coded, numbered tags to bumblebees and was thereby able to map the foraging paths of individual bees among flowers. The behavior of particular individuals at different places and times could be documented and compared. Any method that allows ready identification of individual insects facilitates estimation of the important parameter of behavioral variation among individuals. With many of the observational techniques mentioned here, use of binoculars can facilitate greatly visual location and tracking of an insect.

These are but some of the varied techniques that can aid in behavioral observations of insects. As in behavioral studies of all animals, any such techniques should be used in concert with a primary methodology of careful, discerning observation coupled with patience and thought. Systematic behavioral observations can be and often are accomplished with a mini-

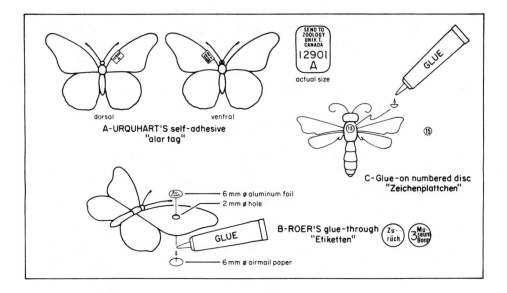

Figure 2. Tehniques for attaching identification tags to insects: (A) Self-adhesive tags: (B) Glue-through tags that require that a hole be made in the insect wing or elytra: (C) Glue-on discs that are attached directly to the insect body. (From Walker and Wineriter, 1981.)

Figure 3. Insect holding and marking techniques: (A,B) Examples on insect ▶ holding techniques that require lightweight net to restrain an insect physically. (C,D) Examples of techniques that require suction to restrain an insect: (E) Insect marking technique using a pin to apply a small spot of paint from a paint pot that will not spill: (F) Similar marking technique using a syringe to apply a small spot of paint: (G) A paper clip used as a paint micro-applicator: (H) Liquid paper has excellent adhesive qualities when applied directly onto an insect, it comes in a variety of colors, and may be overwritten with a symbol using a technical pen or waterproof felt tip marker: (I) Rubber stamp and ink pad to place marks directly on large insects. (From Walker and Wineriter, 1981.)

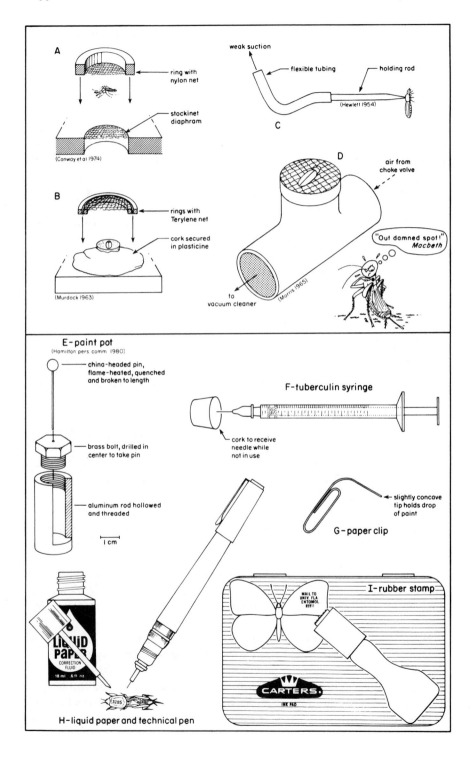

A   ring with
      nylon net

      stockinet
      diaphram

(Conway et al 1974)

B   rings with
      Terylene net

      cork secured
      in plasticine

(Murdock 1963)

weak suction

flexible tubing   holding rod

C

(Hewlett 1954)

D   air from
      choke valve

"Out damned spot!"
Macbeth

to
vacuum cleaner   (Morris 1965)

E- paint pot
(Hamilton pers comm 1980)

china-headed pin,
flame-heated, quenched
and broken to length

brass bolt, drilled in
center to take pin

aluminum rod hollowed
and threaded

1 cm

F-tuberculin syringe

cork to receive
needle while
not in use

slightly concave
tip holds drop
of paint

G-paper clip

LIQUID PAPER
CORRECTION FLUID
18 ml .6 fl oz

MAIL TO
UNIV FLA
ENTOMOL
85911

I-rubber stamp

CARTERS.
INK PAD

H-liquid paper and technical pen

mum of technology or equipment while yielding a great deal of useful information.

V. Methodologies

The techniques used for behavioral observations are tools that facilitate information gathering, storage, retrieval, and analysis. The type of information gained from studies of insect behavior depends on the research goals and the methodologies employed.

We have divided methods of observing plant–insect interactions into three categories: field observations, observations under seminatural conditions, and laboratory observations. Although most behavioral studies may be placed into one of these three categories, often researchers will employ more than one methodology in a study. The categories are not, therefore, necessarily mutually exclusive, and all three methodologies should be evaluated for their strengths and weaknesses when designing a behavioral study.

5.1. Field Observations

Behavioral observations conducted under natural conditions in the field should be the starting point for every behaviorally oriented investigation. A general (Gestalt-type) understanding of the behavior of an insect in its natural surroundings underlies meaningful project conceptualization. Also, it provides a solid foundation of information needed for experimental hypothesis formulation. How much time an insect spends and what kinds of behaviors it exhibits on different structures and types of plants (e.g., hosts vs. nonhosts) in nature, for example, are basic items vital to illuminating the nature of plant–insect interactions.

Field studies should not be considered less useful or less accurate than lab studies but should be recognized for their many strengths. For instance, field studies limit the amount of unnatural disturbance the study insect encounters, and, at the same time, allow the environment to fluctuate naturally so that environmental interactions can be monitored. Thus, the observer can describe the behavioral repertoire of the insect in its natural context, providing valuable information for the development of experimental designs.

Nevertheless, the limitations of field observations and the necessity of careful methodology should be recognized. For example, quantification and analysis of behavior are sometimes difficult in the field because parameters affecting behavior may vary in time and space, and the lack of controls may yield nonreproducible results. In nature, the probability of events correlated with particular behavioral acts being causally related is less than under seminatural or laboratory conditions. In addition, the history of the insect being observed and thus its physiological state cannot be known. Decisions must be made regarding how to select which individual to watch. Individuals easiest to detect may receive disproportionate attention. Tracking certain insects within or among plants can prove especially difficult, reinforcing the desirability of choosing an appropriate study animal for behavioral observation.

An observer should be particularly aware that while in the field, he or she may be in the perceptual world of the observed animal and thus the observer's presence may influence the animal. The less apparent the observer, the less likely the animal will be affected. Because insect senses often differ from those of man, it may be difficult to determine precisely which stimuli affect the behavior of an insect. As so aptly stated by V.G. Dethier (1978, p. 224), "It is clear that different species, different populations of species, and different individuals within a population have different sensitivities (thresholds) to commonly encountered stimuli. . .". It can be presumed, however, that at least three main human factors (rapid movement, darkly or brightly colored clothing, and foreign odor) could affect insect behavior, and that these should be minimized, if possible. The observer should try to blend with the environment or should consider using an observational "blind" to attempt to remove himself or herself from the perceptual world of the insect.

A principal problem faced in field studies is how to go about choosing an individual to observe. One common method is to census locales (selected systematically or randomly) thoroughly at intervals, recording each individual seen and its activities (Altmann, 1974), e.g., the number and activities of pollinators visiting certain flowers (Bawa, 1977), or the number and activities of individuals on specific portions or species of plants (Prokopy et al., 1971; Prokopy and Bush, 1973; Smith and Prokopy, 1981). To illustrate more fully, Malavasi et al. (1983) once per week over a 7-week period carried out observations of the activities of *Anastrepha fraterculus*, a tephritid fruit fly, on 11 trees which were divided into four groups: major hosts with fruit, major hosts without fruit, minor hosts with fruit, and nonhosts without fruit. Once per hour, they examined a sector of each tree for 5 min and recorded the location and type of activity of each fly detected. With such information, a quantitative picture was drawn of how the occurrence of different behaviors changed with time and location. With this and all other insect sampling procedures, emphasis should be placed on an unbiased method in which each individual in a given locale has an equal probability of being chosen and observed.

A second method is to select systematically or randomly individuals for continuous observation over a given period, recording all behaviors and interactions observed. This has been termed focal animal sampling (Slater, 1978). To observe host choice and oviposition by *Battus philenor* butterflies, for example, Rausher (1980) walked a circular path through his study area until encountering a female *B. philenor*, which he then observed for 30-min, recording each plant on which she alighted and each act of oviposition. He then compared the proportion of total egg clusters laid on each plant species with the proportion of total host plants that each species comprised. In several studies of tephritid fruit fly mating and host-searching behavior (eg., Prokopy et al., 1971; Smith and Prokopy, 1982; Malavasi et al., 1983), individuals had been selected for observation by pointing blindly to a portion of a host or nonhost plant and then watching the fly closest to that spot for a given period of time.

In some studies, observers have wished to establish the direction of insect flights within or between plants. Levin et al. (1971) accomplished this simply by scoring the directions of at least two consecutive flights of bee and butterfly pollinators. These workers created eight directional alternatives and then recorded in which of these directions the flights took place. Brussard and Ehrlich (1970), however, were interested not only in the direction of butterfly movements, but also in the behaviors that were exhibited during and between flights. Brussard and Ehrlich needed two observers to record behavior and simultaneously follow the flight track. In this particular case, more detailed observations were accomplished with additional manpower rather than additional technology.

Several early biological investigators did not hesitate to delve into field behavioral observations and subsequently propose varied biological hypotheses from their studies. Consider, for example, the far-ranging implications of the field observations and evolutionary hypotheses that Charles Darwin formulated during the voyage of the HMS Beagle. With a paucity of technology but a great deal of patient observation and thought, Darwin was able to derive a truly amazing amount of useful information concerning behavior and evolutionary relationships. Today, some behaviorists perhaps have a tendency to avoid purely descriptive studies because of a perceived (but in many cases illusory) lack of obtainable conclusive results. A central purpose of unmanipulated systematic field observations of insects is to develop an appreciation for how behavior is shaped by its relationship to and interactions with biotic and abiotic agents in the habitat. The examples cited here show that highly informative observational studies can be conducted in nature, and furthermore, that such studies provide an excellent starting point for experimental behavioral investigations.

5.2. Seminatural Conditions

Conducting behavioral observations under natural field conditions often clarifies the differential importance of various factors affecting the insect's behavior; controlling for or quantifying the effects of those factors is then in order. Use of seminatural situations in which the insect or environment has been manipulated to observe insect behavior allows comparison of behaviors under different circumstances and the determination of proximate factors influencing behavior.

A primary strength of seminatural behavioral studies is that the habitat can be segregated into components for assessing the impact of environmental factors individually or in combination. This allows some degree of experimentation. Yet, the experimental arena is not in an entirely artificial environment so that behaviors exhibited can be expected to approximate at least roughly behaviors under natural conditions. Another great strength is that the history of the individuals under observation may be controlled, yielding greater reproducibility and interpretation of results. Because it is possible to develop controls, greater quantification and more in-depth causation analyses of behaviors are possible. In this setting, the relative importance of component parts of behavioral patterns are more readily analyzed.

A drawback to using seminatural situations is that certain nonnatural conditions, such as cages, greenhouses, or other types of enclosures, may greatly influence the insect's behavior. Thus, studies under seminatural conditions must be based on and compared with observations conducted under natural field conditions. One must know how an insect behaves typically in nature in order to interpret behaviors under less than fully natural circumstances. Mimicking the resource structure and composition of the natural habitat under seminatural situations may also present design difficulties because some natural settings may not be easily duplicated or controlled. Again, the most useful and elegant studies are often simple and based on common sense and extensive natural observation.

As examples of behavioral observations under seminatural conditions, let us turn first to some enlightening studies of the foraging behavior of bees. One of the simplest methodologies is exemplified by the investigation of Heinrich (1979). Individual bumblebees were observed foraging in open fields where densities of flower heads had been manipulated by clipping and where all foragers had been excluded for 2 days prior to the study. In this particular case, Heinrich created various orientations of flower heads by bending inflorescences with string. He then recorded the reactions of individual foragers. In somewhat more complex studies, Waddington and Heinrich (1979) and Waddington and Holden (1979) created artificial inflorescences (Fig. 4) and observed bee visitations within outdoor screen cages, ranging in size up to 3.4 × 3.4 × 2.1 m. This design allowed manipulation of flower type, flower distribution, nectar rewards, and life histories of the bees. Both types of design yielded a great deal of useful information regarding bee foraging behavior.

In a unique study of parasitoid foraging behavior, Waage et al. (unpublished) placed ca. 50–100 potted cruciferous plants in an area surrounded by fields of planted crucifers. The field crucifers were naturally

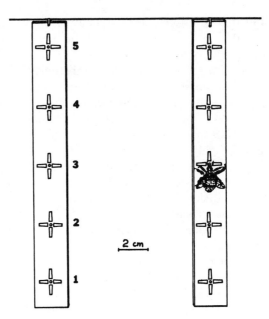

Figure 4. Artificial "inflorescences" used in bee foraging experiments showing spacing and numbering of flowers on each inflorescence and spacing of inflorescences suspended horizontally across foraging cage. (From Waddington and Heinrich, 1979.)

infested by various lepidopterous pests that supported several species of parasitoids, whereas the potted crucifers were manipulated to have one to a few unparasitized lepidopterous larvae per plant. The potted plants were placed in the field every morning and removed every evening. Each day, several persons continuously circled the potted plants, and, using binoculars, observed immigration rate and foraging behavior of parasitoids flying in from the fields of naturally infested crucifers. Thus, these workers manipulated the host plant and insect population densities and structures to observe the foraging behavior of natural populations of parasitoids.

Numerous seminatural behavioral studies have been undertaken to document the foraging behavior of phytophagous insects. For instance, movements of wild and laboratory-reared butterflies among plants were observed in field situations where both the numbers and arrangements of plants were manipulated (Jones, 1977; Zalucki and Kitching, 1982). The techniques used ranged from simple direct observation of individuals and recording by hand (Jones, 1977) to use of a continuously recording tracking device to obtain three-dimensional coordinates of individual insect positions (Zalucki and Kitching, 1982). Some studies have focused more on description of actual behaviors than flight movements exhibited during foraging under controlled conditions (Yamamoto et al., 1969). Still other studies have focused on extrinsic factors that affect foraging behavior. Using wild, laboratory-maintained female apple maggot flies, *Rhagoletis pomonella*, of defined physiological state, Roitberg and Prokopy (1982) and Roitberg et al. (1982) released flies singly into a large field cage, up to 8 m diameter by 2.5 m high, containing host trees with and without fruit clusters. Recording all movements of each fly released (Fig. 5), they were able to determine the effects of host fruit density and quality on the amount of time a fly spent searching for and ovipositing in fruits.

A potential factor influencing insect behavior and plant–insect interactions is the presence of predators. Field cages often do not effectively

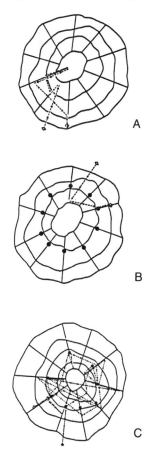

Figure 5. Search paths of *R. pomonella* flies in host trees. Innermost area of the maps represents the highest elevation level in the tree, while the outermost ring represents the lowest elevation level. Lines radiating from the center ring represent the major tree limbs. An *R. pomonella* fly in a tree: (A) devoid of fruit, (B) harboring eight clusters of nonhost fruit, (C) harboring eight clusters of host fruit each marked twice with oviposition deterring pheromone. (From Roitberg, et al. 1982.)

exclude predators, which have been found to influence greatly the behavior of phytophagous insects (e.g., Roitberg et al., 1979; Messina, 1981). Thus, in design of behavioral studies under seminatural conditions, all aspects of the environment (biotic as well as abiotic factors) plus the physiological state of the insect should be taken into account.

The topic of behavioral observations under seminatural conditions covers a wide variety of studies that in some way concern manipulation of the insect or environment. Manipulations may involve alteration of densities, compositions, or structures of host plants, use of field or greenhouse cages, or use of insects reared or handled under varying protocols. The degree of deviation from the natural situation may be great or small. In each instance, however, the methodology employed should in some way approximate the natural situation as closely as possible, while still altering one or more variables.

5.3. Laboratory Studies

For detailed description and causation analysis of insect behavior, investigators have turned often to laboratory studies. In the laboratory, meteorological conditions, the physiological state of the insect, and several external stimuli can be at least partially controlled simultaneously. Because indoor studies often utilize laboratory-reared or field-collected, laboratory-maintained insects, they are less seasonally constrained or influenced than natural field observations. Caution must be exercised when using laboratory-reared insects, however, because they may exhibit behaviors different from those of wild insects. Use of insects in the laboratory often allows greater sample sizes and thus facilitates more detailed statistical treatment of data. This, coupled with the use of internal and external controls, yields greater reproducibility of findings. The end result is often greater discernment of causation from correlation in behavioral pattern analysis.

As in studies under seminatural conditions, however, the relevance of lab behaviors to field conditions may be compromised by unnaturalness. Some animals exhibit aberrant behaviors or will not behave at all normally in the laboratory, for reasons unknown. The experimental apparatus itself may adversely affect behavior. Therefore, caution should be exercised when designing laboratory studies. As for observations under seminatural conditions, information gained from laboratory studies may be severely limited or misleading without knowledge of and reference to field observations. Thus, to maintain applicability to the field situation, it is important to attempt to mimic natural resource structure and components when designing a laboratory observational situation.

Under laboratory conditions, the role of the host plant in the orientation of insects to hosts has been investigated by a number of researchers. Often, however, these studies have involved use of a protocol arrived at after numerous previous protocols failed to elicit "normal" behavioral responses from the study insect. So long as the behavioral repertoire of the insect in the laboratory is substantially comparable to the insect's behavior in nature, an inexperienced behaviorist should not be disillusioned by false starts when designing a laboratory regimen. Many factors, including age, physiological state, nutrition, and sex of the individual (e.g., Hardee et al., 1966; Hawkes and Coaker, 1976), level of illumination (Sparks and Cheatham, 1970), humidity (e.g., Traynier, 1967), and time of day (Hardee et al., 1966), may influence the response of an insect to plants. For example, through comparison of insect behaviors under different meteorological conditions in a 120 × 72 × 5 cm laboratory chamber, Saxena (1978) found that water deprivation, temperature, and humidity greatly influenced the host-searching behavior of the red cotton bug, *Dysdercus koenigii.* Thought and careful use of controls are often the best approaches to account for effects of such variables.

Several investigators have observed insect orientation to host plants via use of olfactometers (e.g., Traynier, 1967) or wind tunnels (e.g., Hawkes and Coaker, 1976; Kennedy, 1977; Brantjes, 1978; Visser and Ave, 1978). Studies such as these normally employ a comparative approach. For example, Traynier (1967) compared the effects of host plant and nonhost plant odor on the activity of gravid female cabbage root flies, *Delia radicum,* in an olfactometer. Hawkes and Coaker (1976), working with this same species, and Visser and Ave (1978), working with the Colorado potato beetle, *Leptinotarsa decemlineata,* utilized wind tunnels to compare the responses of insects to host-plant odors and their individual components.

Not all observations of insect orientation to plants need be conducted in moving air. Brantjes (1978) compared landings of several species of night-flying moths on paper strips containing extracts of flower odors and held in still air. Orientation of moths to flower odors was also studied in a wind tunnel. Other workers have investigated visual orientation of Lepidoptera in still air by utilizing artificial plants, both with and without odor components (Sparks and Cheatham, 1970; Saxena and Goyal, 1978).

One may be interested also in the behavior of an insect after it has located a host. Carlson and Hibbs (1970), in their studies of the leafhopper *Empoasca fabae,* were not concerned with choice of plant for oviposition, but wished to examine more closely the details of ovipositional behavior after leafhopper arrival. To accomplish this, they used a twofold approach: motion pictures were taken of leafhoppers ovipositing into a plant stem, and then, to halt the oviposition at a particular point in the process, the female leafhopper and plant were instantaneously flooded with a hot fixative solution. Histological sections were prepared of the ovipositor and plant stem, yielding detailed information about the location of the ovipositor in the stem during egg-laying. Another important behavior that may be exhibited after host location in phytophagous insects is feeding. Using choice and nonchoice feeding trials coupled with observation of behavior, recording of time until feeding commencement, and assessment of meal size, Boys (1978) was able to assess the acceptability of various grass species for feeding by the grasshopper *Oedaleus senegalensis*.

In holometabolous insects, the adult usually is the more vagile stage, being equipped with the capacity for flight. Thus, components of adult host-seeking behavior may occur in more rapid succession and over a greater area than analogous behaviors exhibited by immature insects. Yet, the immature, from a pest management point of view, is often the more damaging stage, and therefore, its behavioral orientation mechanisms should not be overlooked. Sutherland (1972) observed and traced movements of subterranean New Zealand grass grubs, *Costelytra zealandica*, in 15 × 15 × 0.6 cm glass chambers in the laboratory. The lid of each chamber had fine, vertical black lines painted on it so that larval movements could be followed and traced onto graph paper. Because these grubs normally live underground, all tests had to be conducted with only red light for illumination. To ensure involvement only of olfactory stimuli, larvae were not allowed to contact the odor sources. Saxena and Khattar (1977) wished to test the effects of size, distance, and combination patterns of visual stimuli on orientation responses of *Papilio demoleus* caterpillars. They created sheets of *Citrus* leaves sandwiched between glass plates, thereby allowing natural leaf color and artificial shape and orientation of the sheets, but probably not odor, to influence the behavior of the larvae in the lab. The movement of the larvae, whether along or across the axis of the leaf sheets or toward or away from the stimulus, was then recorded.

Laboratory studies allow greater elucidation of behavioral detail than do field or seminatural studies. From these few examples, it can be seen that laboratory observations do not necessarily require greater technology than field observations. Nevertheless, rather few behaviorists conduct direct laboratory observations because laboratory studies have limited applicability to natural situations unless they have a strong basis in field observations. Most often, researchers turn to laboratory situations to conduct bioassays in which the effects of different stimuli and protocols are compared. The use of bioassays in plant–insect interactions is discussed in the following chapters of this book.

VI. Conclusions

We have attempted to collect a number of examples of techniques and methodologies for direct observation of plant–insect interactions, emphasizing the importance of simplicity in design of observational protocols.

Manipulation of the environment and experimentation can yield a great deal of information regarding factors affecting insect behavior, but observation and thought give rise to the hypotheses that are the basis of sound experimental research. Often, investigators do not publish the direct observations that led to the development of their experimental protocols, but this does not mean descriptive observations did not take place and were not necessary. New, imaginative techniques for behavioral research may be discussed frequently in scientific papers but are not necessarily more important or enlightening than simple direct observational approaches.

References

Altmann J (1974) Observational study of behavior: sampling methods. Behaviour **49**:227–267.

Baker TC, Cardé RT (1979) Courtship behavior of the Oriental fruit moth (*Grapholitha molesta*): experimental analysis and consideration of the role of sexual selection in the evolution of courtship pheromones in the Lepidoptera. Ann Entomol Soc Am **72**:173–188.

Bawa KS (1977) The reproductive biology of *Cupania guatemalensis* Radlk (Sapindaceae). Evolution **31**:52–63.

Bennett SR, McClelland GAH, Smilanick JM (1981) A versatile system of fluorescent marks for studies of large populations of mosquitoes (Diptera: Culicidae). J Med Entomol **18**:173–174.

Boys HA (1978) Food selection by *Oedaleus senegalensis* (Acrididae: Orthoptera) in grassland and millet fields. Entomol Exp Appl **24**:78–86.

Brantjes NBM (1978) Sensory responses to flowers in night-flying moths. In: The Pollination of Flowers by Insects. Richards AJ (ed), Academic Press, New York, pp 13–19.

Brussard PF, Ehrlich PR (1970) Adult behavior and population structure in *Erebia epipsodea* (Lepidoptera: Satyrinae). Ecology **51**:886–891.

Carlson OV, Hibbs ET (1970) Oviposition by *Empoasca fabae* (Homoptera: Cicadellidae). Ann Entomol Soc Am **63**:516–519.

Chabora PC, Shukis AA (1979) The automated recording of insect activity: the house fly. Ann Entomol Soc Am **72**:287–290.

Dethier VG (1978) Other tastes, other worlds. Science **201**:224–228.

Drickamer LC, Vessey SH (1982) Animal Behavior: Concepts, Processes, and Methods. Willard Grant, Boston.

Eisner T, Silberglied RE, Aneshansley D, Carrel JE, Howland HC (1969) Ultraviolet video-viewing: the television camera as an insect eye. Science 166:1172–1174.

Faegri K (1978) Trends in research in pollination ecology. In: The Pollination of Flowers by Insects. Richards AJ (ed), Academic Press, New York. pp 5–12.

Fagen RM (1978) Repertoire Analysis. In: Quantitative Ethology. Colgan PW (ed), John Wiley, New York, Chap 2, pp 25–42.

Finch S (1980) Chemical attraction of plant-feeding insects to plants. In: Applied Biology. Coaker TH (ed), Academic Press, New York, pp 67–143.

Fitzpatrick SM, Wellington WG (1983) Contrasts in the territorial behavior of three species of hover flies (Diptera: Syrphidae). Can Entomol 115:559–566.

Frankie GW, Opler PA, Bawa KS (1976) Foraging behavior of solitary bees: implications for outcrossing of a neotropical forest tree species. J Ecol 64:1049–1057.

Hardee DD, Mitchell EB, Huddleston PM (1966) Effect of age, nutrition, sex, and time of day on response of boll weevils to an attractant from cotton. Ann Entomol Soc Am 59:1024–1025.

Hawkes C, Coaker TH (1976) Behavioural responses to host-plant odours in adult cabbage root fly (*Erioischia brassicae* [Bouche]). Symp Biol Hung 16:85–89.

Heinrich B (1976) The foraging specializations of individual bumblebees. Ecol Monogr 46:105–128.

Heinrich B (1979) Resource heterogeneity and patterns of movement in foraging bumblebees. Oecologia (Berl) 40:235–245.

Hinde RA (1970) Animal Behavior: a Synthesis of Ethology and Comparative Psychology. McGraw-Hill, New York.

Hoy JB, Globus PA, Norman KD (1983) Electronic tracking and recording system for behavioral observations, with applications to toxicology and pheromone assay. J Econ Entomol 76:678–680.

Jackson RR (1983) The biology of *Mopsus mormon*, a jumping spider (Araneae: Salticidae) from Queensland: intraspecific interactions. Aust J Zool 31:39–53.

Jones RE (1977) Movement patterns and egg distribution in cabbage butterflies. J Anim Ecol 46:195–212.

Kareiva P (1982) Experimental and mathematical analyses of herbivore movement: quantifying the influence of plant spacing and quality on foraging discrimination. Ecol Monogr 52:261–282.

Kennedy JS (1977) Olfactory responses to distant plants and other odor sources. In: Chemical Control of Insect Behavior. Shorey HH, McKelvey Jr JJ (eds), Wiley-Interscience, New York, pp 67–91.

Lehner PN (1979) Handbook of Ethological Methods. Garland, New York.

Levin DA, Kerster HW, Niedzlek M (1971) Pollinator flight directionality and its effect on pollen flow. Evolution 25:113–118.

Lew AC, Ball HJ (1979) The mating behavior of the western corn rootworm *Diabrotica virgifera* (Coleoptera: Chrysomelidae). Ann Entomol Soc Am 72:391–393.

Lingren PD, Sparks AN, Raulston JR, Wolf WW (1978) Night vision equipment for studying nocturnal behavior of insects. Paper number 4. Bull Entomol Soc Am 24:206–212.

Malavasi A, Morgante JS, Prokopy RJ (1983) Distribution and activities of *Anastrepha fraterculus* (Diptera: Tephritidae) flies on host and nonhost trees. Ann Entomol Soc Am 76:286–292.

Messina FJ (1981) Plant protection as a consequence of an ant-membracid mutualism: interactions on goldenrod (*Solidago* sp.). Ecology 62:1433–1440.

Owens ED, Hauschild KI, Hubbell GL, Prokopy RJ (1982) Diurnal behavior of plum curculio (Coleoptera: Curculionidae) adults within host trees in nature. Ann Entomol Soc Am 75:357–362.

Prokopy RJ (1976) Feeding, mating, and oviposition activities of *Rhagoletis fausta* flies in nature. Ann Entomol Soc Am 69:899–904.

Prokopy RJ, Bennett EW, Bush GL (1971) Mating behavior in *Rhagoletis pomonella* (Diptera: Tephritidae) I. Site of assembly. Can Entomol 103:1405–1409.

Prokopy RJ, Bennett EW, Bush GL (1972) Mating behavior in *Rhagoletis pomonella* (Diptera: Tephritidae) II. Temporal organization. Can Entomol 104:97–104.

Prokopy RJ, Bush GL (1973) Mating behavior of *Rhagoletis pomonella* (Diptera: Tephritidae) IV. Courtship. Can Entomol 105:873–891.

Prokopy RJ, Roitberg BD, Averill AL (1984) Chemical mediation of resource partitioning in insects. In: The Chemical Ecology of Insects. Bell WJ, Cardé RT (eds). Chapman and Hall, London, pp 301–330.

Rausher MD (1980) Host abundance, juvenile survival, and oviposition preference in *Battus philenor*. Evolution 34:342–355.

Roitberg BD, Myers JH, Frazer BD (1979) The influence of predators on the movement of apterous pea aphids between plants. J Anim Ecol 48:111–122.

Roitberg BD, Prokopy RJ (1982) Influence of intertree distance on foraging behaviour of *Rhagoletis pomonella* in the field. Ecol Entomol 7:437–442.

Roitberg BD, van Lenteren JC, van Alphen JJM, Galis F, Prokopy RJ (1982) Foraging behaviour of *Rhagoletis pomonella,* a parasite of hawthorn (*Crataegus viridis*), in nature. J Anim Ecol 51:307–325.

Saxena KN (1978) Role of certain environmental factors in determining the efficiency of host plant selection by an insect. Entomol Exp Appl 24:466–478.

Saxena KN, Goyal S (1978) Host-plant relations of the citrus butterfly *Papilio demoleus* L.: orientational and ovipositional responses. Entomol Exp Appl 24:1–10.

Saxena KN, Khattar P (1977) Orientation of *Papilio demoleus* larvae in relation to size, distance, and combination pattern of visual stimuli. J Insect Physiol 23:1421–1428.

Scriber JM (1983) The evolution of feeding specialization, physiological efficiency and host races in selected Papilionidae and Saturniidae. In: Variable Plants and Herbivores in Natural and Managed Systems. Denno RF, McClure MS (eds), Academic Press, New York, pp 373–412.

Slater PJB (1978) Data collection. In: Quantitative Ethology. Colgan PW (ed), John Wiley, New York, Chap 1, pp 7–24.

Slater PJB, Ollason JC (1973) The temporal pattern of behaviour in isolated male zebra finches: transition analysis. Behaviour 42:248–269.

Smith DC, Prokopy RJ (1981) Seasonal and diurnal activity of *Rhagoletis mendax* files in nature. Ann Entomol Soc Am 74:462–466.

Smith DC, Prokopy RJ (1982) Mating behavior of *Rhagoletis mendax* (Diptera: Tephritidae) flies in nature. Ann Entomol Soc Am 75:388–392.

Sparks MR, Cheatham JS (1970) Responses of a laboratory strain of the tobacco hornworm, *Manduca sexta,* to artificial oviposition sites. Ann Entomol Soc Am 63:428–431.

Stanton ML (1983) Searching in a patchy environment: foodplant selection by *Colias p. eriphyle* butterflies. In: Herbivorous Insects: Host Seeking Behavior and Mechanisms. Ahmad S (ed), Academic Press, New York, pp 125–157.

Sutherland ORW (1972) Olfactory responses of *Costelytra zealandica* (Coleoptera: Melolonthinae) larvae to grass root odours. New Zeal J Sci 15:165–172.

Thompson JN, Price PW (1977) Plant plasticity, phenology, and herbivore dispersion: wild parsnip and the parsnip webworm. Ecology 58:1112–1119.

Thorsteinson AJ (1960) Host selection in phytophagous insects. Annu Rev Entomol 5:193–218.

Tinbergen N (1963) On aims and methods in ethology. Z Tierpsychol 20:410–433.

Traynier RMM (1967) Effect of host plant odour on the behaviour of the adult cabbage root fly, *Erioischia brassicae*. Entomol Exp Appl 10:321–328.

Visser JH, Ave DA (1978) General green leaf volatiles in the olfactory orientation of the Colorado beetle, *Leptinotarsa decemlineata*. Entomol Exp Appl 24:538–549.

Waage JK (1979) Foraging for patchily-distributed hosts by the parasitoid, *Nemeritis canescens*. J Anim Ecol 48:353–371.

Waddington KD, Heinrich B (1979) The foraging movements of bumblebees on vertical "inflorescences": an experimental analysis. J Comp Physiol 134:113–117.

Waddington KD, Holden LR (1979) Optimal foraging on flower selection by bees. Am Nat 114:179–196.

Walker TJ, Wineriter SA (1981) Marking techniques for recognizing individual insects. Fla Entomol 64:18–29.

Woodell SRJ (1978) Directionality in bumblebees in relation to environmental factors. In: The Pollination of Flowers by Insects. Richards AJ (ed), Academic Press, New York, pp 31–39.

Yamamoto RT, Jenkins RY, McClusky RK (1969) Factors determining the selection of plants for oviposition by the tobacco hornworm *Manduca sexta*. Entomol Exp Appl 12:504–508.

Zalucki MP, Kitching RL (1982) The analysis and description of movement in adult *Danaus plexippus* L. (Lepidoptera: Danainae). Behaviour 80:174–198.

Chapter 2

Assessing Host-Plant Finding by Insects

Stan Finch[1]

I. Introduction

It is now more than 30 years since Dethier (1947) remarked that "no one attractant alone performs the service of guiding an insect to its proper host-plant, food or mate, and that the desired end is achieved only by a complex array of stimuli, such as chemical, light, temperature and humidity, acting in harmony." Even so, it is doubtful whether the full complexity of the stimuli involved has yet been envisaged. Thus, just as sex pheromones are sometimes mixtures of olfactory stimuli, whose relative proportions can vary in space, time, and from population to population, the same is also true of the visual and olfactory stimuli used by insects in host-plant finding (Finch, 1977, 1980; Miller and Strickler, 1984; Miller and Harris, 1985).

Although much work has been carried out in recent years on insect-to-insect (pheromone) communication, insect-to-plant systems have been neglected and the statement of Thorsteinson (1960) that "the literature concerned with host-plant selection is not voluminous" is still appropriate. Because of this neglect, some of the techniques included in this review were developed initially for studies on pheromones. Nevertheless, they are just as relevant here, since they relate to principles basic to the understanding of insect reponses to volatile chemicals.

In this chapter, I have not made specific reference to nocturnal insects, though many of the methods described apply equally well to both diurnal

[1]Entomology Section, National Vegetable Research Station, Wellesbourne, Warwick CV35 9EF, United Kingdom

and nocturnal insects, nor have I included studies on pollinators; these have been reviewed critically by Eickwort and Ginsberg (1980).

II. Types of Movement of Insects During Host-Plant Finding

At some stage during their life, most phytophagous insects move to find a host plant. Some merely reclimb the plant on which they, or their parents, developed; other disperse hundreds of kilometers. If the insect does not respond to "vegetative" stimuli (Kennedy, 1961), such as food or host plant, during this dispersal phase, the movement is regarded as true "migration" (Southwood, 1962). In practical terms, understanding the dispersal phase, whether true migration is involved or not, is extremely important, since it defines the area into which the population will spread. Once the population has dispersed, movements in response to "vegetative" stimuli are classed as "trivial" movements (Thorpe, 1951). Although separating dispersal into "migrating movements" and "trivial movements" in this way clarifies the theoretical situation, it does little to clarify the practical situation, since it is often difficult in the field to determine when migratory movements cease and trivial movements begin. This is particularly so for insects such as Diptera and Lepidoptera, in which flight is not restricted to just one short period of adult life but is the normal means of locomotion. Consequently, most studies have concentrated only on the final stage of host-plant finding, when the insect is 1–10 m away from a suitable source of host-plant material.

III. Mechanisms for Finding an Odor Source

Three mechanisms of behavioral responses assist an insect in finding an odor source (Shorey, 1973). In the first, the insect can align its body in the direction of the odor source (true chemotaxis) as a result of being able to sense directly the gradient of odor molecules. By Koehler's (1950) definition such a "taxis," or turn, includes no progressive movement. Consequently, for the "turned" insect to arrive at the odor source a separate locomotor reaction, or kinesis, must be initiated.

Such kineses belong to Shorey's (1973) second category of behavioral responses. Here the insect does not detect the direction of the odor gradient but instead is stimulated to move at different rates (orthokinesis) or to turn at different frequencies (klinokinesis) by changes in the local concentrations of odorous chemicals. Thus if the insect is stimulated to move rapidly at low concentrations of odor and slowly at high concentrations (inverse orthokinesis) it should eventually arrive at and get stopped near the odor source. Because the insect responds this way, the odor in effect becomes a stronger "arrestant" (Dethier et al., 1960) the closer the insect approaches the source. In certain instances the odor becomes so effective that the insects alight short of the source and, presumably once sufficient adaptation has occurred, then proceed in a series of short "hops" to reach their final goal (Hawkes and Coaker, 1979; Dindonis and Miller, 1980). In contrast, the majority of insects studied fly directly to the source of the odor, mainly because the arrestant effect of the odor is partially nullified by the increase in attraction to visual stimuli the closer the insect approaches its landing site.

A second mechanism that should enable insects to find a source of odor involves turning more frequently or severely when the stimulus is decreasing (inverse klinokinesis) and less frequently or severely when the concentration of odors from host plants is increasing. Results from trapping insects in the field indicate that visual stimuli do not override chemical stimuli at close range; instead the two work in harmony. For example, the cabbage root fly responds visually to flourescent yellow traps from a distance of about 2.5 m (Finch and Skinner, 1974). Despite this strong visual attraction to the yellow color of the trap, fewer flies were caught when a dispenser releasing the attractant, allylisothiocyanate, was moved from the center of the trap to a point only 0.3 m away (Finch and Skinner, 1982a). Hence once the insect is in the plume of odor close to its source, klinokinesis may be so strong that at least a proportion of the insects are unable to deviate from their course, even when offered an adjacent supernormal visual stimulus. Thus, even close to the source, visual stimuli interact with the chemical stimuli rather than substituting for them. The unnatural situation of separating the appropriate chemical and visual stimuli may be unacceptable to a proportion of the responding insects.

In the third category of behavioral response, odor molecules cause insects to orient to some other stimulus. The most commonly quoted example of this mechanism is where molecules of an "attractive" chemical stimulate receptive insects to turn upwind (anemotaxis). It is this mechanism that most workers now consider to be a major means by which insects orientate to a distant source of odor, such as that emanating from an appropriate host plant. Unfortunately, attempts to show that this is the chief mechanism operative in host-plant finding in the field have so far been largely unsuccessful.

The papers of Kennedy (1977, 1978) are recommended for a critical appraisal of the traditional classification of insect locomotor reactions to odors and the earlier review of Shorey (1973) for its lucid exposition of how these terms apply to practical situations.

Once airborne, insects actively engaged in host-plant finding generally fly close to the ground and visually assess the apparent movement of the objects over which they are flying (Kennedy, 1939). During such flights, the speed of upwind flight is governed largely by the speed at which the visual ground images pass the eyes (Kennedy and Thomas, 1974). However, certain insects prefer to fly at a height where they maximize their flight speed rather than the velocity of the ground images passing their eyes (Kuenen and Baker, 1982). Insects flying upwind (optomotor anemotaxis) steer so the ground patterns move approximately parallel to their longitudinal body axes from front to rear. The possible advantages gained by insects flying against, across, or with the prevailing wind during host-plant finding need further study. It should be borne in mind, however, that crosswind flights may be only the resultant displacement of insects trying to fly upwind. Similarly, many insects that disperse by flying close to the ground seem unable to take advantage of wind-assisted flights during host-plant finding, and land as soon as the wind, often a relatively moderate airflow, becomes too strong for them to regulate their own speed of flight.

IV. Direct Observation of Insect Responses to Plants in the Field

The advantages that can be gained during insect behavioral studies in the field by choosing insects that are both clearly visible from a distance and conspicuously marked have been described in considerable detail by Opp and Prokopy (Chapter 1). Unfortunately, insects that fit such descriptions often can be studied only by theoretical ecologists, allowed freedom of choice of study subjects, or by the few practical ecologists lucky enough to work on a pest species that is both conspicuous and amenable to observation. Many practical ecologists are restricted to species whose populations must reach high densities before they cause economic damage to crop plants. The individuals of such species tend to be small, can often by separated from closely related nonpest species only by microscopical examination, and frequently do not exhibit any obvious dimorphism between the sexes. Hence much of the data recorded under field conditions may be collected on male rather than on female activity. To be absolutely certain that observational data pertain to the species being studied, the individuals observed in the field must be caught and positively identified. If this is not done, there is a danger of even confusing certain plant feeding insects (e.g., *Delia* spp.) with their predators (e.g., *Caricea* spp.).

Despite the apparent lack of "cooperation" by many test insects, there is, as Opp and Prokopy stress, a great deal to be gained from good observational data collected under field conditions. To progress at an acceptable speed, however, most ecologists make considerable use of field-cage and laboratory studies in which certain experimental treatments or conditions (see Chapter 1) can be partially controlled.

Although such studies can be criticized as being somewhat artificial, they are often the only way of unravelling some of the complicated be-havioral repertoires that occur during host-plant finding. Laboratory stud-ies are also used frequently to formulate hypotheses that can be tested in the field. Such studies are of paramount importance where either envi-ronmental constraints or univoltinism ensure that the pest insect is active in the field for only a short period each year.

V. Techniques for Studying Insect Flight

5.1. Laboratory Techniques

As it is difficult to obtain data on insect disperal in the field, laboratory studies are usually carried out to determine whether a particular species has the capacity for sustained movement prior to, or during, host-plant finding.

A. Tethered Insects

In early studies sustained flight was assessed by tethering (see Miller, 1979 for current methods) an insect to a pin stuck into a suspended cork so that the insect dangled just above the working surface. The insect was then given an object to hold in its tarsi. When this object was withdrawn, the insect usually started to fly. A disadvantage of this method is that it provides only a crude measure of flight. An advantage, however, is that many insects can be tested at the same time and hence it is possible to measure the variation for flight within a population. Consequently, this method can be used to determine periods in the insect's life when host-plant stimulated flight is most common.

B. Flight Mills

Hocking (1953) devised a rotary flight mill that allowed suspended insects to fly in a circle so that the duration and speed of flight and also distance flown could be measured. The results were difficult to analyze, however, since they included such factors as the change in the work done by the insect to overcome the drag and friction of the mill, the steering efforts of the insect flying a curved path, and the support of the insect's weight by the mill arm. Henson (1962) and Chapman (1982) suggest ways of ov-ercoming these difficulties. Modern flight mills use analogue recorders that are sufficiently sensitive to record the rapid changes that occur in insect flight when volatile plant chemicals are blown over flying insects (Borden and Bennett, 1969; Cullis and Hargrove, 1972). Even with the use of such recorders, however, an inordinate amount of time is still re-

quired to process flight-mill data. This disadvantage has now been eliminated by the development of an interface system allowing data from up
to 12 flight mills to be entered directly into a microcomputer for high-
speed processing (Clarke et al., 1984).

C. Flight Balances

In addition to flight mills, simple balances are used to study the lift component of flight. These consist of a thin rod or wire balanced across a
precision bearing, such as a galvonometer mechanism with the hairspring
removed (Baker et al., 1980), the center of gravity of the rod being just
on the same side as the attached insect. As the insect raises the arm during
flight, its lift is counterbalanced by a restoring couple caused by displacement of the center of gravity. More complicated flight balances in which
the tethered insects fly in an air stream have been used for locusts (Weis-
Fogh, 1956) and bugs (Ward and Baker, 1982) to measure the vertical (lift)
and horizontal (yaw, pitch, and roll) components of displacement.

D. Sustained-Flight Wind Tunnels

The most sophisticated apparatus for measuring flight is the wind tunnel
in which the insects fly freely (e.g., Gatehouse and Lewis, 1973; Miller
and Roelofs, 1978). Usually the flight speed of an insect is balanced by
the air wind speed so that the insect flies mainly in the central portion of
the tunnel. To keep the insect flying, the tunnel floor is patterned (see
David, 1982) and moved on rollers so that the insect receives the optomotor
information appropriate to its apparent movement. By choosing a wind
speed that permits the insect to progress upwind, either visual or olfactory
stimuli can be assessed. The working lengths of most tunnels do not exceed
2–3 m and so their use without movement of the floor is restricted largely
to slow-flying insects such as aphids and small moths.

5.2. Field Techniques

A. Indirect Methods

Experiments on the ability of insects searching for host plants to pass
over inhospitable terrain, such as large stretches of water or grassland,
can be carried out only where the local situation is appropriate. An alternative is to find (Leggett and Roach, 1981), or to spray insecticides to
create, an uninfested zone and then record how quickly the insects reinfest
it. It may not be possible, however, to produce a completely insect-free
zone by spraying, since some individuals may survive the treatment. Furthermore, large "insect-free" zones are required and hence such exper-

iments tend to be expensive. An advantage of this method, however, is that natural populations are used and so the rearing and release of insects is avoided. In addition, the virtual removal of the local population means that any insect caught yields useful information. This contrasts sharply with release/recapture experiments where considerable time is required to separate the marked insects from the feral insects caught in the traps. This separation is particularly tedious when only a few insects can be released. Because trap data are required for conditions appropriate to a particular stage in the insects' life cycle, the marked insects must be released when the feral population is large.

B. Direct Methods

These normally employ some form of mark/recapture method.

(a) Marking methods. The types of marks applied to insects, and the major difficulties in selecting one that will persist under field conditions, have been described by Southwood (1966). The fundamental requirement of any mark is that it does not affect the behavior of the insect.

Individuals can be marked (see also Chapter 1 of this volume) by attaching small labels, by applying spots of different colored paints, or by physically marking the integument, a process often inaptly described as "mutilation." As many as 9999 individual marks can be applied to certain beetles by etching a pattern of dots into the elytra with an insect pin mounted in a high-speed drill (Best et al., 1981).

Most insects, however, are difficult to recapture and it is usual therefore to mark and release as many as possible. Populations of insects can be marked by allowing larvae to feed on a diet containing dye (Brewer, 1982) or by allowing adults to feed on sugar solutions containing radioactive tracers (Hawkes, 1972). In both instances, the insects have to be maintained under laboratory conditions for some time and this can affect their subsequent behavior. In an extreme example, a colony of the brown plant hopper, *Nilaparvata lugens* (Stahl), was almost unable to fly after only one generation in the laboratory (Baker et al., 1980). Other factors that need to be considered when working with radioactive markers are the few marks available and the relatively expensive equipment required for their detection.

Large numbers of insects are usually marked by spraying with a colored dye or dusting with a colored powder (see Southwood, 1966). Insects marked by either method are easy to identify, as dyes can be eluted from the insects using an appropriate solvent and small particles of colored powders can be observed easily using a light microscope.

(b) Recapture methods. Once the test insect has been marked and released, it has to be recaptured. Many suitable traps (see below) have been developed for most of the major pest species.

A frequent limitation of the marking–release–recapture method is that few insects are recaptured, even though many thousands or even millions may be released. It is also generally assumed that the behavior of the few recaptured, often fewer than 1% of the total released, is an unbiased, representative sample of the original population (Johnson, 1969). If released insects are to be recaptured in large numbers, traps must be spaced throughout their dispersal zone. As Southwood (1962) pointed out, although the whole population appears to move during the dispersive phase, some individuals are left behind, at the start and all along the route, so that the end result is a progressive scattering of the original population over a wide area. Furthermore, although traps baited with plant chemicals may attract or arrest more insects than an individual host plant in any recapture area, each trap competes against large numbers of host plants. Hence, even though many of the released insects may be present in the crop, few may be recaptured.

VI. Isolation, Concentration, and Identification of Odors

6.1. Isolation

The volatile chemicals that give a plant its characteristic odor are those that attract and/or arrest insects. Although plant odors are often extremely pungent, particularly in crushed tissues, the chemicals that produce them are present only at very low concentrations (μg/g) in living plants. To be able to extract a sufficient amount of these chemicals for testing is the main chemical challenge in insect–plant relationships. It is essential that the characteristic odor of the plant is not destroyed during the extraction. If the extract does not smell at all like the crushed plant material, then it may be pointless to proceed with any attempt to identify active factors.

A. Distillation

To isolate sufficient volatile material for biological testing, large quantities of plant material are usually macerated, chopped, or pressed. The technique must ensure that the odors are not destroyed by enzymes or by chemical change. Furthermore, as most chemical changes result from interaction with nonvolatile components, it is advisable to separate the volatile from the nonvolatile chemicals as soon as possible (Teranishi et al., 1971).

The separation usually involves some form of distillation, since volatility is the only property that the chemicals of interest have in common. If the volatile substances are not damaged by high temperatures, flash distillation is possible. In most cases, however, flash distillation reduces the volatile chemicals to simpler derivatives and hence the milder treatment of distillation under reduced pressure has to be employed (e.g. Visser et al., 1979). An extremely simple method that can provide concentrated extracts is described by Likens and Nickerson (1964). It is based on the principle that many odorous compounds, e.g., terpenes and other essential oils, are soluble in organic solvents but not in water. The crude plant extracts are placed into the distillation flask and the distillates are collected in an organic solvent such as hexane (Fig. 1a). Both the steam distillate and the organic solvent condense on the cooling column and drop into the U-shaped tube. The hexane containing the volatile extracts is immiscible with water and so floats. Volatile fractions accumulate in the hexane distillate until it siphons via tube A into the small collecting vessel (Fig. 1a). Similarly the excess water that condenses in the tube siphons back via tube B into the distillation flask. This system requires little solvent and the solvent recycles. Once a distillate has been obtained, it is usually concentrated by freezing, adsorption, or further extraction.

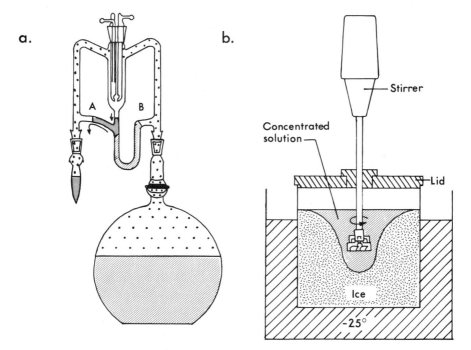

Figure 1. (a.) Distillation-extraction apparatus. (From Likens and Nickerson, 1964.) (b) Freeze concentration apparatus. (From Visser, Van Straten and Maarse, 1979.)

6.2. Concentration

A. Freezing

Freezing is commonly used to remove the water from dilute aqueous distillates of plant odors, as it reduces the rate of chemical reactions and
minimizes the loss of heat-labile compounds. Distillates can be placed in
stainless steel beakers inserted in baths of methanol maintained at about
$-25°C$. While the distillate is stirred, the water is selectively frozen out
around the edges of the beaker, leaving the concentrated volatile materials
in a conical hole around the stirrer blade (Fig. 1b) (Kepner et al., 1969;
Visser et al., 1979). To avoid evaporation of volatile fractions, a lid is
placed on the beaker. Problems arising when the rates of cooling or stirring
are too fast are described by Teranishi et al., (1971).

B. Adsorption

Adsorption on porous compounds is also a common method for removing
volatile substances from aqueous solutions. Powdered charcoal is often
used because it has a large adsorptive capacity and is not deactivated by
water. Adsorption has been used to concentrate volatile material from
onions and apples (see Finch, 1980). The time to reach the adsorption
equilibrium can vary from minutes to hours, depending on the type of
charcoal used, and some components may be irreversibly held. Using
charcoal to extract volatile chemicals dissolved in a range of solvents
gradually becomes less effective across the solvent range from water,
ethanol, esters, and acetone to chloroform (Weurman, 1969). Diethyl ether
is generally used to elute the volatile material from the charcoal.

Adsorption chromatorgaphy often uses silica gel columns (Murray and
Stanley, 1968) instead of charcoal, the separated volatile fractions being
eluted from the column with trichlorofluoromethane (Freon 11), an odor-
less, non flammable chemical with a boiling point of only 23°C.

The adsorbents Porapak Q (an ethylvinylbenzene divinylbenzene co-
polymer) and Tenax GC (a polymer based on 2,6-diphenyl-p-phenylene
oxide) are the two most commonly used at present, as they have a low
affinity for both water and ethyl alcohol. It is more usual, however, to
use these two adsorbents to collect head-space vapors (see below) from
above other extracts.

C. Extraction

Extraction by solvents should not be attempted until the distillates have
been concentrated as much as possible by other methods. This avoids
dilute plant extracts in large volumes of solvent, which add considerably

to costs and which are time-consuming to remove. Furthermore, as most solvents contain some impurities, evaporation of large quantities of solvent also concentrates any impurities and often it is these that react with the odorous chemicals to produce artifacts. It is generally essential, therefore, to redistill purchased solvents to attain the high level of purity required for most successful extractions.

6.3. Identification

To identify small amounts of volatile chemicals, gas chromatography can be used. The relative retention time of the test chemical is compared with that of a known compound, both being measured against an "internal" standard chemical, such as a paraffin. Such retention times are useful for low-molecular-weight compounds but not for large molecules and particularly not for those that readily degrade to produce several different peaks.

To determine the structure of a chemical umambiguously, volatile fractions can be collected from the effluent gas of the gas chromatograph and their structure determined by mass spectrometry, infrared spectroscopy, and/or nuclear magnetic resonance. Teranishi et al. (1971) introduce this subject and provide concise descriptions of the principles underlying the equipment currently available for the identification of plant volatile materials. For those wishing to pursue the subject in greater detail, Macleod (1973) describes many of the pitfalls that can arise both in the preparative work and in the resolution of the analytical equipment.

6.4. Head-Space Analysis

Surprisingly few studies have been carried out to determine whether the volatile chemicals obtained during plant extraction are those normally liberated from plants into the air (Finch, 1980) and the results from such studies have often been conflicting. For example, although the air extracted from around cotton plants contained only 6 of the 58 chemicals found in cotton bud essential oil, it possessed a cotton plant odor (Hedin et al., 1975). Extracting air from around plants is the best indicator of the chemicals most important in host-plant finding by insects, as shown by the attraction of insects to chemicals in the air around flowers of *Rosa* spp (Sirikulvadhana et al., 1975) and around the leaves of brassica plants (Cole and Finch, 1978).

"Head-space collection" is being used increasingly to collect volatile substances from the air around plants. The advantages of this system are that it requires little plant material, it avoids artifacts from impure solvents, and it provides an estimate of the concentration of the volatile chemicals that the insects normally experience during host-plant finding in the field. Its most serious disadvantage, however, is that many of the chemicals are present in such small quantities that the analytical techniques are often operating close to their limits. Hence, even for the analysis of volatile substances collected using the "head-space" technique, some form of concentration if often carried out prior to analysis.

Head-space collection is carried out using either a static or a dynamic system. In the static system, chemicals released by the plants are allowed to reach an equilibrium level in a closed container before a large gas-tight syringe is used to sample the air. As direct injection of such a large-volume sample onto a gas-chromatographic column would result in band broadening and poor resolution, the sample is concentrated before injection using the small-trap method described below. In the dynamic system, an air stream is used to flush the volatile chemicals from around plants through a trap half filled with an adsorbent such as Porpak Q or Tenax GC (Fig. 2a, after Cole, 1980). Heat and a system of valves (Fig. 2b) (Williams et al., 1978; Cole, 1980) are then used to desorb the volatile materials from the absorbent and to move them into the unfilled portion of the trap where they can be condensed by the collar of solid carbon dioxide. After 30 min of desorption, the dry ice is removed and the heating element lowered to transfer the volatile substances onto the front of the gas-chromatographic column prior to starting the temperature programming (Cole, 1980). Head-space collection can also be used to collect volatile chemicals above macerated plant tissues, thereby minimizing some of the difficulties that occur with more conventional methods of extraction. Similarly, Cane and Jonsson (1982) have used the head-space principle to develop an inexpensive suction/adsorption system capable of collecting submicrogram quantities of odors under field conditions.

VII. Bioassays for Attractants

7.1. General

An assay in which the detector is a living organism or part of its sensory system (Frazier and Hanson, Chapter 7) is referred to as a bioassay. Bioassays are the basic method for assessing the biological activity of chemical and visual stimuli used during host-plant finding. When an insect, or part of an insect, responds immediately to a colored object or to a crude plant extract it is easy to design a suitable bioassay. The results of such bioassays are not always easy to interpret, however, since crude plant extracts may not be very reproducible. Hence such extracts often induce different behavioral changes in the test insects depending on how the individual chemicals flux and interact from sample to sample (Finch, 1978). Similarly, it is possible that under natural conditions fluctuations in the incident light could alter the relative attractiveness of colored objects. For critical experiments, therefore, it is usual to deal with quali-

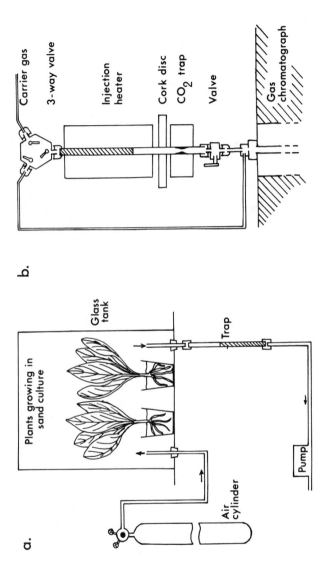

Figure 2. (a) Apparatus for the collection of "head space" volatile chemicals. (After Cole 1980.) (b) Three-way valve arrangement for transferring adsorbed volatile chemicals from the collection trap to the gas chromatograph. (After Cole 1980.)

tatively and quantitatively defined chemical stimuli as well as defined wavelengths and intensities of light. Often the goal is to synthesize a stimulus mixture with an attractancy similar to, or better than, that of the insect's preferred host plant. This is often less difficult for visual aspects of host-plant finding, since many insects attracted to green host plants show a "supernormal" response to yellow.

The proportion of insects that move towards or are arrested in the vicinity of a source of odor and/or color is used to indicate the effectiveness of the test stimulus. Compared to pheromones, however, sterotyped repertoires of insect behavior (e.g., Alberts et al., 1981) associated with either visual or chemical stimuli involved during host-plant finding are less obvious. However, such repertiores may exist but may be overlooked because they are not as overtly ritualized as those induced by sexual stimuli.

7.2. Problems in Bioassays

The response of an insect to a particular color or odor can be evaluated critically only when the stimulus can be controlled qualitatively and quantitatively. Shorey (1970) should be consulted for a list of the potential problems in designing bioassays. The way in which such problems relate to host-plant selection by chemical cues was reviewed by Finch (1980). As most problems are concerned with either standardization of the physical conditions, or with selection of the insect stage most suitable for testing, they also apply equally well to the bioassay of visual stimuli.

7.3. Types of Bioassay

The two main types of bioassay are based on either behavioral or electrophsiological responses. The earliest behavioral bioassays involved placing colored objects or odorous plant material near a group of insects and observing how this affected their activity. As such tests were only qualitative, considerable research has been carried out subsequently to quantify the insect's response to plant stimuli. One major difficulty is that the photons that induce visual stimuli travel in straight lines, whereas the odorus molecules that induce olfactory stimuli can move in all directions, since they are transported mainly in air currents. Consequently, apparatuses used to determine the contributions of visual and olfactory stimuli to host-plant finding are different. For this reason the two aspects are better considered separately.

A. Bioassay of Visual Stimuli

Terms such as "dark green" or "yellowish" are only vague descriptions of the colors of plants. The three main variables used by physicists to describe a color are its *hue, saturation,* and *intensity.*

The hue refers to the principal colors of the visible spectrum—red, yellow, green etc., whereas saturation defines the density of the color. Thus the more white, gray etc. that is mixed with a particular color the less it is saturated. Hence, pastel colors are said to be desaturated whereas strong vivid colors are highly saturated. If part of a colored card reflects directly the incident white light as a shiny area or "flash point," the saturation of color in that area is greatly reduced. Furthermore, if half of a sheet of colored card is in shadow, then hue and saturation remain the same; only intensity (or brightness) has changed (Langford, 1974).

(a) *Colored papers.* These can be selected from a standard series (e.g., Munsell Colors, Baltimore, Maryland) consisting of 100 different hues. It is usual initially to present the test insects with a selection of colored papers to determine which is the preferred of the major hues (five principal and five intermediate) named as red, yellow-red, yellow, green-yellow, green, blue-green, blue, purple-blue, purple, and red-purple. Each hue is represented by a series of papers given numbers to represent their color quality, the highest quality color being saturated. In nature, however, most colors are usually desaturated in some degree by other pigments. Consequently, papers are now available in which the same hue is produced at several different levels of saturation, by incorporating fixed amounts of either white or gray pigments. Once a preferred hue has been found, papers at different levels of saturation are then tested to determine if any are more attractive to the test insect than the saturated hue.

(b) *Neutral papers.* To confirm that the insect sees color and does not select a particular paper just because of its "brightness," a comparable test is carried out using a series of neutral, or "gray," papers. This is the equivalent of watching a color film on a black-and-white television. For this test, leaves are cut from a series of papers (nine in the Munsell series), ranging from black to white, and the insects are either counted or trapped when they visit the various papers. Neutral papers used in this way help to show whether the insects select various plants, or plant parts, on account of their relative brightness.

The contributions of Ilse and Vaidya (1956) and Vaidya (1969) illustrate the basic principles in studies of this type. Vaidya (1969) also showed that although results are obtained easily, their biological significance is not always easily understood. If the insect normally responds to two stimuli either simultaneously or in quick succession, then presenting it with only one type of stimulus may induce a displaced response that could be difficult to interpret. For example, in 1969, Vaidya presented a screened citrus plant outside a test cage and its odoractivated butterflies to select an apparently "inappropriate" colored paper. Whether separating the two sets of stimuli in this way can be expected to produce results comparable to normal host-plant finding is uncertain.

(c) *Colored lights.* Using colored papers can be criticized because such papers reflect more ultraviolet light than gray papers and can therefore be detected more easily by insects sensitive to such wavelengths (see Wigglesworth, 1965). The test insects should therefore also be subjected to restricted spectral bands of light.

The importance of correct timing is illustrated by work that showed that whiteflies were attracted by wavebands in the blue/ultraviolet at about 400 nm and to wavebands in the yellows at about 550 nm. When a mixture of the two wavebands was reflected from a surface, however, it was not attractive (Mound, 1962). The explanation offered was that ultraviolet radiation was a stimulus for migratory behavior and yellow was a stimulus for "vegetative" behavior during host-plant finding. To prove this, Coombe (1981) designed an apparatus to present the two wavelengths separately. It consisted of a cylindrical flight chamber open at one end and closed at the other end with a vertical, ground-glass screen, one half of the screen being back-illuminated with 400 nm light, the other with 550 nm light. The lights were screened so that the rest of the chamber was unlit (Fig. 3a). Whiteflies were introduced into the dark end of the cylinder and the numbers that arrived on the two colors were recorded, either by photography or by being trapped on sticky material applied to the ground-glass screen. The two recording techniques produced different results, illustrating the dangers in making simple interpretations of complex orientation behavior (Coombe, 1982). The sticky method measured attraction towards the lights while the insects were in flight, accumulating them over time as they landed. The photographic method recorded the instantaneous result of a variety of responses, e.g., attraction during flight, landing, and also takeoff, so that at any one instance, the numbers of insects recorded were many fewer than had responded during the experiment.

A simple apparatus for studying the effects of various wavelengths can be constructed from a high intensity microscope lamp, and a combination of Kodak"Wratten" and appropriate neutral-density filters (Fig. 3b, Tanton, 1977). This apparatus produces a range of wavebands, each about 40 nm wide, spanning the visible spectrum (440–680 nm). Cells, 1 cm deep, containing 5% aqueous copper chloride solution placed between the lamps and the filters remove the infrared radiation. The apparatus is arranged so that one, two, or three lamps can be used simultaneously. Lamps 1 and 2 project a 2-cm diameter semicircle of light onto the bottom edge of the two ends of the test chamber. Using this design, it is possible to study the response of insect larvae to wavelength, to brightness, and to the angles that the light sources subtend on the light-sensitive cells of the larvae.

(d) Real plants. Experiments on color discrimination can also be carried out on real plants or leaves. The main approaches are to work on either (1) insects that do not respond to plant odors or (2) host plants that have differently colored leaves, but the same odor. This is a rare combination. To ensure that only visual cues are available, freshly picked leaves are

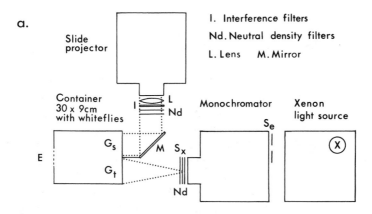

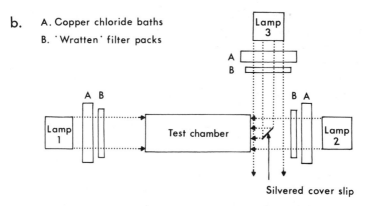

Figure 3. (a) Diagram of the apparatus used to measure the settling response of whiteflies. X, 150W xenon lamp; Se, entrance slit, and Sx, exit slit for light. G_s, ground glass screen illuminated by the standard 400 nm light; G_t, ground glass screen illuminated by test light. E, entrance hole for whiteflies and for taking photographic records. (From Coombe 1981.) (b) Plan of arrangement of lamps, filters and test chamber used by Tanton. (From Tanton 1977.)

usually placed between thin sheets of glass, the edges of which are sealed with silicone grease. The leaves are then tested with light transmitted through or reflected from their surfaces (Vaishampayan et al., 1975). Internal dyes or external sheets of colored cellulose acetate can also be superimposed to show how changes in hue affect the attractiveness of leaves. The possible effects of the glass in filtering the ultraviolet portion of the stimulus are frequently not considered.

(e) Plant mimics. Anyone who works on host-plant selection soon realizes that it is difficult to produce a "standard" (control) plant for ex-

periments conducted at different times of the year, unless relatively so-phisticated growth rooms or cabinets are readily available. In the absence of such facilities, most authors describe the age of their test plants by some arbitrary calendar time scale, such as days or weeks after sowing. Such descriptions are generally not very informative, as, for example, a "6-week old radish plant" could be a plant either in the first true-leaf stage or one ready for harvest depending on the time of year and the place where it is grown. Obviously, unless the "control" plants can be stand-ardized from one experiment to the next, it becomes impossible to make meaningful comparisons between experiments. To overcome this difficulty, Prokopy et al., (1983) advocated the use of artificial plant mimics con-structed from painted cardboard and wire. By mixing paint pigments in specific proportions, "leaves" of artificial plants can be painted to have reflectance patterns similar to those of real leaves (Prokopy et al., 1983). It is now possible, therefore, not only to make comparisons between real and artificial plants but more importantly to obtain reproducibility between experiments. A further advantage of using plant "mimics" of this type is that desaturation can be varied more or less at will to show how other pigments interact with the reflectance of preferred wavelengths and how this subsequently affects insect choice. At present, the technique is re-stricted to reflectance studies because the pigments required to produce the desired "leaf" color are opaque.

In other instances, authors have used simple objects, such as pieces of glass tubing, to provide their test insects with basic physical stimuli and have made no attempt to mimic the real plant. Despite the simplicity of such "surrogate plants" they can be remarkably effective. The elegant series of experiments by Harris and Miller (1982) together with the detailed description (Miller and Harris, 1985) of the reasoning behind each ex-periment are highly recommended reading for anyone considering using surrogate plants in their experiments. With their simple surrogates, the above authors showed clearly that, in addition to color, the onion fly *Delia antiqua* also selects host plants on their shape and on the angle the "plant" subtends to the ground. Interestingly, even though the onion fly must select plants containing specific sulfur chemicals if its larvae are to survive, the data suggested parity between the physical and chemical stimuli during host-plant selection, not supremacy of the chemical stimuli (Miller and Harris, 1985).

(f) *Subtended angle.* Meyer (1975) suggested that the angle subtended by the host-plant or crop at the insect's eye determines whether or not the insect will respond to visual stimuli from the host crop. This effect operated over 35 cm under laboratory conditions, but has not been con-firmed over the larger distances involved in color attraction in the field. Nevertheless, this approach is worthy of further consideration.

B. Bioassay of Olfactory Stimuli

(a) Olfactometers. Before attempting to copy any olfactometer it is advisable to scrutinize carefully the information that has been collected using the apparatus. Too often olfactometers are described with insufficient data to reveal the consistency of the results and this must cast doubt on their effectiveness. Furthermore, even olfactometers that are both expensive and difficult to construct may still fail to provide reliable or appropriate information. There is therefore no substitute for seeing an apparatus being used by someone adept at handling it. There are often relatively insignificant or unwitting steps in techniques that are not mentioned in a written account, but that determine the difference between success and failure. In all cases, simple systems are preferably if they provide the required results.

(b) Y-tubes. The simple Y-tube olfactometer provides the insects with one of two choices. Odor is passed through one arm of the Y and clean air through the other (McIndoo, 1926). The proportion of insects entering the arm containing the odor is used as a measure of the chemical's attractiveness. A modern olfactometer using just a single-tube glass unit and developed for the bioassay of pheromones (Giannakakis and Fletcher, 1978) seems suitable for assaying host-plant volatile chemicals. The units are small and sealed and hence a large number of replicates can be tested in a short period of time. This is a particularly important consideration for insects that have a strong diurnal periodicity and are active for only short periods during each day.

(c) Small arenas. These are commonly used to screen chemicals as larval attractants (Matsumoto, 1968; Matsumoto and Thorsteinson, 1970), since they require no expensive equipment and yet are capable of showing whether the test chemical is an attractant or repellent. Although such systems may be designed specifically to test larval attractants, larvae are often used just for convenience, since larval attractants also usually "attract" (see Section VIII) adults, a behavioral association recognized to be common to many insects (Thorsteinson, 1960). Bioassays with larvae of anthomyid flies can involve no more than placing a 5 mm diameter absorbent disc containing the test chemical, and another with a similar quantity of solvent, at opposite sides of a 9-cm Petri dish and releasing freshly emerged larvae at the center of the dish (Matsumoto, 1968).The percentage of larvae responding to the test chemical, within as little as 5 min, can provide a measure of its attractiveness or repellency.

Tests using larvae require little space or time and so are easy to replicate. In contrast, tests involving adult insects generally require the adults to be maintained in large cages at something approaching field conditions before they will respond appropriately. As larvae in small closed chambers use mainly chemotaxis to move up the odor gradient diffusing from the source (Jander, 1963), the larvae can be tested in the dark. This is a further advantage, since it removes the competition from visual stimuli and aids consistency of response. However, the main advantage of using larvae is their rapid response since, unless they find a host-plant shortly after hatching, many phytophagous larvae die of starvation or desiccation. Hence newly hatched larvae are generally highly responsive to any directional cues given off by host-plant materials (Finch, 1980).

(d) Coupled electrophysiology/gas chromatography. This bioassay is usually concerned with utilizing the antennal sensory organs to detect chemicals and involves linking the outlet from a gas chromatograph (GC) to the insect's sense organs, prepared for the recording of electroantennograms (EAGs) (see Frazier and Hanson, Chapter 7).

A simple method for preparing antennae for EAG recordings, together with several recent improvements, is described by Roelofs (1977). The improvements eliminate the need for a Faraday cage and enable the equipment to be easily set up in laboratories or classrooms. As described, it is possible within a few minutes to remove the insect's head or antenna, position it in wax, and record responses to olfactory chemicals. After the antenna is prepared, the volatile constituents of the test material are separated on a chromatographic column and the the effluent is split between a flame ionization, or other, detector and the antenna. Simultaneous recordings of the GC signal and the insect's responses enable the retention times of the effluent components to be correlated with the EAG response (Moorhouse et al., 1969). Since electroantennographic sensitivity often does not translate directly into overt behavioral responses in the intact insect, additional bioassays must be carried out with intact insects to determine whether an EAG-active fraction elicits attraction, repellency, or neither.

Misunderstandings and misinterpretations of using the EAG technique to identify the stimulatory volatile chemicals usually arise when investigators use only part, rather than the whole, of the technique (Roelofs, 1977). If the host-plant chemical system is analogous to that of pheromones, then the complete study involves: (1) determining the GC retention times of the plant chemicals; (2) obtaining EAG profiles of closely related naturally occurring chemicals; (3) obtaining as many chemical analyses as possible of the GC-collected EAG active fractions: (4) synthesizing the proposed chemical; (5) comparing the GC retention times and EAG activity of the synthetic chemical with those of the natural chemical; and (6) testing the chemical in behavioral tests in the laboratory and field. Conclusions regarding the attractive components should be made only when the data from all steps are in agreement (Roelofs, 1977). Methods for using this system with only limited numbers of insects are described by Greenway et al. (1977).

(e) Wind tunnels. Responses of insects to host-plant odors have been studies in a variety of "wind tunnels" designed for either walking (Haskell et al., 1962; Visser, 1976) or flying insects (Kennedy and Moorhouse, 1969; Hawkes and Coaker, 1979).

Basically each tunnel consists of three parts: 1) an *effuser* or entrance zone in which the air is accelerated and the flow is "smoothed," 2) a *working section* usually under 1 m in diameter, and 3) a *diffuser* or exhaust zone where the air is deccelerated (Vogel, 1969). The fan can be placed either in the effuser (Fig. 4a) or in the diffuser (Fig. 4b). Often there is a large area referred to as a *settling chamber* joined to the working section by a *contraction* in which the cross-sectional area of the tunnel is reduced. Contracting the air in this way produces a more laminar flow through the working section. Wind tunnels can be classified according to whether the air in the tunnel is recirculated (closed-circuit tunnel) or whether it is taken from the experimental room and exhausted back into the same room (open-circuit tunnels). In open-circuit tunnels, particularly in those utilizing odors (e.g., Hawkes and Coaker, 1979), fresh air is usually taken from outside the room and the odor and exhaust air are expelled from the room through an exit port remote from the inlet port. Alternatively, tunnels can be classified by whether the working section is enclosed (closed throat) or open to the atmosphere of the room (open jet).

The advantages and disadvantages of each system are described in detail by Vogel (1969). It is notable that to be useful a tunnel does not have to be sophisticated and expensive. Very useful tunnels can be built at little cost from materials readily available at most research laboratories (Miller and Roelofs, 1978). Before starting construction, it is advisable whenever possible to visit a laboratory where this type of work is done routinely and observe a tunnel during an actual bioassay. Invariably this will involve watching a preconditioned moth fly upwind to a pheromone source. To an ecologist it is a never-to-be forgotten euphoric experience of what can be achieved. The outstanding problem, however, is to design a tunnel suitable for bioassaying host-plant odors. To date, there has been a pronounced lack of success in this endeavor. This could be because inappropriate species have been used in the bioassays, since most studies have concentracted on a narrow range of pest Diptera, and species within this order seem generally to be unresponsive in tunnels. The lack of success could also be the result of our incorrectly assuming that since both pheromones and plant odors are volatile, insects should respond similarly to both. If the majority of pheromones are attractants and the host-plant odors arrestants,then the present system of trying to stimulate insects to fly upwind to sources of host-plant odor may be totally inappropriate. Perhaps the appropriate bioassay for host-plant odors is to fly insects either in wind tunnels or on flight mills and then test specifically for odors that terminate, rather than initiate, insect flight.

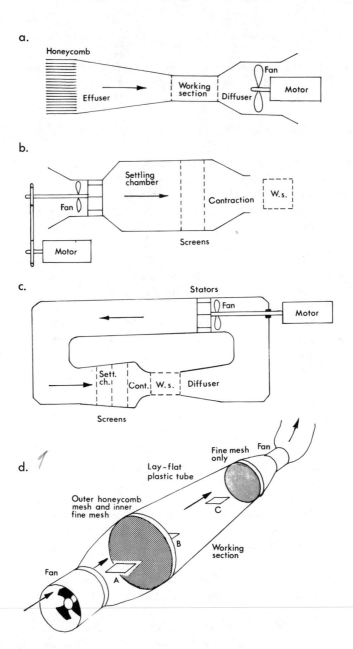

Figure 4. Examples of several types of wind tunnel. (a) Open curcuit, closed throat. (b) Open circuit, open jet. (c) Closed circuit, open jet, or closed throat. (With permission from Vogel S. In: Experiments in Physiology and Biochemistry, Vol 2, p. 297. Kerkut GA [ed]. Copyright: Academic Press Inc. (London) Ltd.) (d) Diagram of wind tunnel of Jones et al. (From Jones OT, Lomer RA, Howse PE. Physiol Entomol **6**:176, 1981. Reproduced with permission of Blackwell Scientific Publications.) Arrows indicate the air flow; A and B, the positions for placing the test chemicals; and C, the point of release of insects.

A major problem in using wind tunnels to assess the effectiveness of volatile chemicals is that the chemicals may impregnate or coat the fabric of the tunnel. To overcome this, Jones et al. (1981) constructed small wind tunnels from inflated "lay flat" plastic-film tubing (Fig. 4d), which was cheap enough to be discarded when contaminated. In larger wind tunnels, the working area is often made of glass as this can be readily decontaminated using appropriate solvents. In addition, the patterned moving floors (Kellogg and Wright, 1962; David, 1982) that produce the optomotor stimuli are now situated outside the working area and so do not have to be replaced regularly because of contamination (Sanders et al., 1981).

Smoke or a clearly-visible chemical vapor is normally used to reveal the path of the odor plume through the tunnel. Separate point sources of concentrated ammonium hydroxide and hydrochloric acid placed at the chemical release point will form a smoke of ammonium chloride particles. Alternatively, Miller and Roelofs (1978) soaked coarse muslin (mesh size 5 mm) in 0.02% crystal violet and then suspended it in the tunnel. Where the vapors from a concentrated hydrochloric acid source passed through the muslin, the crystal violet turned yellow.

The air speed in the working area is usually measured with a hot-wire anemometer. If the air-flow characteristics of the tunnel are to be mapped, the anemometer is coupled to some form of storage oscilloscope and/or tape recorder to integrate the signal.

To stimulate certain insects to perform part of their normal behavioral repertoire in the relatively close confines of a tunnel, the physical conditions within the tunnel must simulate closely those of the natural environment of the insect. Apart from a suitable air flow, light, temperature, and humidity may also need to be closely controlled. The paper by Visser (1976) illustrates clearly the equipment required just to produce a laminar flow of air over beetles walking on a flat surface in a large tunnel; but, whether airflow in the working section of the tunnel needs to be exceedingly laminar as in Visser's study is doubtful. With other insects, e.g., locusts, the physical conditions may be less critical (Kennedy and Moorhouse, 1969). For those investigating behavioral mechanisms, choosing the "right"insect is the most important decision. In this aspect of host-plant finding, data collection is generally a lengthy process and involves relatively few insects, even with "co-operative" species. Studies on uncooperative species in which only a small portion of individuals may respond in a "characteristic" way at some unpredictable time during the test period are especially difficult.

Farkas and Shorey, Kennedy et al., and Vogel (for references, see Kuenen and Baker, 1982) should be consulted for detailed descriptions of the apparatus used and the mechanisms involved in insect flight in an odor plume.

VIII. Trapping Devices for Measuring Attraction Stimulated Chemically and/or Visually

The chemical and visual stimuli that attract insects to their host plants have been incorporated into a wide range of insect traps. A specific color, chemical, or a combination of the two can be used to attract insects to traps (Finch and Skinner, 1974; Prokopy, 1977). Most current studies of trapping concentrate on improving the efficiency of a specific trap rather than on obtaining information about host-plant finding. Unlike many laboratory studies involving host-plant stimuli (see Miller and Strickler, 1984), improvements in trap effectiveness in the field are generally based on the previously most effective trap and hence improvements tend to be significant rather than sensational. When traps are based on both visual and olfactory stimuli, it is tempting to think that they are of use in directly controlling the pest (e.g., Hamilton et al., 1971). In most instances, however, they are not sufficiently effective. Their main use is for monitoring pest populations.

Finding attractants from among the natural colors and chemicals of plants, or from a large range of laboratory materials, comprises the initial step. However, the number of insects caught in a trap is a function of the attractiveness of the trap, the activity and responsiveness of the insects, and their population density (Southwood, 1966). Consequently as trap type, shape, size, alignment, height, and positon all affect trap efficiency, it is difficult to optimize a trap, or trapping system, to suit a range of circumstances or pest species.

8.1. Types of Trap

Sticky traps (Fig. 5a, b, and f), water traps (Fig. 5d and e), and inverted cone traps (Fig. 5c) are used most commonly for catching insects in the field. Sticky traps are made by covering objects, such as colored boards, with a thin coating of an adhesive so that insects that settle on or are blown onto the surface are retained (Fig. 5a, b, and f). A range of suitable adhesive resins and greases is described by Southwood (1966), the choice depending on the insect being studied. A highly sticky adhesive may clog the trap with insects, whereas a less sticky one will let some insects escape and, if chosen carefully, may selectively retain those species being studied. The removal of insects from sticky traps is extremely time-consuming and the adhesives can also make the insects difficult to identify unless they are first cleaned in a solvent. Nevertheless, sticky traps are used widely because the final aim of most trap development is to kill insects, their condition being immaterial. In contrast to sticky traps, water traps (Fig. 5d and e) retain the insects in good condition and, for many studies on host-plant finding, they are preferable to sticky traps, particularly if the

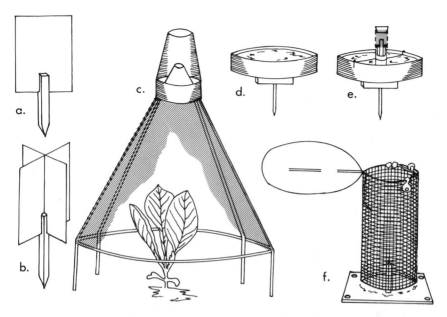

Figure 5. Examples of insect traps. (a, b) Sticky traps. (c) Inverted cone trap. (After Eckenrode and Chapman 1971.); (d, e) Water traps without and with attractants. (From Finch, Freuler and Stadler 1980.) (f) Directional sticky trap. (From Dindonis and Miller 1980.)

presence of water vapor itself is also important in attracting insects to the trap (Finch and Skinner, 1982a).

8.2. Visual Traps

Under field conditions, it is easy to test the attraction of insects to various visual stimuli by painting boards of different shapes and sizes with colors differing in either hue, saturation, or intensity. Attempts can also be made to simulate the shape of an object. Hence Prokopy (1968) has shown that apple maggot flies *(Rhagoletis pomonella)* are attracted to red and black spheres because they simulate apples. The spheres are thought to be preferred not on true color discrimination but because they contrast strongly with the background (Prokopy, 1968). Nevertheless, because the absence of transmitted light is crucial in fruit finding, it does not necessarily follow that transmitted light is unimportant in studies on host-plant finding based on leaf color.

A problem is relying solely on color, e.g., yellow, as the attractant for an insect trap is that color attraction generally lacks specificity and hence a wide range of different species are attracted to the trap. It would be interesting to study: (1) whether a particular species does clearly prefer a green characteristic of its favored host plant to other greens of slightly differing spectral composition, and (2) for how long it maintains this preference in the presence of competing visual stimuli. If, however, there is no preference and the test insect responds similarly to a range of greens, this may indicate that the particular test species is less discriminating with respect to visual stimuli.

8.3. Visual/Chemical Traps

In contrast to traps relying solely on color as the attractant, neutral-colored traps relying solely on host-plant volatile chemicals usually catch few insects (Finch and Skinner, 1982a). Nevertheless they catch more insects than similar traps lacking the attractant, confirming that insects in the field are capable of finding an odor source by chemical cues alone. It is more usual in studies on chemical attractants to paint a trap an attractive color and then try to see how much more attractive it can be made by the addition of a chemical attractant. As the added chemical is required to improve the performance of an already visually effective trap, most results tend to be significant rather than sensational. If water vapor itself attracts a particular insect, a colored watertrap is already making use of combined visual/olfactory stimuli.

Care is required in the interpretation of trapping data, as the terminology used commonly by the practical rather than the theoretical ecologist can be misleading. For example, it a trap catches more insects following the addition of a host-plant chemical, then in practical terms the added chemical is described as an "attractant." In reality, however, the chemical may only have arrested the insects in the vicinity of the trap, more insects being caught as a result of subsequent trivial movements. It is easy in such situations to understand why Kennedy has consistently (1939–1978) stressed the importance of describing accurately the response of an insect to each particular stimulus, rather than just using an all embracing term, such as "attractant," to describe the overall result, culminating from a series of different behaviors. If host-plant chemicals released from traps arrest certain insects within the effective zone of the trap then, instead of trying to improve trap efficiency against such species by searching for more effective mixtures of host-plant odors, searching for a chemical to increase the pest insects' trivial activity near to the trap could be more rewarding.

It is easy to test under field conditions the relative "attractancy" of chemicals extracted from plants or obtained commercially. Unfortunately, most volatile plant chemicals are either not yet known or they are not yet available commercially. Consequently, many of the available chemicals tested in the field may at best be only related, though not necessarily closely, to the naturally occurring chemicals. When traps are compared in areas that exclude the host crop, they rarely catch the desired insect species. When comparing the efficiency of various "attractants" in the field, one can move traps daily to ensure that each trap occupies every position at least once (see Southwood, 1966). It is better, however, to check whether this is so before undertaking any extensive screening program, because interchanging traps is time-consuming and is not always necessary (see Finch and Skinner, 1982a). Although screening may reveal one chemical that is more attractive than the others, host-plant finding is likely to be based on the insects' response to mixtures of volatiles, rather than to individual, host-plant chemicals (Finch, 1980; Miller and Strickler, 1984).

When an attractant or mixture of attractants has been found, the most difficult part of designing an efficient trap is to determine how best to release the attractive chemicals. Although numerous dispensers have been developed for the daily release of nano- or microliter quantities of pheromones from traps, they are not suitable for the release of host-plant chemicals which at present are highly effective only when released in larger (ml) amounts/day (Finch, 1980). By releasing such quantities, the chemicals may become repellent in the immediate vicinity of the trap. Experiments designed to determine whether this occurs were described by McGovern and Beroza (1970).

Sometimes mixtures of chemicals with differing volatilities can be released in the desired proportions from a dispenser consisting of only a wick dipping into a reservoir of the mixture (McGovern and Beroza, 1970; Beroza, 1970). The mixtures tested to date, however, have generally involved groups of closely related chemicals. Other mixtures may not behave similarly in this respect. It is possible that a final blend will include volatile, secondary plant chemicals to provide the specific attractants, plant alcohols for their "green plant" odor (Visser and Ave, 1978), and water to provide sufficient humidity. In water traps, the alcohols could be placed directly into the water of the trap. In addition to acting as part of the attractive mixture, such alcohols would also help to preserve the trapped insects and, by killing bacteria, keep the water in the trap clean.

8.4. Trap Variables

Many factors have to be considered during the development of an effective trap. The necessary experiments may be simple, but there are often complications.

A. Trap Shape

A cylindrical trap is subjected to fewer effects from eddies and catches more insects than a flat card or lamina of the same surface area (Heathcote, 1957). However, a cylindrical trap is also equally visible from all directions whereas a flat card is not. Hence, square tubular traps have to be included in the experiments to determine that trap shape is the important variable.

B. Trap Size

Trap size has often been studied (e.g., Heathcote, 1957; Prokopy, 1977). Although more insects are usually caught as trap size increases, catch is not proportional to size, the smallest traps catching the largest number of insects per unit area of trap. Hence the trap size chosen is usually that most easy to handle or to service.

C. Trap Alignment

Trap alignment is not important in round, omnidirectional traps but is important when the attractant is released from a covered trap and emitted only from its downwind edge. Such traps catch more insects when they are aligned parallel to the prevailing wind than at right angles to the wind (Lewis and Macaulay, 1976).

D. Trap Height

Trap height is invariably important and often the optimum heights for catching males and females differ (Finch and Skinner, 1974). Whenever possible, therefore, insects caught during any trap development program should be sexed. When insects are associated with herbaceous plants the effects of trap height are easily assessed, since most of these pests stay close to the ground. It is not so easy with pests such as bark beetles infesting tall trees, since traps for such insects can be effective 20 m above ground (Birch et al., 1981). With other species optimum trap height may depend on the attractant. For example, plant volatile chemicals are most effective in trapping Japanese beetles when the trap is 56–84 cm above ground level whereas the trap becomes most effective at heights of 28–56 cm (Ladd and Klein, 1982) when the female sex attractant (Japonilure) is added.

E. Trap Background

Trap background can also influence the numbers of insects entering a trap. Traps placed against the green background of the crop canopy are most effective for the bollweevil (Johnson et al., 1982) whereas traps within the crop and against a soil background are most effective for female cabbage root flies (Finch and Skinner, 1974). The effectiveness of a trap is also influenced by the presence of host plants to arrest the insects in the immediate vicinity. For example, approximately 20% of the bark beetles released in an isolated area containing elm trees were recaptured, whereas only 1% were recaptured in a comparable area without trees (Birch et al., 1982).

F. Trap Position

When traps are used to reduce an insect population, they are usually placed within the crop (Finch and Skinner, 1975) and referred to as "infield" traps. An advantage of placing traps in the crop rather than alongside hedgerows or other natural boundaries is that they do not become filled with nontarget species. Blow flies are particularly troublesome when traps using yellow as the visual stimulus are placed alongside flowering hedgerow plants. In the case of weevils the catch probability can be increased 10–20% by adjusting infield traps away from a uniform spacing and placing more of them around the borders of the crop where the weevils are most likely to land first (Witz et al., 1981). Trap deployment suitable for fields of various shapes and sizes is worthy of further study, since there is often a maximum size of field above which traps are no longer required in the center.

G. Directional Traps

Directional traps have been developed that catch slow-flying insects using an interception barrier to make them fly upwards while heavier fast-flying insects are stunned on impact with the barrier and drop downwards (Wilkening et al., 1981). Other traps utilize wind vanes (Fig. 5f) so that the same sector is always in the downwind quarter (Dindonis and Miller, 1980). It is then possible to show the preferred direction of movement towards the trap, albeit from close range.

H. Timing Traps

Timing traps, based on a rotating sticky drum with only one segment exposed during a set time interval, are useful for studying not only the diurnal flight activity of insects to visual and olfactory stimuli but also the effects of weather on the responses to such stimuli (e.g., Riedl and Croft, 1981).

8.5. Other Considerations

The devices used for trapping a particular species are often compared directly with one another. Unfortunately, most are compared after, rather than during, development. Hence comparisons are not usually made on the numbers of insects caught per unit area of trap. However, when this is done, water traps are generally found to be more effective then sticky traps. Although this generalization seems universally true, there is still a lack of basic information comparing water traps and similar traps coated, only up to the water level, with adhesive. Whereas water traps must remain horizontal, placing equivalent sticky traps both vertically and horizontally should indicated the relative effectiveness of the two trapping systems.

As most trapping systems are based on visual and olfactory stimuli many authors have tried to "assess" the relative contributions of the two types of stimuli in host-plant finding. This is frequently inappropriate, however, since in the field the insect normally encounters the two types of stimuli simultaneously. Furthermore, the two stimuli often act complementarily and in such quick succession that they are for all intents and purposes inseparable.

When an insect is trapped, the trap effectively ends any further search for a host plant. Under field conditions, insects that land on a plant are not forced to stay and hence those that do not find the plant acceptable take off again. Of a bark beetle population that landed on experimental trees in Texas, 22% eventually entered the tree, 43% flew away, 32% dropped off, and 2% were eaten (Bunt et al., 1980). Furthermore, there is no guarantee that traps sample a population at random. With certain species the trapped insects may be the weak individuals most likely to perish in any case.

A further difficulty in the development of an effective trap is that of ensuring that the trap is developed under appropriate conditions. A water trap releasing ANCS will catch 20 times as many cabbage root flies as a standard water trap during September in England, when the cabbage root fly populations are high and the early morning and evenings are balmy (Finch and Skinner, 1983). In April, however, when crops are most at risk to this fly, it is unusual to obtain more than a fourfold difference in trap catch. The efficiency of a trap must therefore be determined at the time of year when it will be used, since it may not remain equally effective throughout the year.

IX. Distances Insects Respond to Host-Plant Stimuli in the Field

9.1. Response to Visual and Chemical Trap Stimuli

The simplest way to determine the distance that insects will respond to a trap, often called the traps' "effective zone", is to assume that the insect population is distributed evenly throughout an area of host plants. The test area is then divided into equal plots and different numbers of traps are placed in each, usually following a logarithmic progression (Finch and Skinner, 1974), to allow the traps to be evenly spaced in each plot. Each plot is then replicated several times to conform to a "randomized block" or "Latin square" design. When the number of flies caught per trap ceases to increase with trap density, half the distance between the traps is assumed to be the radius of the effective zone of the trap.

A second method is to vary trap spacing between the rows and columns of several rectangular blocks (Lin and Morse, 1975), so that plots of different size and shape form a compact layout (Fig. 6). Trap spacings of *n, 2n,* and *4n* etc. can then be chosen to fit the experimental site, so that each block of six traps at one spacing is separated from adjacent treatments by guard traps. An advantage of this design is that, if the insects are dispersed evenly through the host patch, each test trap can be used as a replicate. The plots do not then need to be replicated in space. It is still necessary, however, to replicate in time. By using the design in Fig. 6, 76 guard and 54 treatment traps will provide nine different combinations of trap spacing. The effective zone of the trap is determined as above.

In a third method, Wolf et al., (1971) proposed releasing marked coded insects at 24 of the intersections of a 5 × 5 grid and trapping the released insects at a central point. The numbers of coded insects captured from each release point are used to compute the probability of capture from that point. Lines are then drawn between points of equal probability to define the effective zone of the trap.

All of these methods are suitable for measuring the effective zones of

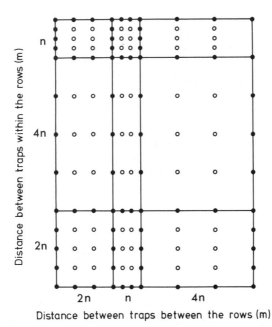

Figure 6. Layout of trap spacing experiment used to determine the effective zone of a trap for *Delia radicum,* showing the relative positions of the guard (●) and treatment (○) traps. (From Finch and Skinner 1974.)

visual, chemical, and visual/chemical traps. The results, however, are only an indirect estimate of the distance of response as trivial movements will inevitably bring some insects close to the traps from time to time during the trapping period. The true distance of response could therefore be closer than the estimated distance.

9.2. Responses Solely to Chemical Plant Stimuli

The methods described in this section apply only to olfactory stimuli.

A. Indirect Method (Laboratory)

Because of the difficulties in devising direct methods for determining the distances over which insects respond to olfactory stimuli, several authors (e.g., Wright, 1958; Bossert and Wilson, 1963) developed indirect mathematical methods, based on Sutton's (1953) gas diffusion models, to determine the distances over which certain odorous chemicals could be attractive. Such calculations were based on the fact that within the odor gradient emanating from plants there is a zone where the odor molecules are at, or above, the concentration required to produce a behavioral response in receptive insects. When insects move into such a zone they respond accordingly and, as a consequence, Bossert and Wilson (1963) referred to the zones as "active spaces". In still air, the active spaces around plants are considered to be hemispherical whereas in moving air calculations based on Sutton's model produce a semiellipsoidal shape whose long axis is aligned downwind. Such a shape was generated mainly because entomologists based their estimates on the average number of odor molecules passing though an area in a fixed (3 min) period and because they expressed the amount of odor likely to influence responding insect in terms of odor concentration (e.g., $g\ cm^{-3}$). As insects respond to odors within milliseconds rather than minutes and odor perception is more closely related to odor flux ($g\ cm^{-2}\ s^{-1}$) than odor concentration, Sutton's model has now been superceded (Mankin et al., 1980; Cardé, 1984).

Currently the active space is believed to be an instantaneous odor plume (Murlis and Jones, 1981; David et al., 1982), consisting of a narrow swath of disjunct odor filaments that meanders downwind along with the large-scale turbulent eddies (Elkington and Cardé, 1984). To verify that the odor plume meanders in this way, Murlis and Jones (1981) showed that when ionized air, used to simulate odor molecules, was released at a constant rate from a point source, the ions arrived at points up to 15 m downwind not as a constant signal but as a series of bursts of variable strength and duration. Hence, much of the time stationary insects close to and downwind of an odor source experience little or no odor interspersed by bursts of high concentration. The change in chemical flux along the plume depends on the changing values of the dispersion coefficients which are now derived, not from theoretical estimates (e.g., Sutton, 1953), but from empirical tests using visible tracer gases. Experiments to determine the mechanisms flying insects use to locate the source once within an odor plume are currently being debated. One hypothesis is that within the "snaking" odor plume the wind is aligned on the source of any windborne odor wherever that odor can be detected (David et al., 1982). Consequently, by flying upwind on meeting an appropriate odor, and holding station against the wind with or without casting across it on losing the odor, is a mechanism that seems likely to be highly adaptive. The alternative hypothesis of flying along the plume, which David et al., (1982) consider unlikely, as it frequently takes the insects in "wrong" directions, is supported by Cardé (1984) because it allows continued progress towards the source at those times when the wind suddenly ceases. Doubtless, some insects will use one or other of the two extreme alternative mechanisms whereas, if past experience is anything to go by, many will probably use both mechanisms to some degree, or some compromise of the two.

Furthermore, the premise that insects do not go in "wrong" directions (David et al., 1982) needs to be considered carefully. For example, recent attempts to produce a budget, based on a physiological time scale (day-degrees above 6°C), of the field behavior of the cabbage root fly have been unable to account for a period of approximately 1 week duration in the spring between the end of the pre-oviposition period and the predicted state of oviposition. This is not just a consequence of spring temperatures being sufficiently high for egg maturation but too low for ovipositioin, since a corresponding delay was also recorded at the time of the mid summer generation. An unaccounted delay of this duration suggests that even "specialist feeders" (Rhoades and Cates, 1976) have sufficient time available during their sequence of host-plant finding to make "wrong" turns.

B. Direct Methods (Field)

Little is known about how far insects are "attracted" by odors of host plants in the field. None of the great distances cited in the literature has been supported by experimental proof (Shorey and Gaston, 1967; Finch, 1978).

(a) Caged insects. At present, the maximum distance for insect orientation to naturally produced host-plant volatile chemicals in the field appears to be the 15 m recorded for the cabbage root fly by Hawkes (1974) and confirmed recently by Holmes and Finch (1983). This distance of orientation was obtained, however, by noting the response of flies in cages placed at varying distances away from a small plot of cabbage plants, a source more attractive than a single plant but not necessarily comparable to a large cabbage field. By comparing how many insects moved to the upwind side of the cage when it was placed at a known distance upwind or downwind of the crop it was possible to determine the proportion of the population moving upwind in response to the odor from the crop. Clearly if the sun was shining from the same direction as the prevailing wind, flies in the upwind and downwind cages would not be subjected to the same visual images. Flies looking with the sun (downwind) would "see" illuminated plants whereas those looking into the sun would see only silhouettes. This experiment must therefore be repeated on several days in different wind directions or only on those days, or parts of the day, when the sun is more or less at right angles to the direction of the prevailing wind.

(b) Coned insects. A method developed for pheromone studies by Baker and Roelofs, (1981) may also be suitable for studies on insect attraction to host-plant volatile chemicals. It involves placing three test insects in a 8 cm high × 10 cm basal diameter screen cone completely open at the base and then, holding the cone base forwards, walking upwind towards an odor source. While walking, the observer notes the behavior of the insects through the back of the cone. Maximum distance of response is recorded by dropping a marker flag whenever any of the three test insects flies out of the cone. As with most methods, once the test insects take off it is generally not possible to follow them for more than a few meters upwind.

(c) Marked insects. The distances over which free-flying insects are influened by volatile host-plant chemicals can be determined by releasing marked insects at the center of two circular areas, surrounding one area by a band of host crop and the other by a band of a nonhost crop. In this way the test insects are subjected to either a host crop or nonhost crop odor. A circular experimental design enables the dispersal of the insects to be determined in differing wind directions and intensities, without interrupting the flow of plant odor passing over the central release point.

For example, by comparing the distribution of recaptures in a nonhost-plant circle with that in a host-plant circle, it was clear that female cabbage root flies flew upwind without prior stimulation from either a host crop or from a large number of traps each releasing up to 3 ml/day of the attractant allylisothicyanate (Finch and Skinner, 1982b). Systems of this type could be useful for many species. It would be a major breakthough if an insect shown to disperse at random in the presence of odors from a nonhost crop was influenced to fly upwind by odors from a host crop at a distance of 100 m. Although it would be acceptable to surround the release field by four similar shaped fields of host crop, or four fields of the same nonhost crop, a circular design was used to avoid hedgerows and other barriers, since these greatly influence the relative effectiveness of neighboring insect traps (Lewis and Stephenson, 1966). Mardia (1972) is recommended for a description of the methods currently used to analyze directional data.

X. Conclusions

Present proposals for utilizing host-plant visual and chemical stimuli in crop-protection systems are either still untried or highly speculative. To develop them further will require collaboration between entomologists, physicists, meterologists, and biochemists whose combined efforts will have to be intensified considerably before these complex systems can be better understood and used to greater advantage in agricultural systems.

Acknowledgments. I thank my colleagues Mr. G.A. Wheatley and Drs. G.H. Freeman, A.R. Thompson, and P.R. Ellis for their constructive comments on this chapter.

References

Albert SA, Kennedy MK, Cardé RT (1981) Pheromone-mediated anemotactic flights and mating behavior of the sciarid fly *Bradysia impatiens*. Environ Entomol **10**:10–21.

Baker PS, Cooter RJ, Change PM, Hashim HB (1980) The flight capabilities of tropical field populations of the Brown planthopper, *Nilaparvata lugens* (Stal) (Hemiptera: Delphacidae). Bull Entomol Res **70**:589–600.

Baker TC, Roelofs WL (1981) Initiation and termination of oriental fruit moth male response to pheromone concentrations in the field. Environ Entomol **10**:211–218.

Beroza M (1970) Current usage and some recent developments with insect attractants and repellents in the U.S.D.A. In: Chemicals Controlling Insect Behaviour. Beroza M (ed), Academic Press, New York and London, pp 145–163.

Best RL. Beegles CC, Owens JC, Ortiz M (1981) Population density, dispersion and dispersal estimates for *Scarites substriatus, Pterostichus chalcites,* and *Harpalus pennsylvanicus* (Carabidae) in an Iowa cornfield. Environ Entomol **10**:847–856.

Birch MC, Miller JC, Paine TD (1982) Evaluation of two attempts to trap defined populations of *Scolytus multistriatus*. J. Chem Ecol **8**:125–136.

Birch MC, Paine TD, Miller JC (1981) Effectiveness of pheromone mass-trapping of the smaller European elm bark beetle. Calif Agric **35**:6–7.

Borden JH, Bennett RB (1969) A continuously recording flight mill for investigating the effect of volatile substrances on the flight of tethered insects. J Econ Entomol **62**:782–785.

Bossert WH, Wilson EO (1963) The analysis of olfactory communication among animals. J Theor Biol **5**:443–469.

Brewer FD (1982) Development of food utilization of tobacco budworm hybrids fed artificial diet containing oil soluble dyes. Georgia Entomol Soc **17**:248–254.

Bunt WD, Coster JE, Johnson PC (1980) Behavior of the southern pine beetle on the bark of host trees during mass attack. Ann Entomol Soc Am 73:647–652.

Cane JH, Jonsson T (1982) Field method for sampling chemicals released by active insects. J Chem Ecol 8:15–21.

Cardé RT (1984) Chemo-orientation in flying insects. In: Chemical Ecology of Insects. Bell WJ, Cardé RT (eds), Chapman and Hall, London and New York, pp 111–124.

Chapman MG (1982) Experimental analysis of the pattern of tethered flight in the Queesland fruit fly, *Dacus tryoni*. Physiol Entomol 7:143–150.

Clarke JL III, Rowley WA, Christiansen S, Jacobson DW (1984) Microcomputer-based monitoring and data acquisition system for a mosquito flight mill. Ann Entomol Soc Am 77:119–122.

Cole RA (1980) The use of porous polymers for the collection of plant volatiles. J. Sci Food Agric 31:1242–1249.

Cole RA, Finch S (1978) Studies on pest biology: cabbage root fly. Vapours from intact plants. Rep Natn Veg Res Stn 1977: p. 84.

Coombe PE (1981) Wavelength specific behaviour of the whitefly *Trialeurodes vaporariorum* (Homoptera: Aleyrodidae). J Comp Physiol 144:83–90.

Coombe PE (1982) Visual behavior of the greenhouse whitefly, *Trialeurodes vaporariorum*. Physiol Entomol 7:243–251.

Cullis NA, Hargrove JW (1972) An automatic device for the study of tethered flight in insects. Bull Entomol Res 61:533–537.

David CT (1982) Competition between fixed and moving stripes in the control of orientation by flying *Drosophila*. Physiol Entomol 7:151–156.

David CT, Kennedy JS, Ludlow AR, Perry JN, Wall C (1982) A reappraisal of insect flight towards a distant point source of wind-borne odor. J Chem Ecol 8:1207–1215.

Dethier VG (1947) Chemical Insect Attractants and Repellents. Lewis, London.

Dethier VG, Browne LB, Smith CN (1960) The designation of chemicals in terms of the responses they elicit from insects. J Econ Entomol 53:134–136.

Dindonis LL, Miller JR (1980) Host-finding behavior of onion flies, *Hylemya antiqua*. Environ Entomol 9:769–772.

Eckenrode CJ, Chapman RK (1971) Observations on cabbage maggot activity under field conditions. Ann Entomol Soc Am 64:1226–1230.

Eickwort GC, Ginsberg HS (1980) Foraging and mating behaviour in Apoidea. Annu Rev Entomol 25:421–446.

Elkington JS, Cardé RT (1984) Odour dispersion. In: Chemical Ecology of Insects. Bell WJ, Cardé RT (eds), Chapman and Hall, London and New York, pp 73–91.

Finch S (1977) Effect of secondary plant substances on host-plant selection by the cabbage root fly. In: Comportement des Insects et Milieu Trophique; Colloques Internationaux du C.N.R.S. No. 265. Paris, C.N.R.S., pp 251–268.

Finch S (1978) Volatile plant chemicals and their effects on host plant finding by the cabbage root fly. Proceedings of the 4th Insect/Host Plant Symposium. Entomol Exp Appl 24:350–359.

Finch S (1980) Chemical attraction of plant-feeding insects to plants. Appl Biol 5:67–143.

Finch S, Freuler J, Stadler E (1980) Trapping Hylemya brassicae adults. S.R.O.P./ W.P.R.S. Bull 1980/III/1: Int. Control "Brassica Corps": 11–17.

Finch S, Skinner G (1974) Some factors affecting the efficiency of watertraps for capturing cabbage root flies. Ann Appl Biol 77:213–226.

Finch S, Skinner G (1975) An improved method of marking cabbage root flies. Ann Appl Biol 79:243–246.

Finch S, Skinner G (1982a) Trapping cabbage root flies in traps baited with plant extracts and with natural and synthetic isothiocyanates. Entomol Exp Appl 31:133–139.

Finch S, Skinner G (1982b) Upwind flight by the cabbage root fly, *Delia radicum*. Physiol Entomol 7:387–399.

Finch S, Skinner G (1983) Studies of pest biology. Cabbage root fly. Monitoring fly numbers. Rep Natn Veg Res Stn 1982: p.33.

Gatehouse AG, Lewis CT (1973) Host Location behaviour of *Stomoxys calcitrans*. Entomol Exp Appl 16:275–290.

Giannakakis A, Fletcher BS (1978) An improved bioassay technique for the sex pheromone of male *Dacus tryoni* (Diptera: Tephritidae). Can Entomol 110:125–129.

Greenway AR, Lewis T, Mudd A, Scott GC, Wall C (1977) Some chemical and entomological problems in the investigation and use of behaviour-controlling chemicals. In: Crop Protection Agents—Their Biological Evaluation. Proceeding of the International Conference on the Evaluation of Biological Activity. McFarlane NR (ed), Academic Press, London, pp 167–185.

Hamilton DW, Schwartz PH, Townshend BG, Jester CW (1971). Traps reduce an isolated infestation of Japanese beetle. J Econ Entomol 64:150–153.

Harris MO, Miller JR (1982) Synergism of visual and chemical stimuli in the oviposition behaviour of *Delia antiqua*. Proceedings of the 5th International Symposium of Insect-Plant Relationships, Wageningen 1982. Pudoc, Wageningen, pp 117–122.

Haskell PT, Paskin MWJ, Moorhouse JE (1962) Laboratory observations on factors affecting the movement of hoppers of the desert locust. J Insect Physiol 8:53–78.

Hawkes C (1972) The estimation of the dispersal rate of the adult cabbage root fly *(Erioischia brassicae* [Bouché]) in the presence of a brassica crop. J Appl Ecol 9:617–632.

Hawkes C (1974) Dispersal of adult cabbage root fly *(Erioischia brassicae* [Bouché]) in relation to a brassica crop. J Appl Ecol 11:88–93.

Hawkes C, Coaker TH (1979) Factors affecting the behavioural responses of the adult cabbage root fly, *Delia brassicae,* to host plant odour. Entomol Exp Appl 25:45–58.

Heathcote GD (1957) The optimum size of sticky aphid traps. Plant Pathol 6:104–107.

Hedin PA, Thompson AC, Gueldner RC (1975) A survey of the air space volatiles of the cotton plant. Phytochemistry 14:2088–2090.

Henson WR (1962) Laboratory studies on the adult behaviour of *Conopthorus coniperda* (Coleoptera: Scolytidae). III Flight. Ann Entomol Soc Am 55:524–530.

Hocking B (1953) On the intrinsic range and speed of flight of insects. Trans R
    Entomol Soc Lond **104**:223–347.
Holmes A, Finch S (1983) Studies of pest biology. Cabbage root fly. Bioassay of
    volatile chemical. Rep Natn Veg Res Stn 1982: p. 34.
Ilse D, Vaidya VG (1956) Spontaneous feeding response to colours in *Papilio
    demoleus* L. Proc. Indian Acad Sci **XLII**:23–31.
Jander R (1963) Insect orientation. Annu Rev Entomol **8**:95–114.
Johnson CG (1969) Migration and Dispersal of Insects by Flight. Methuen, London.
Johnson WL, Mitchell EB, Huddleston PM, Cross WM, Heiser RF (1982) Boll
    weevil capture efficiency: position and density of traps and Grandlure dosage.
    J Econ Entomol **75**:446–448.
Jones OT, Lomer RA, Howse PE (1981) Response of male Mediterranean fruit
    flies, *Ceratitis capitata,* to trimedlure in a wind tunnel of novel design. Physiol
    Entomol **6**:175–181.
Kellogg FE, Wright RH (1962) The olfactory guidance of flying insects. III. A
    technique for observing and recording flight paths. Can Entomol **94**:486–493.
Kennedy JS (1939) The visual response of flying mosquitoes. Proc Zool Soc Lond
    (Series A) **109**:211–242.
Kennedy JS (1961) A turning point in the study of insect migration. Nature (Lon-
    don) **189**:785–91.
Kennedy JS (1977) Olfactory responses to distant plants and other sources. In:
    Chemical Control of Insect Behaviour: Theory and Application. Shorey HH,
    McKelvey JI, Jr (eds), John Wiley and Sons, New York and London, pp 67–91.
Kennedy JS (1978) The concepts of olfactory 'arrestment' and 'attraction'. Physiol
    Entomol **3**:77–89.
Kennedy JS, Moorhouse JE (1969) Laboratory observations on locust response
    to wind-borne grass odour. Entomol Exp Appl **12**:473–486.
Kennedy JS, Thomas AAG (1974) Behaviour of some low-flying aphids in wind.
    Ann Appl Biol **76**:143–159.
Kepner RE, Van Straten S, Weurman C (1969) Freeze concentraction of volatile
    components in dilute aqueous solutions. J Agric Food Chem **17**:1123–1127.
Koehler O (1950) Die Analyse der Taxisanteile instinktartigen Verhaltens. Symp
    Soc Exp Biol **4**:361–384.
Kuenen LPS, Baker TC (1982) Optomotor regulation of ground velocity in moths
    during flight to sex pheromone at different heights. Physiol Entomol **7**:193–
    202.
Ladd TL Jr, Klein MG (1982) Japanese beetle (Coleopters: Scarabaeidae): effect
    of trap height on captures. J Econ Entomol **75**:746–747.
Langford MJ (1974) Advanced Photography. Focal Press, London and New York,
    pp 435.
Leggett JE, Roach SH (1981) Boll weevil: movement into an uninfested area and
    detection with Grandlure-baited traps. Environ Entomol **10**:995–998.
Lewis T, Macaulay EDM (1976) Design and elevation of sex-attractant traps for
    pea moth, *Cydia nigricana* (Steph.) and the effect of plume shape on catches.
    Ecol Entomol **1**:175–187.
Lewis T, Stephenson JW (1966) The permeability of artificial windbreaks and the
    distribution of flying insects in the leeward sheltered zone. Ann Appl Biol
    58:355–363.

Likens ST, Nikerson GB (1964) Detection of certain hop oil constituents in brewing produces. Proc Am Soc Brew Chem pp 5–13.

Lin C-S, Morse PM (1975) A compact design for spacing experiments. Biometrics **31**:661–671.

MacLeod AJ (1973) Instrumental Methods of Food Analysis. Elek Science, Londen, 802 pp.

Mankin RW, Vick KW, Mayer MS, Coffelt JA, Callahan PS (1980) Models for dispersal of vapour in open and confined spaces. J Chem Ecol **6**:929–950.

Mardia KV (1972) Statistics of Directional Data. Academic Press, London.

Matsumoto Y (1968) Volatile organic sulphur compounds as insect attractants with special reference to host selection. In: Control of Insect Behaviour by Natural Products. Wood DL, Silverstein RM, Nakajima M (eds), Academic Press, New York and London, pp 133–160.

Matsumoto Y, Thorsteinson AJ (1970) Olfactory response of larvae of the onion maggot, *Hylemya antiqua* Meigen (Diptera: Anthomyiidae) to organic sulphur compounds. Appl Entomol Zool **3**:107–111.

McGovern TP, Beroza M (1970) Volatility and compositional changes of Japanese beetle attractant mixtures and means of dispensing sufficient vapour having a constant composition. J Econ Entomol **63**:1475–1479.

McIndoo NE (1926) An insect olfactometer. J Econ Entomol **19**:545–571.

Meyer JR (1975) Effective range and species specificity of host recognition in adult alfalfa weevils, *Hypera postica*. Annu Entomol Soc Am **68**:1–3.

Miller JR, Harris MO (1985) Viewing behavior-modifying chemicals in the context of behavior: lessons from the onion fly. In: Semiochemistry: Flavors and Pheromones. Acress TE, Soderlund DM (eds), Walter de Gruyter, Berlin and New York.

Miller JR, Roelofs WL (1978) Sustained flight wind tunnel for measuring insect responses to wind-borne sex pheromones. J Chem Ecol **4**:187–198.

Miller JR, Strickler KL (1984) Finding and accepting host plants. In: Chemical Ecology of Insects. Bell WJ, Cardé RT (eds). Chapman and Hall, London and New York, pp 127–157.

Miller TA (1979) Insect Neurophysiological Techniques. Springer-Verlag, New York.

Moorhouse JE, Yeadon R, Beevor PS, Nesbitt BF (1969) Methods for use in studies of insect chemical communication. Nature (London) **223**:1174–1175.

Mound LA (1962) Studies on the olfaction and colour sensitivity of *Bemisia tabaci* (Genn.) (Homoptera, Aleyrodidae). Entomol Exp Appl **5**:99–104.

Murlis J, Jones CD (1981) Fine-scale structure of odour plumes in relation to insect orientation to distant pheromone and other attractant sources. Physiol Entomol **6**:71–86.

Murray KE, Stanley G (1968) Glass separation of flavour volatiles by liquid chromatography on silica gel at 1°. J Chromatogr **34**:174–179.

Prokopy RJ (1968) Visual responses of apple maggot flies, *Rhagoletis pomonella* Diptera: Tephritidae): orchard studies. Entomol Exp Appl **11**:403–422.

Prokopy RJ (1977) Attraction of *Rhagoletis* flies to red spheres of different sizes. Can Entomol **109**:593–596.

Prokopy RJ, Collier RH, Finch S (1983) Leaf color: a character used by cabbage root flies to distinguish among host plants. Science **221**:190–191.

Rhoades DF, Cates RG (1976) Towards a general theory of plant herbivore chemistry. In: Biochemical Interaction Between Plants and Insects. Recent Adv Phytochem 10:168–213.

Riedl H, Croft BA (1981) A timing trap for segregating catches of insects by discrete intervals. Can Entomol 133:765–768.

Roelofs WL (1977) The scope and limitations of the electroantennogram technique in identifying pheromone components. In: Crop Protection Agents—Their Biological Evaluation. Proceedings of the International Conference on the Evaluation of Biological Activity. McFarlane NR (ed), Academic Press, London, pp 147–165.

Sanders CJ, Lucuik GS, Fletcher RM (1981) Responses of male spruce budworm (Lepidoptera: Tortricidae) to different concentrations of sex pheromone as measured in a sustained-flight wind tunnel. Can Entomol 112:943–948.

Shorey HH (1970) Sex pheromones of Lepidoptera. In: Control of Insect Behaviour by Natural Products. Wood DL, Silverstein RM, Nakajima M (eds). Academic Press, New York and London, pp 249–284.

Shorey HH (1973) Behavioural responses to insect pheromones. Annu Rev Entomol 18:349–380.

Shorey HH, Gaston LK (1967) Pheromones. In: Pest Control: Biological, Physical and Selected Chemicals Methods. Kilgore WW, Doutt RL (eds), Academic Press, New York and London, pp 241–265.

Sirikulvadhana S, Jennings WG, Vogel G (1975) Collection of flower aroma concentractes for gas chromatographic analysis. Flavour Ind 6 (2):126–128.

Southwood TRE (1962) Migration of terrestrial arthropods in relation to habitat. Biol Rev 37:121–214.

Southwood TRE (1966) Ecological methods: with Particular Reference to the Study of Insect Populations, 3rd edit, Methuen, London.

Sutton OG (1953) Micrometeorology. McGraw-Hill, London.

Tanton MT (1977) Behavioural response to different wagelength bands and attraction to colour by larvae of the mustard beetle *Phaedon cochleariae*.Entomol Exp Appl 22:35–42.

Teranishi R, Hornstein I, Issenberg P, Wick EL (1971) Flavour Research—Principles and Techniques. Marcel Dekker, New York.

Thorpe WH (1951) The definition of terms used in animal behaviour studies. Bull Anim Behav 9:34–40.

Thorsteinson AJ (1960) Host selection in phytophagous insects. Annu Rev Entomol 5:193–218.

Vaidya VC (1969) Investigations on the role of visual stimuli in the egg-laying and resting behaviour of *Papilio demoleus* L. (Papilionidae Lepidoptera). Anim Behav 17:350–355.

Vaishampayan SM, Waldbauer GP, Kogan M (1975) Visual and olfactory response in orientation to plants by the greenhouse whitefly, *Trialeurodes vaporariorum* (Homoptera: Aleyrodidae). Entomol Exp Appl 18:412–422.

Visser JH (1976) The design of a low-speed wind tunnel as an instrument for the study of olfactory orientation in the Colorado beetle *(Leptinotarsa decemlineata)*. Entomol Exp Appl 20:275–288.

Visser JH, Avé DA (1978) General green leaf volatiles in the olfactory orientation of the colorado beetle, *Leptinotarsa decemlineata*. Proceedings of the 4th Insect/Host Plant Symposium. Entomol Exp Appl 24:738–749.

Visser JH, Van Straten S, Maarse H (1979) Isolation and identification of volatiles in the foliage of potato, *Solanum tuberosum,* a host plant of the Colorado beetle, *Leptinotarsa decemlineata.* J Chem Ecol **5**:11–23.

Vogel S (1969) Low speed wind tunnels for biological investigations. In: Experiments in Physiology and Biochemistry. Kerkut GA (ed), Academic Press,London, pp 295–325.

Ward JP, Baker PS (1982) The tethered flight performance of a laboratory population of *Triatoma infestans* (Klug) (Hemiptera: Reduviidae). Bull Entomol Res **72**:17–28.

Weis-Fogh T (1956) Biology and physics of locust flight. II. Flight performance of desert locust *(Schistocerca gregaria).* Philos Trans R Soc (Series B) **239**:459–510.

Weurman C (1969) Isolation and concentration of volatiles in food odor research. J Agric Food Chem **17**:370–384.

Wigglesworth VB (1965). The Principles of Insect Physiology, 6th edit; Methuen, London.

Wilkening AJ, Foltz JL, Atkinson TH, Connor MD (1981) An omnidirectional flight trap for ascending and descending insects. Can Entomol **113**:453–455.

Williams AA, May HV, Tucknott OG (1978) Observations on the use of porous polymers for collecting volatiles from synthetic mixtures reminiscent of fermented ciders. J Sci Food Agric **29**:1042–1054.

Witz JA, Hartstack AW, LLoyd EP, Mitchell EB (1981) Effects of infield trap spacing on potential catch of adult boll weevils entering cotton: a computer simulation. Environ Entomol **10**:454–457.

Wolf WW, Kishaba AN, Toba HH (1971) Proposed method for determining density of traps required to reduce an insect population. J Econ Entomol **4**:872–877.

Wright RH (1958) The olfactory guidance of flying insects. Can Entomol **90**:81–89.

Chapter 3

The Definition and Measurement of Oviposition Preference in Plant-Feeding Insects

Michael C. Singer[1]

I. Introduction

For many plant-feeding insects, the selection of an ovipositon site is a critical stage in their choice of host. This is especially true when the newly hatched offspring are not capable of searching for additional hosts until they have fed on the individual chosen by their mother (but see Wint, 1983; Futuyma et al., 1984). An understanding of the physiological, behavioral, ecological, or evolutionary interactions between these insects and their hosts requires precise representation of the events and factors influencing oviposition. Ecologists and entomologists have tried to simplify the complex processes involved by using shorthand terms to represent abstract concepts that summerize the most pertinent variables. For example, insect preference and host acceptability summarize the array of factors determining the immediate outcome of a particular encounter between insect and plant, whereas larval performance, host suitability, and plant response to attack describe the subsequent consequences of oviposition. These and other concepts have been used in a variety of ways by different authors, many of whom have failed to define and distinguish among the terms they have used.

Although much of this chapter concerns the development of techniques for assessing preference, its first goal is to seek clarity by distinguishing among the terms applied by past workers, discarding synonymies and ambiguities where possible, and defining those terms that represent concepts

[1]Department of Zoology, University of Texas at Austin, Austin, Texas 78712.

useful in furthering our understanding of insect ovipositon. These concepts fall into two classes, one describing the way an individual insect interacts with a variable population (or community) of plants, and the other summarizing the manner in which an individual plant is treated by a variable population (or community) of insects. The two classes are often confused, but it is critical to a meaningful understanding of plant–insect relationships that they be distinguished when it is possible to do so. The following sections attempt this task.

II. Definitions of Terms

2.1. Properties of a Single Insect Faced with a Variable Population of Hosts: Preference and Performance

The influences on oviposition of sensory physiology, functional morphology, and development operate through modifications of behavior, so it is appropriate that behavioral abstractions are used to describe them. Insects have accordingly been described as choosing, selecting, preferring, and discriminating among their host plants. These behavioral terms are often used as synonyms to describe an individual insect making active choices among a range of plants. The term "preference" embodies these notions, and is defined by deviations from random behavior (Mackay and Singer, 1982), where "random" refers to the situation in which variation in insect behavior (ovipositing on some plants and not on others) is not related to variation among the plants encountered. If, for example, an insect's likelihood of ovipositing on plants that it encountered varied with its motivational state rather than with plant phenotype, its behavior would be described as random in this sense. In contrast, if an encounter with plant A is more likely to result in oviposition than an encounter with plant B by the same insect at the same time, the insect prefers A over B. Preference can be measured as the relative likelihoods of accepting plants that are encountered (for definitions of "encounter" and "acceptance" see below). If the likelihoods are equal, the insect shows no preference. Preference can be used to refer to the behavior of an insect towards parts of an individual plant, particular plants of the same species, or towards a *specified* set of plant species.

It is important to stress that oviposition preference tells us nothing about the fates of eggs or larvae on the host. Larvae on different plants may survive with different probabilities and grow at different rates to produce adults with different fertilities and fecundities. They may have different efficiencies of host digestion and conversion to insect biomass. All of these traits can be measured by well-established techniques (Scriber and Slansky, 1981) and are often described as different aspects of insect "performance". Just as preference describes the set of probabilities that particular plants will be accepted, we need a term to describe the set of probabilities that particular performances would result from feeding on specific plants. In this paper I use "intrinsic performance" to describe the performance *characteristics* of the insect in the same way that "preference" describes its host acceptance characteristics. Strictly speaking, experimental results provide *measurements* of acceptance and performance, from which *estimates* of preference and intrinsic performance can be made.

This section has introduced preference and intrinsic performance as characteristics of an insect faced with a variety of plants. The next section considers the description of properties of a plant interacting with a number of different insects.

2.2. Plant Characters: Acceptability and Suitability

I shall describe the properties of whole plants rather than plant parts, and refer the reader to Whitham (1983) for discussion of the importance of within-plant variation. I use the term "acceptability" to describe the likelihood that a plant will be accepted if it is encountered and "suitability" to encompass the various aspects of host quality that affect insect performance. Acceptability may refer to the likelihood that a single plant will be accepted by a single insect, or to the pattern of acceptance probabilities shown by specified members of an insect population or community. The acceptability of a plant to an insect population or community summarizes for the plant in question the effects of all of the preferences of all the insects that encounter it.

There is some overlap between the meanings of acceptability and preference. If an insect prefers plant A over plant B, then A is more acceptable than B to that insect. Even when we can separate plant and insect characters conceptually, we may not be able to do so in practice from available experimental data. However, acceptability and preference are *not* synonyms because variation among plants occurs independently of variation among insects, and we need both terms if we are to describe both sets of variables. If we offer different plants to members of an insect clone, we can measure variation in plant acceptability to this insect genotype. Conversely, offering one or two plant clones to different insects generates data that can be used to show variation in insect acceptance (of one clone) or preference (for two). Note also that a plant that is acceptable to most insects may still be rejected by a individual with unusual preference.

I have discussed only those situations in which a single insect interacts Cwith many plants or a single plant interacts with many insects, but in the field insect populations and communities interact with plant populations and communities. In this circumstance it is possible to investigate variation among plants by testing many plants with few insects, and then to examine variation among insects by offering few plants to many insects. The collective properties of groups of individuals should be described as completely as possible. For example, the "mean preference" of a group of insects has much less predictive value than the proportions of insects with specific preferences.

Plant suitability relates to insect intrinsic performance in the same way that plant acceptability relates to preference. The suitability of an individual plant is defined with respect to the survival and growth of an individual, population, or community of insects using that plant. Although both the suitability of a plant and the intrinsic performance of an insect are estimated by measuring insect growth and survival, suitability and intrinsic performance are not synonomous, for the same reasons that preference and acceptability are not synonyms.

Excellent summaries of the large (mainly agricultural) literature on variation of all these characters are given by Gould (1983) and Rausher (1983), and other papers in the same volume. However, it seems that most of the genetic work referred to by Gould concerns plant suitability and acceptability to feeding stages. Effects of plant *genotype* on acceptability to ovipositing adults have apparently received scant attention, although many effects of plant *phenotype* have been documented. A few of these phenotypic characters known to be important are: phenology (e.g., Courtney, 1982, Messina, 1984), color (e.g., Harris and Miller, 1982), leaf shape (Rausher, 1978; Rausher and Papaj, 1983b), size (Prokopy and Owens, 1978), nutritional status (Myers, 1985), stress (Lewis, 1984), and secondary chemistry (Feeny et al., 1983). The distributions and identities of neighboring plants are also influential (Stanton, 1983; Root and Kareiva, 1984).

Once we have separated the concepts of preference, intrinsic performance, acceptability, and suitability, and defined the level (individual, population, or community) at which they are being applied, we can ask more detailed questions. For example, if an insect species shows host use patterns that differ between areas with similar vegetational compositions, we can ask whether intersite variation of plants or of insects is responsible. If we ascertain that the variation is principally among the insects, we can then ask whether it is variation of preference, intrinsic performance, or both.

2.3. Encounter, Acceptance, and Host Use

I follow Miller and Strickler (1984) in using "encounter" to refer to the "point where sensory information from the resource is first received." This is equivalent to defining encounter as the first perception of the plant by the insect; perception of chemical stimuli can be investigated with neurophysiological techniques (Frazier and Hanson, Chapter 10, *this volume*). Perception of visual stimuli is usually more difficult to ascertain: we often cannot tell whether an insect that passes close by a plant without overt response does so because it fails to perceive the plant or because it does perceive it but rejects it on the basis of its appearance or odor. However, occasionally we can come close to answering this question. For example, Roitberg (personal communication) has described tephritid fruit flies hopping from leaf to leaf in a tree in their search for fruits in which to oviposit. From time to time a fly will conspicuously turn its head towards an individual fruit, then either fly to the fruit or hop to another vantage point. Here we have good circumstantial evidence that head rotation indicates visual perception, and that it is possible in this exceptional case to tell which visual encounters are followed by acceptance (flying to the fruit) and which by rejection. It is still possible, however, that some classes of fruit are perceived but do not elicit head rotation. As behavioral knowledge and expertise in physiology of visual perception increase, these questions will become less problematical.

"Acceptance" is used to denote a positive response to an encounter. each oviposition may be preceded by a series of encounters and acceptances as the insect proceeds through a sequence of behaviors (Kennedy, 1965). For example, my own study organism, the butterfly *Euphydryas editha,* first encounters plants by perceiving them visually and accepts one by alighting on it. After alighting (a second level of encounter) it tastes the plant with its foretarsi and responds to chemical stimuli. Rejection of plant chemistry causes the insect to resume flight (eventually), while acceptance does not produce oviposition, but further examination; the butterfly extrudes its ovipositor and searches for appropriate tactile stimuli. At this stage the plant may still be rejected on the basis of its physical characteristics; if, however, it is accepted, actual oviposition is the result.

The sequence of decisions is important. Within any single stage of the ovipositional sequence, acceptance or rejection may depend on the relative magnitudes of positive and negative influences acting synchronously (see "rolling fulcrum" model of Miller and Strickler, 1984). However, interactions between stages are quite different. For example, the response of *E. editha* to physical stimuli requires prior acceptance of plant taste. If the response to chemical stimuli is rejection, the physical characteristics of the plant are simply not investigated. Because these stimuli act in sequence in this way, a plant could be accepted on the basis of its chemistry and rejected because of its physical traits but *not* vice versa. This insect could be described as possessing several different oviposition preferences that operate in sequence. Each time a decision to accept or reject is made, a preference could be measured. In practice, a readily quantifiable subdivision is that between prealighting and postalighting preferences (Rausher et al., 1981; Stanton, 1982; Mackay, 1985). Insects sometimes show prealighting preference for one plant and post alighting preference for another (Stanton, 1982; Papaj and Rausher, 1983), in which case the measurement of these separate components is more informative than an overall preference measure. Rausher (1983a) expects these differences between pre- and postalighting preferences, predicting that the former should maximize rates of host encounter whereas the latter should rank hosts in order of suitability.

"Host use" and "host utilization" are used to describe patterns of insect attack on different hosts in the field. For example, Singer (1983) estimated the proportion of eggs allocated by a single insect population to four host species, and described the result as a "pattern of host use." I believe that these terms are not controversial.

2.4. Apparency, Findability, and Perceptual Ability

A plant that is likely to be perceived has been called "apparent" by Feeny (1975). Variation of plant apparency should interact with variation of insect perceptual ability in a manner analogous to the relationships already described between acceptability and preference or between suitability and intrinsic performance. If chemical perception can be measured (see above), then so can chemical apparency; However, the technical difficulties of ascertaining when a plant is visually perceived make it difficult to distinguish in practice between plants that are visually unapparent to an insect and those that *are* apparent, but are rejected after initial visual encounter (see definition of "encounter" above). With few exceptions, notably the work of Roitberg described above, any attempt to measure visual apparency is likely to be, in effect, an estimate of some combination of apparency and prealighting preference (cf. Rausher et al., 1981; Wiklund, 1984). Perhaps for this reason, Miller and Strickler (1984) offer "findability," the likelihood that a plant will be found, as a substitute for apparency. The most important difference between the two concepts stems from the the definitions of "find" and "perceive." If encounter is followed by "establishing and maintaining proximity with" the plant, the plant has been "found" (Miller and Strickler, 1984). Hence findability is more easy to measure than apparency, since the results of high findability (proximity and in many cases physical contact between insect and plant) are readily quantified by the behavioral end results. However, findability contains variable components of both apparency and acceptance, and will therefore vary among hosts of equal apparency but differing acceptabilities. For example, a nonhost may be apparent, but not likely to be found according to Miller and Strickler's definition, as many insects are able to reject nonhosts from a distance after cursory examination. In summary, apparency is conceptually a property of a plant that is, in practice, difficult to measure without being confounded by insect preference. Findability is conceptually a property of the insect–plant interaction but is easier to measure. In the light of the bias of this paper towards the conceptual separation of insect and plant properties wherever possible, I recommend the use of apparency and insect perceptual ability as interacting plant and insect characters, and I hope that future advances will render them more easily separated in practice then thay can be at present.

III. Comparison with Other Usage

3.1. Insect Characters

"Preference" provides a plethora of problems. It has been extended to include not only the various factors that generate an instantaneous insect distribution over diverse plants, but even rates of population increase over unstated numbers of generations (Zehnder and Trumble, 1984). The distributions of adult insects, and hence those of their eggs, are affected by the disposition of their own resources, such as nectar (Murphy et al., 1984). Larval distributions result from combinations of the effects of initial adult distribution and density, oviposition preference and plant acceptabilities, egg survival patterns, larval movement, preference and performance, and host suitabilities, including differential predation on different hosts. These distributions should not be described simply as "preference." as is frequently done (e.g., Cates, 1981; Zehnder and Trumble, 1984). Such usage obscures the process that generates the distribution.

The concept of preference tends to be used differently by ecologists than by agricultural biologists, as described below.

A. Ecological Usage

Ecologists tend to use consumption-based rather than behavioral concepts of preference. It is often both defined and measured in terms of nonrandom patterns of feeding. If different plant species are not consumed in proportion to their relative abundances, the insects are said to show host preference (Hassell and Southwood, 1978; Crawley, 1984).This is not equivalent to the behavioral definition used in this paper. The distinction is important because the two approaches generate quite different interpretations of the same experimental results. I shall give three examples.

(a).   Kareiva (1982) worked with flea beetles feeding on and moving among collards of two different qualities,planted in small, homogeneous patches.The difference in beetle density between high-quality (lush) and low-quality (stunted) hosts was much less when patches were 11 m apart then then they were closer (3 m). If, at some greater distance still, the densities on the two patch types became equal, the insects would have no preference according to a consumption-based definition, but I would

argue that the plant spacing prevented them from expressing their preference.

(b). A second example also concerns insect responses to plant spatial patterns, recently summarized in detail by Stanton (1983). One frequently observed effect is an increased risk of insect asttack on isolated plants compared to those that grow in denser patches (Jones, 1977; Mackay and Singer, 1982; Root and Kareiva, 1984). An insect that initiates its search at a randomly chosen point, lays a single egg on the first host that it encounters, and then starts its next search at another random point will, if presented with equal numbers of plants arranged singly and in groups, lay more eggs on the isolated plants (see also Hamilton, 1971 for a general description of the phenomenon). Even though every plant that is encountered receives an egg, there is an "apparent preference" for isolated plants (Mackay and Singer, 1982). What happens if we present such an insect with two host species that differ in their degree of aggregation? It will lay more eggs on the plant species that is less clumped. This insect would show no preference among the two host species by the behavioral definition of preference used here, but preference would be detected if a consumption-based definition were used.

(c). Finally, suppose that an insect that avoids conspecific eggs (Rothschild and Schoonhoven, 1977; Williams and Gilbert, 1981; Prokopy, 1981) is presented with several plants of different quality. If egg avoidance is strong, it may lay only one egg on each plant, starting with the most preferred individual plant. When this process is complete, one might infer from the egg distribution that the insect had no preference among the plants. Repeated trials would be needed to demonstrate that the sequence in which eggs were laid was not random, and preference could be detected. The final egg distribution alone would not be a very good predictor of the behavior of the same insect in other circumstances, for example in the presence of a superfluity of hosts (see Thomas, 1984).

These scenarios convince me that oviposition preference should not be defined in terms of egg distribution patterns, nor should preference be measured from these patterns in the field. However, oviposition preference *can* be measured by recording the distributions of eggs laid by captive insects under carefully controlled experimental conditions.

B. Agricultural Usage

Gould (1983) and others describe host resistance to attack as being based on some combination of antibiosis and nonpreference. "Antibiosis" is the inverse of suitability, while "nonpreference" refers to low acceptability in the terminology developed here. I disagree with this use of "nonpreference" on these grounds: (1) It is not clearly a trait of either plant or insect—since it is described as a source of plant resistance it is perhaps viewed more as a plant character. (2) It could be confused with lack of preference (i.e., random behavior). (3) It is used whether or not comparison with other plants is implied. A plant can be described as having an acceptability in some absolute sense, but if it is said to be "not preferred," this should imply that it is less acceptable than particular other plants.

Because ecologists, applied and evolutionary biologists have borrowed the term "preference" from behavioral biology and used it in such different ways it seems that confusion over the meaning of this term is likely to persist. I hope that authors who use the term will define it for their own use or refer to a published definition. No one should assume that the meaning of "preference" is obvious. For example, Miller and Strickler (1984) offer an alternative definition to that used here. They argue that "preference" should be restricted to "cases where weighing of alternatives is known to occur." This seems to define preference as an active choice among alternatives, all of which are perceived. In most circumstances it is reasonable to assume that accepted plants are perceived, though insects very highly motivated to oviposit may do so at random without regard to external stimuli. A problem arises in determining whether plants presented as alternatives are all perceived, or whether some are rarely or never accepted because they are rarely or never perceived. In practice, Miller and Strickler recommend use of "preference" when plants are presented simultaneously or when presentation is sequential but the insect can be shown to remember previous encounters and use this information in its decision-making. In other cases, they recommend "acceptability" in lieu of "preference." Here there is a second difficulty: even if presentation is simultaneous, encounter in an experimental arena is likely to be sequential, so the difference between simultaneous and sequential presentation may often not be real from the insect's point of view. Singer (1982, 1983 and *this chapter*) has used "preference" to describe the results of sequential trials performed on insects that show no evidence of learning. This violates Miller and Strickler's criteria.

3.2. Plant Characters

Miller and Strickler (1984) use "acceptability" as a property of an insect rather than a plant trait, referring to the tendency of an insect to accept a plant. However, they and others write that plants are "accepted" by insects. It is consistent with this second usage that "acceptability" should be a plant character, as used here.[2]

There is no conflict between the terminology developed here and the

---

[2]Editor's note: J. Miller agrees that acceptability is a property of plants, while acceptance is a property of insects; thus he now recommends host-plant "acceptance" in lieu of host-plant "preference" when it is unclear that active choice is being exercised.

use of plant "resistance" to indicate effects of insect attack on yield or on plant fitness (i.e., Gould, 1983). If we know a plant's acceptability to an insect and suitability for it, we still have insufficient information to calculate its resistance. For example, two plants that are attacked by equal numbers of identical insects may differ in resistance because of differences in insect consumption rates (which may vary even if plant suitability is constant) or differences in plant tolerance to insect feeding.

IV. The Uses and Importance of Preference

Applied entomologists, ecologists, evolutionary and behavioral biologists have all worked on insect preferences.The practical value of understanding the fine details of host finding and host preference is realized when these behaviors are used to evaluate resistance in plant breeding programs (Gould, 1983) or manipulated to divert insects to inappropriate hosts or traps (Prokopy and Owens, 1978; Shapiro and Masuda, 1980). Ecologists ask how host-specialist insects differ from generalists (Scriber and Feeny, 1979; Futuyma and Wasserman, 1981) and how the diet of an insect is affected by its interactions with host plants, predators, parasitoids, and competitors (Gilbert, 1979; Price et al., 1980).

Behaviorally inclined biologists (Jones, 1977; Courtney, 1982; Kareiva, 1982; Stanton and Cook, 1983 or 1984, Root and Kareiva, 1984; Mackay, 1985) evaluate the efficiency of host-search behavior. Using the general approach of foraging theory, one can ask how an insect decides to spend a particular length of time searching a habitat patch and how the rate of host encounter should affect the decision to leave the patch. When we understand the behavioral rules that insects follow we should be able to predict both immediate and evolutionary changes in patterns of host use that result from altered plant quality, density, diversity, and distribution (Jones, 1977; Stanton, 1983).

Evolutionary biologists contemplate the role of host preference in speciation (Jaenike, 1981; Bush and Diehl, 1982) and observe evolutionary changes in diet, including the addition of novel hosts that may generate a new pest species from a formerly innocuous insect. They have particular requirements of a definition of preference. The reason is that any *evolutionary* change in the relative frequency with which particular plants are used requires a change in the frequencies of insects with different preferences (unless it occurs solely as a result of host-plant evolution). Such a change depends on several factors (Singer, 1983), including the presence of heritable preference variation among individual insects in a population. Much of the information presented as "preferences" of populations is not sufficient for this purpose. For example, the observation that an insect population deposits 40% of its eggs on host species A and 60% on B does not necessarily indicate variation in either use or preference, since each insect may lay 40% of its eggs on A and 60% on B. If use does vary, and some insects lay most of their eggs on A, while otheres lay mainly on B, it is still possible that all insects are following the same behavioral rules. The variation could result from the different experiences (e.g., host encounters) of individual insects (Rausher, 1983c). If this were so, there would be no real behavioral variation on which selection could act, and no response to selection would be expected, even if one of the hosts were totally unsuitable. For evolutionary purposes we need information on the causes of variation in individual behavior (Rausher, 1983a). We cannot use definitions of preference that measure the collective behaviors of groups of insects, such as the rates of accumulation of eggs on different hosts in the field. We must also clearly separate performance from preference since relative performance on different hosts provides the selection pressure in response to which both preference and performance may evolve. For these reasons, behavioral rather than ecological concepts of preference are most useful to the evolutionary biologist.

V. Measurement of Preference

Components of preference have been tested by a variety of techniques. A stimulus may be presented and the tendencies of insects to fly (Kennedy and Booth, 1963) or walk (Visser, 1982) towards it may be measured (see Chapters 1 and 2, this volume). Preferences for different plant species have been estimated from rates of alighting (Rausher, 1979; Stanton, 1982), fecundities in no-choice trials (Robert et al., 1982; Lofdahl, 1985), rates of emigration from particular hosts (Wint, 1983; Futuyma et al., 1984), and relative oviposition on plants presented simultaneously (Tabeshnik et al., 1981) or sequentially (Singer, 1971). Preference quantified by one technique may be undetected or reversed (Hoffman, 1985) with another method.

Testing the effects of single stimuli may be misleading because of the interaction between insect responses to different modalities (Singer, 1971; Harris and Miller, 1982; Thibout et al., 1982). Singer found low rates of oviposition on plants that were chemically acceptable but presented inappropriate tactile stimuli. Harris and Miller found that the presentation of visual, chemical, or tactile stimuli along to onion flies generated little response: the flies needed all three forms of stimulation to oviposit normally.

I have already listed (page 68) many factors that affect plant acceptability. All of these may influence the outcome of preference trials. Two additional factors deserve mention: plant density and temperature:

A. Host Density

Rausher (1983b) showed changes in insect behavior with host density. As density increased, rates of host encounter by *Battus philenor* butterflies increased more rapidly than rates of oviositon. Rausher suggested that these insects became more discriminating (within host species) in their post-alighting behavior as their rates of host finding increased.

B. Temperature

When insects require high body temperature to perform fine discriminations, a rise in temperature may be expected to increase the strength of preference shown in trials involving simultaneous presentation of hosts. My own experience (Singer, unpublished observations) shows the opposite effect: if preference is measured in terms of lengths of time between first acceptances and hosts presented sequentially (see below), a rise in temperature reduces the apparent strength of preference ("specificity") because the motivational state of the insect changes more rapidly and the time interval between first acceptances of different hosts is reduced as temperature rises.

The following sections discuss in more detail some of the components of preference that have been tested,and some of the techniques that have been used.

5.1. Estimates of Pre- and Postalighting Preference Combined

A. Measurement from Natural Egg Distributions in the Field

As a rule the eggs of plant-feeding insects are nonrandomly distributed, even within the appropriate habitats. Species that drop eggs to the ground without reference to the distribution of host plants are by far the exceptions (Wiklund, 1984). When egg distributions are not closely tied to those of host plants, they are likely to be influenced by physical features of the microhabitat (Thomas, 1985). However, egg distribution is usually strongly correlated with the presence of host plants. I have already discussed my objections to the measurement of preference from egg distribution in the field, and in particular from the relationship between egg distribution and the relative abundances of plants.

B. Measurement from Egg Distributions in Field Experiments

Field experiments may give different results when conducted at different scales, i.e. with different sizes of host plant patches. This occurs partly because large patches are more likely to be encountered than small ones

(Stanton, 1983) and partly because of the effects of scale on the ability of the insects to exert their preference (Kareiva, 1982). An insect may find a broader range of plants acceptable as its search for a oviposition site proceeds (Singer, 1982). Such an insect, initiating its search in a large patch of moderately acceptable hosts, is likely to oviposit before it leaves the patch. Thus it is quite possible for the relative insect abundance among species or cultivars to be influenced primarily by preference if host patches are small and by performance if they are large. It could easily be reversed in direction as one proceeds from small-scale to large-scale experiments. The moral here is that large-scale effects cannot be predicted from small-scale experiments.

C. Measurement from Egg Distribution in the Laboratory

(a) No-choice trials. In no-choice trials insects are allowed contact with only one host, and their times to first oviposition and/or subsequent rates of oviposition are measured. Lofdahl (1985) has used such trials in studies of heritability of preference in cactiphilic *Drosophila*. Two techniques were used: flies were scored binomially according to whether or not they laid eggs and the numbers of eggs laid were counted. The first technique measures acceptance, the second a combination of acceptance and fecundity. By these means many thousands of insects in 1100 families were classified and a heritability estimated at about 0.1 differed from zero with high significance. Such scale would be impractical with more labor-intensive methods.

(b) Sequential choice trials. In these techniques insects are allowed contact with more than one host, but only one is present at any given time. This is realistic because many insects encounter plants sequentially in their natural host-search behavior. However, complexity is introduced by the possibility that an insect's experience with one host may affect its subsequent response to another. Several such effects are known. *Callosobruchus maculatus* beetles are highly unlikely to accept a less-preferred plant after they have encountered a higher ranked host (Mark, 1982). These beetles behaved as though they could make comparisons among plants encountered sequentially, though this was not necessarily the real mechanism causing "exaggerated aversion" to the less-preferred host after exposure to the more-preferred species. Tephritid fruit flies become less ready to accept plants on which oviposition has been attempted and has failed (Prokopy et al., 1982). Some species accept more readily those hosts they encountered as young adults (Jaenike, 1982; Rausher, 1983c), but Hoffmann (1985) has shown the reverse effect . . .increased preference for a host resulting from *lack* of encounter with it by young adults. He used the same *Drosophila* as Jaenike, but a different preference testing technique.

(c) Simultaneous choice experiments. Laboratory choice experiments may use potted plants, plant parts, or filter paper soaked in plant extracts (Stanton, 1979; Renwick and Radke, 1982; Mark, 1982). Insects that lay single eggs are presented with a simultaneous choice and their preferences are judged from the relative numbers of eggs laid. In theory, these experiments are different from those preceding because an insect may make a simultaneous comparison among hosts presented to it in an array. In practice, many insects cannot perceive more than one host at a time and encounter them sequentially even when they are presented simultaneously. If this is the case, a sequential choice design may be preferable because the sequence of host encounters is not a random variable in the experiment.

Discrimination revealed by these techniques is likely to be relevant to the field situation, though this should be checked whenever possible. Captive insects are often less discriminating than they would be at liberty. This may result from: (1) low body temperatures in insects that would normally bask in the sun prior to oviposition search; (2) lack of natural continuity between flight search and postalighting behaviors, (3) high oviposition motivation reached by captive insects that do not search for hosts at times when they would do so if at liberty. Such insects are less discriminating when they finally do encounter hosts. On the other hand, if very few individual plants are used in the trials, spurious preferences may be generated from effects of plant position.

The design used by Stanton (1979) controls perfectly for position effects, since the test arena has only four positions, at the corners of a square. With such a simple design each oviposition substrate can be placed at each position in the arena with equal frequency. However, most published studies fail to mention whether or not precautions against position effects were taken. It is absolutely not adequate simply to select positions at random. The fact that a particular plant may have been randomly chosen for a position favored by the test insect does not alter that fact that a spurious preference for this plant will appear unless the experiment is run repeatedly until each plant has been tested at each position. If an experiment tests one cultivar against another, and many individual plants of each cultivar are available, serious position effects can be avoided by arranging the cultivars in alternation. Even then, it would be wise to use at least two arrangements, e.g.:

        BABABABA   ABABABAB
        ABABABAB   and   BABABABA
        BABABABA   ABABABAB

Even in the field, where effects of habitat boundaries or arena edges can be minimized, position effects are likely to be important. For example, Root and Kareiva (1984) found that "the composition and magnitude of the herbivore load on individual collards varies widely even though the plants have similar genetic background, are the same age, and grow within 10 m of one another in an extremely homogeneous environment."

The fact that insect responses to eggs previously laid may obscure preference has already been discussed. Conspecific eggs often repel searching females (Prokopy 1981) but are also known to attract them in onion flies (J. Miller pers. comm.), *Drosophila* (K. Lofdahl pers. comm.) and some *Heliconius* butterflies (J. Mallet & J. Longino pers. comms.). Occasionally there may be no such effects at all in either direction (Singer and Mandracchia, 1982) but more often there is a tendency to avoid eggs that may be quite sophisticated; Messina and Renwick (1985) have shown that *Callosobruchus maculatus* females can assess egg densities on bean seeds accurately enough to achieve "a nearly uniform dispersion of eggs." Experiments in which egg distributions are allowed to build up before eggs are counted run the risk of being confounded by these effects, unless the sequence in which eggs are laid has been noted.

5.2. Separate Estimation of In-Flight or Prealighting Preference

The most satisfactory preference-testing technique would involve presentation of arrays of hosts to insects under conditions as natural as possible so that normal combinations of pre- and postalighting behaviors remain intact. This is not often feasible with captive insects because natural flight behaviors are not duplicated. Two methods have been used to combat this problem. Wiklund (1974) used an extremely large insectary in which segments of natural flight occurred. Kennedy and Booth (1963) developed a chamber in which a flying aphid was held immobile, balanced on air jets, viewed through an eyepiece. The tendency of the aphid to move towards or away from particular stimuli was measured by the adjustment of air currents needed to keep it in place. However, in-flight preferences have usually been studied in the field (Chew, 1977; Rausher, 1978; Stanton, 1982; Courtney, 1982; Root and Kareiva, 1984; Mackay, 1985; Myers, 1985). Stanton (1983) recommends and describes field experiments with "artificial communities of plants where host-plant density, patch size, and species diversity can be manipulated independently." Field studies in completely natural situations usually involve recording alightings during host search and comparing them to the composition of the vegetation. (Stanton, 1982, Rausher and Papaj, 1983*b*; Mackay, 1985.)

The flight paths used in these studies are approximate paths constructed by joining points of alighting (or points flown over) with straight lines. For field studies, plotting of exact paths for small, fast-moving insects does not seem to be technically feasible yet. Even when this can be done some problems will remain since the alightings should ideally be compared with the vegetation in a corridor around the exact flight path. This corridor should encompass that part of the insect's visual field within which it is prepared to alight or move towards the plants that it perceives. The likelihood that on insect will deviate from its track in response to a particular plant may depend in complex fashion on the nature of the plant and its vegetational background, and on the amount of deviation necessary to fly over it. The technical difficulty of plotting exact flight paths often precludes ascertaining whether individual insects differ in their tendencies to alight on particular plants. However, both Rausher and Papaj (1983b) and Stanton and Cook (1983 or 1984) have tackled this problem. Rausher and Papaj found that *Battus philenor* butterflies in the same population differed in their frequencies of alighting on nonhosts with different leaf shapes, and that this variation was not associated with variation in flight paths. Stanton and Cook, on the other hand, found that individuals with high rates of alighting on particular hosts did tend to fly over those hosts with high frequency.

These estimates of preference from rates of alighting in the field require some measure of relative abundance of hosts. Some authors have done this in terms of ground cover (Stanton and Cook 1983 or 1984, Mackay, 1985), whereas others use leaf area (Cates, 1981). Since leaves of different species overlap to different extents when viewed from above, these two methods do not give the same result. A different technique has been to consider plants or leaves of different species as equivalent and measure insect preference from differences in the numbers of insects per plant or per leaf (Zehnder and Trumble, 1984).

Some insects learn to associate chemical and visual stimuli (Traynier, 1984); these effects influence subsequent search behavior and in-flight preferences.

5.3. Separate Estimation of Postalighting Preference

Postalighting preference can be measured in captive insects by manipulating artificial "alightings," placing the insects in contact with hosts (Singer, 1971) or host extracts (Feeny et al., 1983) and recording actual or attempted ovipositions.

If many natural oviposition events can be observed and all the plants of interest are relatively abundant, postalighting preferences can also be measured in the field (Chew, 1977; Wiklund and Ahrberg, 1978). Stanton (1982) followed *Colias philodice* butterflies and recorded both alightings and ovipositions. She then calculated the proportion of alights on different leguminous hosts that were followed by oviposition. Courtney (1982) has done the same with *Anthocharis cardamines* and its cruciferous hosts, and Makay (1985) has used this technique for *Euphydryas editha*.

VI. A Manipulative Technique

My own methods of preference testing have been devised specifically for use with *E. editha* in order to investigate its extensive variation of host use both within and among populations (Singer, 1982, 1983). Its docility and ease of handling, coupled with its readiness to oviposit when placed on a host, render it suitable for preference-testing techniques that manipulate its encounter with particular plants, including rare species. The manipulative technique that I use has the advantage that it can be applied equally well in the greenhouse or in the field. Preference testing in the field allows the use of plants whose quality has not been influnced by disturbance or by greenhouse conditions. I do not suggest that this procedure will be directly applicable to other study organisms; however, I hope that my experiences with *E.editha* will be informative to others who are developing techniques relevant to natural behavior patterns of their own study insects.

6.1. Study Organism

E.editha is a univoltine butterfly that is widely distributed in the Western U.S.A., but often occurs in local populations of 200–3000 individuals (Ehrlich et al., 1975). Hosts are normally members of the plant families Scrophulariaceae or Plantaginaceae. Some populations of the butterfly are monophagous, and others are oligophagous, ovipositing on as many as four host genera (Singer, 1983). Two of the oligophagous populations have been the subjects of detailed studies. At one of these, near General's Highway, Sequoia National Forest, California, the two principal hosts are *Pedicularis semibarbata* and *Collinsia torreyi*. Each year since 1979 the preferences of *E.editha* individuals for these two hosts have been measured. The following procedure has been used.

6.2. Procedure in Preference Trials

A. *Selection of Test Plants*

Natural egg distributions are used as guides to select plants. Acceptable *Collinsias* have been compared with acceptable *Pedicularis*. Since *Pedicularis* are perennial, the same individuals can be used each year provided that they survive. It is possible to examine variation of preference among insects by testing many insects on few plants, or to measure variation of acceptability among plants (of the same or different species) by testing few insects on many plants. By the latter method one can investigate the ranges of acceptability to particular insects within each plant species and estimate the overlap between the ranges for different plant species.

B. Selection of Test Insects

The insects used fall into various categories depending on the hypothesis being tested. Preference testing in the field facilitates experiments that relate variation of preference to other aspects of the insects' behavior. Insects can be captured performing particular behaviors and their preferences tested. For example, by comparing preferences of insects captured in the act of ovipositing on different host species one can relate observed oviposition to tested preference (Singer, 1983). It would also be feasible to capture, test, release, recapture, and retest in order to examine changes of preference that occur during the lives of individual insects. Thomas and Singer (submitted for publication) have marked and released female butterflies, then preference-tested the recaptured individuals. These experiments have shown relationships between postalighting preference and local dispersal; insects tend to leave areas containing only their second-preferred host and move to patches where their most preferred host grows.

C. Preference Trials

When the test plants and insects have been selected, each insect is placed gently on each of the plants in turn and its responses recorded. This can be done either by temporarily releasing the insect completely or by placing it under a mesh cage. The mesh mush be sufficiently coarse that enough sunlight is admitted to allow the butterfly to reach normal body temperature. If the insect responds by flying away, it is recaptured and replaced on the plant at lease three times. If it merely basks, it is gently lifted and replaced at lease three times. Acceptance of the plant is recorded if the abdomen is curled and the ovipositor extruded for about 3 sec. If no such behavior occurs the plant is deemed to be rejected. The insect is not allowed to oviposit; it is removed from an accepted plant before the first egg has been laid. Trials are restricted to sunny weather and appropriate times of day for ovipopsition (11a.m. to 4 p.m. at many sites).

6.3. Interpretation of Results

Comparisons are made between responses to single *Pedicularis* plants and clumps of the much smaller *Collinsia*. Because the insects are not allowed to oviposit, the range of acceptable plants becomes broader as time passes and motivation to oviposit increases. Preference rank is measured as the order in which the test plants become acceptable. Different plants are first accepted at different times, but once a plant has become acceptable it almost always remains so until oviposition is allowed by the experimenter (Singer, 1983). The length of time during which an insect will accept only its most preferred host may be so short as to be undetectable, or may last a week. This time interval (shown by the dotted lines in the table below) is an estimate of the insect's host specificity, at least in its postalighting preference.

If the test plants are abbreviated as "P" and "C", the results of a set of trials on one insect will take one of the following forms:

(1) Reject "P", reject "C", reject "P", **accept "C"**, reject "P", ...**accept "C", accept "P"**.

(2) Reject "P", reject "C", **accept "P", accept "C"**.

(3) Reject "P", reject "C", **accept "P"**, reject "C", ...**accept "P", accept "C"**.

Butterflies that show pattern (1) are recorded as preferring "C" over "P", since "P" was rejected after "C" became acceptable. Likewise, those showing pattern (3) are described as "P" preferring. Pattern (2) has been classed as "neutral" or "no preference," since such insects do not discriminate between the test plants under the specific conditions used. Clearly, some insects recorded as neutral have slight preferences that would be revealed if the intervals between trials were very short. However, I do not use intervals shorter than 5 min, because apparent "carryover" effects are occasionally observed, similar to those seen by Feeny et al. (1983). An insect that is transferred rapidly from a highly acceptable host to one that is normally unacceptable may behave as though it had not been moved and accept the second plant. It the insects are caused to fly between trials this carryover effect is not evident, even with very short intervals between presentations.

By this technique a preference for "P" relative to "C" can be measured as a continuous variable using lengths of discrimination (Singer, 1982). A curve can be drawn to represent the frequencies in a population of insects with different strengths of preference (Singer, 1983). In practice, one person must either work with very few insects (5–10) at one time or test each insect infrequently. When frequency of encounter is sacrificed in order that many insects can be tested at once it is less meaningful to present the results as a continuous variable.

6.4. Assumptions Implicit in the Technique

The following assumptions are implied by the use of the technique and by the mechod of interpretation of the results:

(1) An insect placed upon a plant in bright sunlight between 11 a.m. and 4 p.m. will behave as though it had alighted naturally on the same plant at the same time. The principal evidence supporting this assumption relates to within-species discrimination. It was gathered by Rausher et al. (1981), who followed individual *E.editha* alighting on *Pedicularis* plants. When a plant was accepted the insect was captured and replaced (in randomized order) on the accepted plant and on the last *Pedicularis* that it had rejected. The correlation between the behavior of undisturbed insects and their behavior after manipulation was significant but by no means absolute.

(2) Once a plant has become acceptable, it remains so until the next oviposition occurs. This assumption has been justified (Singer, 1982, 1983). For most host species it is violated in fewer than 5% of trials.

(3) The current preference of an insect is not affected by previous encounters with particular plants. This is a critical assumption. If it were not true the preference-testing technique would itself affect the behavior that it purports to measure, since it involves manipulated encounters with particular plants at particular times. In view of the evidence from other insects (reviewed above) the assumption seems unlikely to be strictly correct. The effects documented in these other studies are changes of preference that result from encounters with hosts of varying quality. If these effects are important in *E.editha* the preference testing technique should be modified accordingly. Three methods are being used to look for effects of encounter on preference:

(a) Comparison of choice tests with no-choice tests. Teneral (freshly emerged) insects are offered only "P," only "C," or "P" and "C" in alternation. The times of first acceptance of each plant are noted. If Mark's (1982) finding applied to *E.editha,* the mean time to first acceptance of "C" (the less preferred plant for most insects at Generals' Highway) should be greater for those insects that have previously encountered "P" than for those offered only "C." If such an effect exists, it is not very strong; the data show a very slight (nonsignificant) trend in this direction (Thomas and Singer, submitted for publication).

(b) Insects that have been exposed to "P" or "C" alone for the first 5 days of adult life are then preference-tested in a routine manner with exposure to both plants. This should check for adult conditioning of the kind known in *Drosophila.*

(c) Preference-tested insects are allowed to oviposit on either "P" or "C," then retested. No effects of allowed oviposition on subsequent preference have been found. Repeated preference tests performed on butterflies from Schneider's Meadow have given a high estimate of repeatability (0.85) whether or not oviposition occurred on the preferred host species.

6.5. Problems of Interpretation

Although many of the results of preference trials on *E.editha* accord with expectation (Singer, 1982, 1983) there are difficulties with the interpretation of others. Here are three examples:

A. Nonsimultaneous Comparisons

At Generals' Highway the proportion of *Pedicularis*-preferring butterflies was lower in 1984 than in previous years. At Schneider's Meadow the proportion of insects preferring *Plantago* has risen from 1980 to 1984. It would be interesting to know whether these data show microevolutionary changes in insect preference, since directional selection associated with early larval survival would tend to generate both of these trends (Singer, 1984). However, some of the limitations of field work apply here. The changes could result from year-to-year fluctuations in climate or plant quality rather than insect preference. Short-term evolutionary changes in insect behavior could be revealed by testing each year's butterflies on cloned plants in the lab under standard conditions. Unfortunately, they cannot be deduced from the available field data. The moral is that the only insects that can be compared in the field are those that can be compared simultaneously (i.e., on the same day) so that plant condition and weather are controlled variables.

B. Accuracy of Preference Tests

In both the oligophagous populations of *E.editha* where preference trials have been conducted there is a strong relationship between observed oviposition and (subsequently) tested preference. I have used these data to claim that, at Generals' Highway, the insects that oviposit on "C" are a nonrandom sample of the population with respect to their postalighting preferences (Singer, 1983). Although this conclusion is reasonable, the estimated relationship between preference and use is only a minimum value. The reason lies in the role of insects that preferred the host on which they did not oviposit. Some insects that oviposited on "C" were subsequently tested as preferring "P." Did these insects change their preferences between the time of oviposition and the preference trials? Did they prefer "P" at both times and oviposit on the less preferred host? Or did they prefer "C" both times and were mistakenly identified as "P" preferring because the preference trials give the "wrong answer" 10% of the time? I cannot yet answer these questions.

A similar problem arises in repeatability trials. These are performed by recording the orders in which plants become acceptable prior to two or more ovipositions by the same individual insect. When the data show apparent changes of preference, are these changes real, or do they result from inaccuracy of measurement? This question cannot be answered with available data, so the estimate of repeatability must be regarded as a minimum value. The same would apply to heritability data. However, the identification of factors that can cause apparent changes of preference in individuals or differences in repeatability among populations may prove helpful in this respect.

C. Experimenter Bias

How much can the experimenter influence the results of a preference trial? Insects that do not perform as expected cause frustration, and may risk being roughly handled. While an insect that is not prepared to accept a plant cannot be forced to do so, except perhaps by starvation (poorly nourished insects sometimes become more ready to oviposit), a butterfly that would accept under ideal conditions may reject if it is mistreated, given insufficient time or tested under marginal conditions of light or temperature. For this reason trials shoud be performed blind whenever possible, though this cannot be done by one person working alone. The insects in our heritability studies have been numbered by one person and tested by another who did not know the identities or preferences of the parents of the tested butterflies. When more data are available it may be possible to quantify experimenter bias by comparing repeatabilities in blind and nonblind data sets.

6.6. Concordance Among Preference Testing Techniques

There are several examples to show the sensitivity of results to the identity of the technique used. Hoffmann (1985) was able to duplicate Jaenike's (1982) result with a similar technique but found that it was reversed if a larger testing arena was used. Hoffman suggested that this reversal was due to the inclusion of behaviors that take place at greater distances from the host.

If so, this result is related to those of Stanton (1982) and Papaj and Rausher (1983), who found that their study organisms ranked host species in different orders before and after alighting.

In a general sense, the technique used should depend on the specific questions being asked and on the biology of the study organism. For example, the changes in acceptability with passing time that have proven useful in testing *E.editha* may be irrelevant to the oviposition behavior of many species. One may imply this from the work of Ives (1978) on the cabbage butterfly, *Pieris rapae*. These insects showed the same rate of acceptance of less preferred hosts at the beginning and at the end of a bout of oviposition. If an adult insect emerges on a large host individual or in a dense population of small hosts, the most important question may be: how likely is it that the insect will leave the host patch before ovipositing? If it lays eggs quickly in this no-choice situation, behavior in choice tests may be irrelevant.

If behavioral thresholds do change with time, choice trials will produce greater differences among responses to test plants than will no-choice trials. Wasserman and Futuyma (1981) showed, however, that variation of behavior in choice trials was not independent of variation in behavior in no-choice trials. Artificial selection applied in one context produced responses in both. Understanding of preference behavior should be improved by investigating the same insect with several different techniques. The following paragraph gives an example, using the relationships between different aspects of preference in *E.editha*.

Manipulative trials performed at Generals' Highway have given the following result: about 70% of the insects prefer "P" over "C," about 20% show neutral preference, and about 8% prefer "C" over "P." How would this result change if other ways of testing preference were used? Using time to first acceptance in no-choice trials would also show that "P" is preferred over "C" in a general sense (Fig. 1). If reproduction in no-choice trials were used the result would depend on time scale (Tallon and

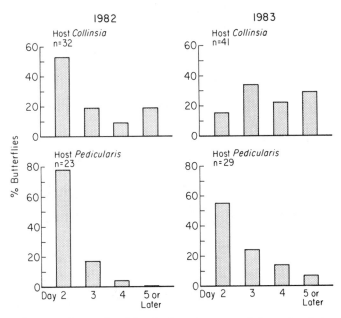

Figure 1. Times to first oviposition in insects exposed from day 1 to only one host species.

Singer, unpublished observations; Fig. 2). "P" received more eggs early in the life of an insect (days 2 and 3) but not subsequently. However, on day 3 of adult life larger clusters were laid on "C" than on "P." This cluster size effect is in the reverse direction from other measures of preference. It probably stems from the fact that most insects wait longer before their first oviposition if they are offered only "C," and the delay in oviposition results in larger cluster size. In spite of the difference in cluster size, mean fecundity on day 3 was the same on both hosts, because the likelihood of oviposition occurring on "C" was lower than that on "P."

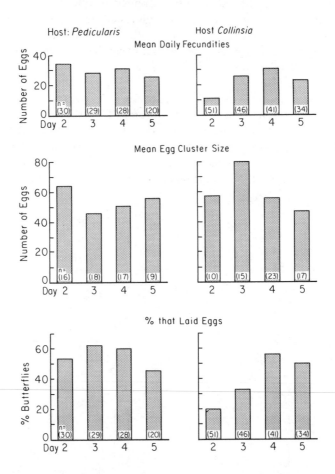

Figure 2. Fecundity, cluster size, and probability of oviposition on days 2–5 of life for butterflies offered only "P" or only "C." Mean fecundity includes ALL insects, whether or not they laid eggs. Cluster sizes are the means for eggs that were laid, zeros are not included.

VII. Conclusions

I have tried to discuss the meaning of ovipositional "preference" in a manner that will help others to use (or avoid) this term. Its relevance to workers in different disciplines is briefly described. This paper recommends that techniques for testing preference be developed in the light of detailed knowledge of the behavioral patterns of specific study insects. Even then, there will still be problems of experimental design and interpretation of results. In spite of difficulties of definition and measurement, preference (a trait of insects) and acceptability (a trait of host plants) will continue to be used in studies of insect–plant relationships.

Acknowledgments.   I am grateful to K. Lofdahl, D. Papaj, C. Parmesan, K. Tallon, C.Thomas, and M. Turelli for permission to cite unpublished data, to Brackenridge Field Lab for computer use, and to the University of Texas and NSF grant BSR 8407701 for partial financial support of this work. Comments, mostly helpful and all welcomed, were made by F.S. Chew, P.P. Feeny, L.E. Gilbert, M.Kirkpatrick, J. Miller, R.A. Moore, D.R. Papaj, C. Parmesan, D. Ng, J.E. Rawlins, M.J. Ryan, M.E. Saks, N. Shah, T.R.E. Southwood , C.D. Thomas, and P. Whittaker.

References

Bush GL, and Diehl SR (1982) Host shifts, genetic models of sympatric speciation and the origin of papasitic insect species. In: *Insect-Plant Relationships*. Visser FH, Minks AK (eds.), PUDOC, Wageningen, pp 297–306.

Cates RG (1981) Host plant predictability and the feeding patterns of monophagous, oligophagous and polyphagous herbivores. Oecologia (Berl) **48**:319–326.

Chew FS (1977) Coevolution of Pierid butterflies and their cruciferous foodplants II. The distribution of eggs on potential foodplants. Evolution 31:568–579.

Courtney SP (1982) Hostplant apparency and *Anthocharis cardamines* oviposition. Oecologia (Berl) **52**:258–265.

Crawley MJ (1984) Herbivory. Blackwells, Oxford.

Ehrlich PR, White RR, Singer MC, McKechnie SW, Gilbert LE (1975) Checkerspot butterflies: a historical perspective.Science **176**:221–298.

Feeny P (1975) Biochemical coevolution between plants and their insect herbivores. In: Coevolution of Animals and Plants. Gilbert LE, Raven PH (eds), University of Texas Press, pp 3–19.

Feeny PP, Rosenberry L, Carter M (1983) Chemical aspects of oviposition behavior in butterflies. In: Herbivorous Insects: *Host-Seeking Behavior and Mechanisms*. S Ahmad, (ed), Academic Press, New York.

Futuyma DJ, Cort RP, Noordwijky IV (1984) Adaption to host plants in the fall cankerworm *(Alsophila pometaria* and its bearing on the evolution of host affiliation in phytophagous insects. Am Nat **123**:287–296.

Futuyma DJ, Leipertz SL, Mitter C (1981) Selective factors affecting clonal variation in the fall cankerworm *Alsophila pometaria* (Lepidoptera: Geometridae). Heredity **47**:161–172.

Futuyma DJ, Wasserman SS (1981) Food plant specialization and feeding efficiency in the tent caterpillars, *Malacosoma disstria* Hubner and *M. americanum* (Fabricius). Entomol. exp. appl. **30**:106–110.

Gilbert LE (1979) Development of theory in insect-plant interactions. In: Analysis of Ecological Systems. Horn EJ, Mitchell RD, Stairs GR (eds), Ohio State University Press, Columbus, pp 117–154.

Gould F (1983) Genetics of plant-herbivore systems: interactions between applied and basic study. Ch. 17 In: Variable plants and herbivores in natural and managed systems. Denno RF, McClure MS (eds), Academic Press, New York, pp 599–654.

Guérin PM, Stadler E (1982) Host odour perception in three phytophagous Diptera—a comparative study. In: Insect-plant relationships. Visser JH, Minks AK (eds), PUDOC, Wageningen, pp 95–106.

Guérin PM, Stadler E (1984) Carrot fly cultivar preferences: some influencing factors. Ecol Entomol **9**:413–420.

Hamilton WD (1971) Geometry for the selfish herd. J Theor Biol. **31**:295–311.

Harris MO, Miller JR (1982) Synergism of visual and chemical stimuli in the oviposition behavior of *Delia antiqua*. In: Insect-plant relationships. Visser JH Minks AK (eds), PUDOC, Wageningen, pp 117–122.

Hassell MP, Southwood TRE (1978) Foraging strategies of insects. Annu Rev Ecol Syst **9**:75–98.

Hoffmann AA (1985) Effects of experience on oviposttion and attraction in *Drosophila:* comparing apples and oranges. Am Nat **126**:41–51.

Ives PM (1978) How discriminating are cabbage butterflies? Aust. J Ecol **3**:261–276.

Jaenike J (1982) Environmental modification of oviposition behavior in *Drosophila*. Am Nat **119**:784–802.

Jones RE (1977) Movement patterns and egg distribution in cabbage butterflies. J Anim Ecol **46**:195–212.

Kareiva P (1982) Experimental and mathematical analyses of herbivore movements: quantifying the influence of plant spacing and quality on foraging discrimination. Ecol Monogr **52**:261–282.

Kennedy JS (1965) Mechanisms of host-plant selection. Ann Appl Biol **56**:317–322.

Kennedy JS, Booth CO (1963) Free flight of aphids in the laboratory. J Exp Biol **40**:67–85.

Lewis AC (1984) Plant quality and grasshopper feeding: effects of sunflower condition on preference and performance in *Melanoplus differentialis*. Ecology, **65**:836–843.

Lofdahl K (1985) Quantitative genetic analysis of habitat selection in the cactus-breeding insect *Drosophila mojavensis*. PhD thesis, University of Chicago.

Mackay, DA (1985) Within-population variation in pre-alighting search behavior and host plant selection by ovipositing *Euphydryas editha* butterflies. Ecology **66** (1):142–151.

Mackay DA, Singer MC (1982) The basis of apparent preference for isolated plants by ovipositing *Euptychia libye* butterflies. Ecol Entomol 7:299–303.

Mark GA (1982) Induced oviposition preference, periodic environments and demographic cycles in the bruchid beetle *Callosobruchus maculatus*. Entomol Exp Appl **32**:155–160.

Meisner J, Ascher KRS, Lavie D (1974) Factors influencing the attraction to oviposition of the potato tuber moth, *Gnorimoschema operculella*. Z Ang Ent **77**:179–189.

Messina F (1984) Influence of cowpea pod maturity on the oviposition choices and larval survival of a bruchid beetle, *Callosobruchus maculatus*. Entomol Exp Appl **35**:241–248.

Messina FJ, Renwick JAA (1985) Ability of ovipositing seed beetles to discriminate between seeds with differing egg loads. Ecol Entomol in press.

Miller JR, Strickler KL (1984) Finding and accepting host plants. In: *Chemical Ecology of Insects*. Bell WJ, Carde RJ (eds), Chapman and Hall Ltd., pp 127–157.

Murphy DD, Menninger MS, Ehrlich PR (1984) Nectar Source distribution as a determinant of oviposition host species in *Euphydryas Chalcedona*. Oecologia **62**:269–271.

Myers JH (1985) Effect of physiological condition of host plant on the ovipositional choice of the cabbage white butterfly, *Pieris rapae*. J Anim Ecol **54**:193–204.

Papaj DR, Rausher MD (1983) Individual variation in host location by phytophagous insects. In: *Herbivorous insects: host-seeking behavior and mechanisms*. Ahmad S (ed), Academic Press, New York.

Price PW, Bouton CE, Gross P, McPheron BA, Thompson JN, Weis AE, (1980) Interacions among three trophic levels: influence of plants on interactions between herbivores and natural enemies. Annu Rev Ecol Syst **11**:41–65.

Prokopy RJ (1981) Epideictic pheromones that influence spacing patterns of phytophagous insects. pp. 181–213 In: *Semiochemicals: Their Role in Pest Control*. Nordlund DA, Jones RL, Lewis WJ, (eds), Wiley & Sons, New York.

Prokopy RT, Averill AL, Cooley SS, Roitberg BA (1982) Associative learning in egglaying site by apple maggot flies. Science **218**:76–77.

Prokopy RJ, Owens ED (1978). Visual generalist–visual specialist phytophagous insects: Host selection behavior and application to management. Entomol Exp Appl **24**:409–420.

Rausher MD (1978) Search image for leaf shape in a butterfly. Science **200**:1071–1073.

Rausher MD (1983a). Ecology of host-selection behavior in phytophagous insects. In: *Variable Plants and Herbivores in Natural and Managed Systems*. Denno RF, McClure MS, (eds), Academic Press, New York, pp 223–258.

Rausher MD (1938b) Alteration of oviposition behavior by *Battus philenor* butterflies in response to variation in host-plant density. Ecology **64**:1028–1034.

Rausher MD (1938c). Conditioning and genetic variation as causes of individual behavior in the oviposition behavior of the tortoise beetle, *Deloyala guttata*. Anim Behav **31**:743–747.

Rausher MD (1985) Alteration of oviposition behavior by *Battus philenor* butterflies in response to variation in host plant density. Ecology, in press.

Rausher MD, Mackay DA, Singer MC (1981). Pre- and post-alighting host discrimination by *Euphydryas editha* butterflies: the behavioral mechanisms causing clumped distributions of egg clusters. Anim Behav **29**:1220–1228.

Rausher MD, Papaj DR (1983a) Demographic consequences of discrimination among conspecific host plants by *Battus philenor* butterflies. Ecology **64**:1402–1410.

Rausher MD, Papaj DR (1939b). Host plant selection by *Battus philenor* butterflies: Evidence for individual differences in foraging behavior. Anim Behav **31**:341–347.

Renwick JAA, Radke CD (1983) Chemical recognition of host plants for oviposition by the cabbage butterfly *Pieris rapae*. Envir Entomol **12**:446–451.

Robert P, Huignard J, Nuto Y (1982) Host plant changes in several population of *Carydeon serratus*. In: *Insect-Plant Relationships*. Visser JH, Minks AK (eds), PUDOC, Wageningen, pp 441–442.

Root RB (1973) Organization of plant-arthropod associations in simple and diverse habitats: the fauna of collards *(Brassica oleracea)*. Ecol Monogr **43**:95–124.

Root RB, Kareiva PM (1984) The search for resources by cabbage butterflies *(Pieris rapae)*: ecological consequences and adaptive significance of Markovian movements in a patchy environment. Ecology **65**(1):147–165.

Rothschild M, Schoonhoven LM (1977) Assessment of egg load by *Pieris brassicae* (Lepidoptera: Pieridae). Nature (Lond) **266**:352–355.

Scriber JM, Feeny P (1979) Growth of herbivorous caterpillars in relation to feeding specialization and to the growth form of their plants. Ecology **60**:829–850.

Scriber JM, Slansky F (1981) The nutritional ecology of immature insects. Ann Rev Entomol **26**:183–212.

Service P (1984) The distribution of aphids in response to variation among individual host palnts: *Uroleucon rudbeckiae* (Homoptera: Aphididae) and *Rudbeckia laciniata* (Asteraceae) Ecol Ent **9**:321–328.

Shapiro AM, Masuda KK (1980) The opportunistic origin of a new citrus pest. Calif Agric **34**:4–5.

Singer MC (1971) Evolution of food-plant preference in the butterfly *Euphydryas editha*. Evolution **25**:383–389.

Singer MC (1982) Quiantification of host preference by manipulation of oviposition behavior in the butterfly *Euphydryas editha*. Oecologia (Berl) **52**:224–229.

Singer MC (1983) Determinants of multiple host use by a phytophagous insect population. Evolution **37**:389–403.

Singer MC (1984) Butterfly-hostplant relationships. In: The Biology of Butterflies. Vane-Wright RI, Ackery P (eds), Symp R, Entomol Soc Lond, XIII.

Singer MC, Mandracchia J (1982) On the failure of two butterfly species to respond to the presence of conspecific eggs prior to ovipositon. Ecol Entomol **7**:327–330

Stanton ML (1979) The role of chemotactile stimuli in the oviposition preferences of *Colias* butterflies. Oecologia (Berl) **39**:79–91.

Stanton ML (1982) Searching in a patchy environment: foodplant selection by *Colias philodice* butterflies. Ecology **63**:839–853.

Stanton ML (1983) Spatial patterns in the plant community and their effects upon insect search. In: Herbivorous Insects: Host-Seeking Behavior and Mechanisms. Ahmad S (ed), Academic Press, New York.

Stanton ML (1984) Short term learning and the searching accuracy of egg-laying butterflies. Anim Behav 32:33–40.

Stanton ML, Cook RE (1983 or 1984) Sources of interspecific variation in the hostplant seeking behavior of *Colias* butterflies. Oecologia 60:365–37 or 61:265–270.

Szentesi A (1981) Antifeedant-treated potato plants as egg-laying traps for the Colorado beetle *(Leptinotarsa decemlineata* Col, *Chrysomelidae)*. Acta Phytopath Acad Sci Hung 16:203–209

Szentesi A, Greany PD, Chambers DL (1979) Oviposition behavior of laboratory-reared and wild Caribbean fruit flies *(Anastrepha suspensa);* Diptera: Tephritidae: I. Selected chemical influences. Entomol Exp Appl 26:227–238.

Tabashnik BE, Wheelock H, Rainbolt JD, Watt WB (1981) Individual variation in oviposition preference in the butterfly, *Colias eurytheme*. Oecologia (Berl) 50:225–230.

Thibout E, Auger J, Lecomte C (1982) Host plant chemicals responsible for attraction and oviposition in *Acrolepiopsis assectella*. In: Insect-Plant Relationships. Visser JH, Minks AK (eds), PUDOC, Wageningen, pp 107–116.

Thomas CD (1984) Oviposition and egg load assessment by *Anthocharis cardamines* (Lep.; Pieridae) Entomol Gaz 35:145–148.

Thomas CD (1985) Specialisations and polyphagy of *Plebejus arqus* (Lep; Lycaenidas) in North West Britain. Ecol Entomol 10:325–340.

Thomas CD Singer MC (submitted for publication) Host preference affects movements in an oligophagous butterfly population. Ecology.

Traynier RM (1979) Long-term changes in the oviposition begavior of the cabbage butterfly, *Pieris rapae,* induced by contact with plants. Physiol Entomol 4:87–96.

Traynier RM (1984) Associative learning in the ovipositional behavior of the cabbage butterfly, *Pieris rapae*. Physiol Entomol 9:465–472.

Visser JH (1982) Olfaction at the onset of host plant selection. In: Insect-Plant Relationships. Visser JH, Minks AK (eds), PUDOC, Wageningen, pp 367–368.

Wasserman S, Futuyma DJ (1981) Evolution of host-plant utilization in laboratory populations of the Southern Cowpea Weevil, *Callosobruchus maculatus*. Evolution 35:605–617.

Whitham TG (1983) host manipulation of parasites: within-plant variation as a defense against rapidly evolving pests. In: Variable Plants and Hervivores in Natural and Managed Systems. Denno RF, McClure MS (eds), Academic Press, New York, pp 15–42.

Wiklund C (1974) Oviposition preferences of *Papilio machaon* in relation to the host plants of the larvae. Entomol Exp Appl 17:189–198.

Wiklund C, Ahrberg C (1978) Host plants, nectar source plants and habitat selection of males and females of *Anthocharis cardamines* (Lepidoptera). Oikos 31:169–183.

Wiklund C (1984) Egg-laying patterns in butterflies in relation to their phenology and the visual apparency and abundance of their host plants. Oecologia (Berl) 63:23–29.

Williams KS Gilbert LE (1981) Insects as selective agents on plant vegetative morphology: egg mimicry reduces egg laying by butterflies. Science 212:467–469

Wint W (1983) The role of alternative host-plant species in the life of a polyphagous moth, *Operophtera brumata* (Lepidoptera: Geometridae). J Anim Ecol **52**:439–450.

Zehnder GW, Trumble JT (1984) Host selection of *Liriomyza* species (Diptera: Agromyzidae) and associated parasites in adjacent plantings of tomato and celery. Environ Entomol **13**:492–496.

Chapter 4

Assays for Insect Feeding

A. C. Lewis[1,3] and H. F. van Emden[2]

I. Introduction

The vital role of plant chemicals in herbivorous insect host choice has long been recognized. Indeed, much of the research on plant–insect interactions has centered on the identification of chemicals important in insect feeding. This chapter summarizes methods of testing nonvolatile chemicals that either elicit or inhibit feeding.

There are four major goals of studies employing bioassay techniques: (1) to study the role of naturally occurring plant chemicals in the host choice of insects; (2) to determine mechanisms of resistance in crop plants; (3) to find antifeedants for pest control; and (4) to increase feeding on artificial diet or trap plants. These goals strongly influence bioassay decisions. Some techniques described here were developed for testing chemicals not of plant origin (Ascher, 1979) and for studying choice between two or more plant species (e.g., Jermy et al., 1968). Certain of the observation methods were developed to study chemoreception and regulation of feeding (Blaney et al., 1973).

Chemicals that elicit feeding have been designated feeding stimulants or phagostimulants; those that inhibit feeding are referred to as antifeed-

[1]Tropical Development and Research Institute, College House, Wright's Lane, London, England.
[2]Departments of Agriculture & Horticulture and Pure & Applied Zoology, University of Reading, Reading, England.
[3]Present address: Department of EPO Biology, University of Colorado, Boulder, Colorado 80309, U.S.A.

ants, deterrents, gustatory repellents, or feeding inhibitors (Dethier et al., 1960). Beck (1965) proposed that a chemical responsible for the first phase of feeding—biting and piercing—be termed a feeding incitant, while a suppressant would be a stimulus that suppresses that activity. However, Chapman (1974) and Schoonhoven (1982) suggest that too little is known about the point at which feeding is affected for these terms to be applied. Recent work on pine-feeding insects indicates that such designations may be valid when experiments can be designed to separate clearly the phases of feeding and the chemicals responsible for them (Raffa and Berryman, 1982). In practice, phagostimulant, feeding deterrent, and antifeedant are the most commonly used terms, with antifeedant more commonly encountered in studies concerned with pest control. These terms do not refer to exclusive entities even for one insect species; the same chemical may be a deterrent, a phagostimulant, or neutral, depending on (1) its concentration, (2) the conditions of the test, and (3) the physiological state of the insect.

These terms refer to feeding itself and not to the effects of feeding. Thus, phagostimulant should not be considered synonymous with nutrient nor antifeedant with antibiotic. Experimental designs must take this into account so that the appropriate behavioral measurements are made. The terms attractant and repellent imply, respectively, oriented movement to or from an odor source and are not used when discussing nonvolatile chemicals.

The basic design used to study phagostimulants and deterrents is to present an insect a substrate with the applied chemical and to measure the response of the insect in one of several ways. A phagostimulant (at the concentration tested) will increase feeding over the control or other test chemical and a deterrent will decrease feeding as compared to the control. The control in a test for phagostimulants is a natural or artificial substrate lacking phagostimulants, or containing deterrents, and not normally ingested. For testing deterrents, the control is a palatable substrate.

The references cited below for a particular technique are not necessarily for its original use. The following papers and reviews can be consulted for the historical development of a particular method: Dethier (1947), Jermy (1966), Ma (1972), Ascher and Meisner (1973), Cook (1976), Schoonhoven and Jermy (1977), and Smith (1978). In addition, Dethier's (1976) classic work on fly chemoreception presents many general principles of bioassay design.

II. Substrate and Application

Ideally, the effect of a plant chemical on insect feeding would be tested with plant lines isogenic except for the allele(s) responsible for the chemical (Bright et al., 1982). Otherwise, the chemical must either be selectively removed from the plant (Bernays et al., 1974; Rowell-Rahier and Pasteels, 1982), which is not an option in most cases, or be applied to a plant or artificial substrate. This latter procedure is appropriate for screening compounds with pest control potential as the bioassay mimics the use of the chemical in the field; but, it is less desirable for studies on the role of chemicals encountered by the insect in a natural situation. The chemical in the leaf may be activated by enzymes released by insect feeding, or may be complexed with inorganic ions, proteins, or other chemicals that may affect its perception by the insect. Moreover, the same chemical may act as neutral, phagostimulant, or deterrent at different concentrations; the response curve may shift according to the other background compounds present, particularly other phagostimulants (Ma, 1972; van Emden, 1978). All this may explain, in part, a lack of correspondence between tests on individual chemicals and tests on either whole plants or crude whole plant extracts (Jermy, 1966; Meisner and Ascher, 1972). Infiltration of a chemical into the plant does not completely avoid these problems if the test plant is not a normal host.

2.1. Substrate Choice and Presentation

Observations of insects feeding in the field with a variety of plant tissues available can provide information about the condition, age, and texture of the preferred tissue that may be useful in choosing a substrate. Both natural and artificial substrates are used, depending on the goal of the experiment. Artificial substrates offer uniformity, evenness of application, and ease of preparation and measurement. However, recent work using both kinds of substrates suggests that an artificial substrate may not be a completely adequate substitute for a natural one. Jermy et al. (1982) found that locusts habituate to nicotine presented on leaf surfaces but not on an artificial substrate. The reason for this is not known but it suggests that the stimulus situation is important. Schoonhoven (1982) remarks that thresholds for the same deterrent may vary as much as 1000 times between natural and artificial substrates, perhaps due to differences in porosity or uptake rates by the insect and therefore differences in the effective concentrations of the chemical presented to the insect. Meisner et al. (1981) suggest that differences in activity of azadirachtin applied to leaves and to an inert substrate are caused by phagostimulants in leaves that mitigate the effects of the deterrent.

For sucking insects, the principal artificial substrate used has been a chemically defined liquid presented between natural (Hamilton, 1930) or artificial membranes (Mittler and Dadd, 1962).

Whatever substrate is chosen, chemical application must not result in textural differences between the control and test substrates that might affect feeding rate. Color differences may also influence insects tested in light (Meisner and Ascher, 1973). The substrate should be presented so that the insect's chemoreceptors encounter it in the usual way; e.g., for edge feeders, substrates are placed above the cage floor.

2.2. Chemical Application

For studying a chemical normally occurring in a potential host, the concentration found in or on the appropriate tissue can be tested in the following ways. (1) For whole extracts, the substrate may be considered to represent the inert part of the plant. After extraction of the plant tissue, the remainder is dried and weighed, and the appropriate volume of the extract is applied to a weight of substrate equal to that of the dried remains. (2) The concentration of particular chemicals may be determined from a sample of the test plants by standard methods and test chemicals applied at those concentrations (Woodhead and Bernays, 1978). (3) Published sources may be consulted (e.g., Alfaro et al., 1981). van Drongelen (1979), when using this method, tested coumarin at a concentration lower than that published, as the chemical is released when the plant is dried or damaged, making published concentrations higher than those found in living plants. Widespread genetic and environmentally induced variation in the concentrations of most plant chemicals requires that a range of concentrations be tested.

A wide range of concentrations should also be tested when searching for thresholds, as responses can be bimodal (Beck, 1956) or reversed at high and low concentrations. This is a particular problem with sucking insects on artificial diets, as uptake is usually very low compared with uptake from plants. For example, Honeyborne (1969) found it necessary to present some insecticidal compounds in diet at 10–20 times the effective dose when applied to plants. Artificial diets for aphids need to contain absolute concentrations, especially of amino acids, very much higher than found in plants, although relative amounts of compounds may be based on analyses of plants (Mittler and Dadd, 1962).

When testing antifeedants for pest control, no attempt need be made to test at natural concentrations as the goal is simply to find effective concentrations for crop protection (Jermy et al., 1981). Higgins and Pedigo (1979) used a foliar phytotoxicity threshold based on the presence of leaf necrosis to determine maximum acceptable concentrations.

Concentrations to test need not be based on any assumptions about the sensitivity of the insect's chemosensilla as very little is known about these sensitivities. Concentrations of some chemicals much lower than those found in plants are perceptible by some insects (Nielson et al., 1979; Ascher et al., 1981).

By chromatographing chemicals re-extracted from the test substrate, some authors have confirmed that the test chemicals have not been altered after application (Erickson and Feeny, 1974; Nielson et al., 1979; Ascher et al., 1981).

Chemicals obtained from commercial sources may differ in their behavioral effects from ones obtained from plants. Ascher et al. (1981) found that grades of a common solvent, methanol, could differ in their effects as well. Natural control substrates with and without solvent should be tested to verify that the solvent has not altered the palatability of the control. If electrophysiological studies are anticipated, the test solutions should include the appropriate electrolyte (Blaney, unpublished observations).

2.3. Natural Substrates

A. Whole Plants

Jermy and Matolcsy (1967) pointed out the advantages of systemic antifeedants for crop protection. Test chemicals are applied either by standing plants or shoots in solutions of the chemical (Erikson and Feeny, 1974) or by soaking the roots in the compound (Meisner et al., 1978). Reed et al. (1982) found that the method of infiltration was critical: azadirachtin applied to the seeds or as a drench to the potting soil of dry seedlings did not show systemic activity, but immersing the roots was effective.

More commonly, the test chemicals are applied to the external surface of the plants by dipping or spraying until runoff. Reed et al. (1982) found that concentrations may differ with these two application methods and suggest that deposits be quantified. A surfactant such as Tween™ or Triton™ may be needed. Darwish and Matolcsy (1981) added 2% potato starch to their solutions as well. They also checked for translocatability and phytotoxic effects of the plant growth regulators tested as antifeedants by surface application.

B. Leaves

Leaves may be infiltrated under pressure (Schoonhoven and Jermy, 1977), but are more commonly treated with the chemical on the surface by dipping, spraying, or painting (e.g., Wensler, 1962; Wyatt, 1969). Leaves should be carefully selected. Damage, disease, wilting, rehydration, and general vigor may all be sources of differential feeding (Lewis, 1979, 1982). Excision may initiate wound responses that could interact with the perception of test chemicals. Raina et al. (1980), however, found that excision did not affect leaf palatability if leaf parts were kept turgid.

C. Leaf Discs

Leaf discs are used more frequently than whole leaves, mainly to normalize foliage area. However, it should be recognized that wound reactions may be more severe for leaf parts and that in certain cases discs can give results completely counter to those obtained for whole leaves (Barnes, 1963).

Rembold et al. (1980) and Ascher et al. (1981) avoided foilage size differences without cutting discs out of leaves by constructing test cages with two bottoms. A plastic petri dish is covered with a petri dish bottom from which a hole is punched with a hot die, exposing an area of specified size to the insect. Other drawbacks of discs can be uneven thickness of discs taken from different leaves and differential uptake (at the cut edge) of chemicals applied by dipping (Antonious and Saito, 1981). Jermy et al. (1981) avoided the latter problem by applying chemicals with a brush or by spraying from a calibrated spray tower. Alternatively, leaves can be sprayed in the field and discs removed for later laboratory testing. In certain cases, cut discs may be entirely appropriate as in Waller's (1982) study of the role of leaf hardness and chemistry in leaf-cutting ant host choice.

Leaf discs are not recommended for work with sucking insects. Physiological changes seem to affect the phloem contents very rapidly (e.g., Müller, 1966; van Emden and Bashford, 1976) and may well induce a generally deterrent condition.

D. Other Natural Substances

Specialized substrates include twigs for sawflies (All and Benjamin, 1976; Ohigashi et al., 1981) and bark beetles (Ascher et al., 1975). Blocks of wood (Lenz and Williams, 1980), board, logs, and paper towel discs (Rust and Reierson, 1977) have all been used for termites. Bark beetle response to phloem extracts of different sources was tested with extract-coated cellulose packed into gelatin capsules (Elkinton et al., 1981).

2.4. Artificial Substrates

Phagostimulants are added to unpalatable artificial substrates for testing deterrents. Sugars, alone or in combination with other chemicals, are the most widely used. If there are no reported studies on phagostimulants for a particular insect, likely candidate chemicals can be assayed (Ascher et al., 1976; Cook, 1977). A particular test chemical may be perceived as neutral, stimulatory, or deterrent, depending on the concentration of the background sucrose (Ma, 1972). To interpret the role of the chemical in the natural feeding responses, the phagostimulant should usually therefore be applied at concentrations found in the host, though this may be too low with assays using artificial diets and sucking insects (see above).

Agar and agar cellulose, presented in petri dishes or as discs, have been widely used (e.g., Hsiao and Fraenkel, 1968). Test chemicals are applied to the surface or mixed into hot solution. To test chemicals in both the surface and matrix of a leaf or stem, a modification of the oviposition substrate used by Szentesi et al. (1979) might be devised. A dome of agar is covered by a wax coating; test and control chemicals can be added to both agar and wax.

Diet, in cups or discs, has been used to test chemicals against gypsy moth (Trial and Dimond, 1979; Doskotch et al., 1980), cotton bollworm (Klocke and Chan, 1982), armyworm (Meisner et al., 1982), and other lepidopteran larvae (Kaethler et al., 1982). Freedman et al. (1979) found that larvae penetrating into the diet received insufficient exposure to the test chemical applied to the surface only, and suggest mixing chemical throughout. For chemicals not soluble in water the solvent should be removed first by coating the chemical onto Alphacel or similar carrier under vacuum (Chan et al., 1978).

Artificial diets are a principal substrate for work with sucking insects. Developing an adequate fully defined diet for a given species is a speculative and long-term program. Even where a diet has been published, this usually involves chance success of a particular genotype of the species and may not be successful in another laboratory. The question then arises, how typical are the responses of a specially diet-adapted strain? For the purposes of testing phagostimulants and deterrents, a simpler diet should suffice. Such a simple less defined diet has been published for *Myzus persicae* (Mittler and Koski, 1974), or interpretable results can often be obtained by substituting a simple sucrose solution or even water for the complex diet between Parafilm℠ membranes.

Thus for short-term tests of both phagostimulants and deterrents, a carrier with no nutritive value may be more suitable. Filter paper has been used in the past (e.g., Marek, 1961 for testing aphid probing responses to various chemicals), but Millipore filters (Bristow et al., 1979) and glass fiber discs (Whatman GF/A) (Städler and Hansen, 1976; Adams and Bernays, 1978) are chemically pure, absorbent, and readily eaten by many insects when phagostimulant is added (Fig. 1). Test chemicals are usually applied with a pipette; rapid drying aids even spread of the chemical. Woodhead (personal communication), using ^{14}C-labeled glucose in ethanol and ^{14}C-labeled lauric acid in hexane, found labeled chemical uniformly distributed throughout the disc with no accumulation on either surface. However, 35% more activity was discovered in the 2-mm peripheral section of a 7-mm disc. This figure is an improvement over similar tests with labeled glucose on elder pith discs (Cook, 1976), and might be reduced with other methods of application. Similar measurement of chemical distribution would be useful for other substrates.

The "styropor method" developed by Ascher and Meisner (1973) has been used successfully on several insects (Ascher and Gurevitz, 1972, Meisner et at., 1974; Meisner and Skatulla, 1975; Ascher and Nemny, 1979). The development of the technique is a good model study for similar

Figure 1. A binary choice test in which a locust is offered a choice between two glass fiber discs, one saturated with sucrose only and one with sucrose and the test chemical. (Photo courtesy of the Tropical Development and Research Institute.)

attempts, as important variables were systematically studied. Test chemical is applied to lamellae or discs of foamed polystyrene, styrofoam (Ma and Kubo, 1977), or polyurethane (Ascher and Nemny, 1979), the latter being used for solutions in nonpolar solvent.

In testing contact chemoreception and oviposition by *Acanthoscelides obtectus,* Jermy and Szentesi (1978) used glass slides and glass beads coated with test chemical in 2% starch solution. Woodhead and Bernays (1977) avoided substrates by presenting chemicals directly to the mouthparts with a syringe. Metcalf et al. (1980) bypassed substrates by simply allowing beetles to feed directly on thin-layer chromatography plates containing cucurbitacin phagostimulants. This technique works well when studying species-specific phagostimulants to which the insects are very sensitive, but its general applicability has yet to be explored.

III. Conditions of the Test

Food deprivation may be desirable to shorten the time of the test (Albert and Jerrett, 1981), but should be undertaken with caution. Thresholds for phagostimulants may or may not alter as hunger increases. As these vary among insects and chemicals, pilot tests can be run to determine if deprivation affects behavior. Jermy et al. (1968) and others terminate tests when 50% of either substrate in a choice trial is consumed so that insects will not lose discrimination due to hunger. Water deprivation before and during the test may lead to reduced feeding. Water is sometimes provided during the test, but may not be necessary for short tests (Clark, 1981).

Ingestion of tannins and other chemicals may be reduced in the absence of water. Mittler (1967) found that *Myzus persicae* were reluctant to settle on artificial diet choice chambers when transferred direct from radish seedlings. Starvation did not overcome this reduction but a period of 24 hr on 20% sucrose solution did. van Emden (1977) showed that starvation of aphids on water for more than 4 hr caused abnormal later feeding behavior.

Accumulating laboratory evidence indicates that previous experience may affect insect diet choice by one of the following: induction (Jermy et al., 1968), habituation (Jermy et al., 1982), food aversion learning (Dethier, 1980), and associative learning (Blaney and Winstanley, 1982). Using an insect only once and for a short test will avoid some of these problems. The diet previous to testing should not include test chemicals if possible, but should contain a balance of nutrients to prevent increased activity or specific hungers. Rearing on artificial diet does not ensure a naive insect. Städler and Hanson (1978) found that *Manduca* larvae were induced to feed preferentially on diet constituents, and thus "only freshly emerged, first instar larvae can be considered uninduced or naive." Smith (1978) suggests some conditioning to the test plant may be necessary when insects are reared on another species.

A few studies indicate that insects lose selectivity in the later instars or after the adult molt, although this is not thoroughly documented (Jermy et al., 1981; Wege and Bernays, unpublished observations; Isman and Duffey, 1982). Periods of low or irregular feeding, either in diurnal cycles (eg., Cull and van Emden, 1977) or before a molt, will give poor results. Pilot observations may therefore be necessary to establish times of maximum feeding. Aphids are increasingly restless and less willing to settle on any substrate (even a preferred one) as each day progresses after the scotophase (van Emden, unpublished observation).

Sex-related differences in feeding rates or preferences have been found in various insects (Reed et al., 1982). These may not be great enough to warrant separate testing (Alfaro et al., 1980).

Temperature, humidity, light levels, and the number of insects to be tested simultaneously can be determined by field observations. Group-feeding insects are best tested in groups lest reduced feeding due to isolation obscure effects due to a deterrent. Individually feeding insects are often tested in groups to ease measurement of feeding activity, but this violates assumptions of independence necessary for statistical tests. Crowding may also lead to behavioral interference: Szentesi (1981) found that insect-derived substances deposited on leaves affected feeding by subsequent insects.

Insect behavior in cages may be profoundly affected by lighting and temperature, to the extent that feeding will not occur unless conditions

Figure 2. Choice test using leaf discs for lepidopteran larva. (Photo courtesy of the Tropical Development and Research Institute.)

are suitable. Beck (1956) tested animals in the dark to avoid overriding positive phototaxis. Testing in the dark is also useful if visual cues are suspected. Most workers with sucking insects have performed their short-term tests in the dark (e.g., Kennedy and Booth, 1951). Uniformity of lighting and temperature are also necessary to prevent position effects. To test for the existence of position effects in binary choice tests, a double control can be run. Caged mobile insects may develop routes that lead to differential encounter with one substrate (Lewis and Harris, unpublished observations). Use of circular cages may help prevent this (Jermy et al., 1968; Weston and Miller, 1985).

Cages should be large enough to permit locomotion as there may be a physiological requirement for locomotion before or after feeding or contact with a deterrent (Lewis, unpublished observations). Insects may not, however, find food even in moderately large cages. The insect may not even search, possibly due to insufficient levels of normal feeding cues, such as plant volatile chemicals. Extraneous odors may also influence the amount of food ingested (Mordue [Luntz], 1979) and possibly host choice (Whelan, 1982). For laboratory testing, a common cage design is a clear polyethylene box or petri dish with appropriate ventilation, extra moisture in the form of saturated filter paper (Reed et al., 1981), and a roost if necessary (Blaney et al., 1973) (Figs. 1 and 2). Insects may also be caged on entire plants, branches, or leaves in the field (Bernays et al., 1977).

IV. Measurement and Observation

4.1. Chewing Insects

In most bioassays for chewing insects, the weight or area of substrate ingested is used to determine the effect of a chemical on feeding. This can be misleading, particularly for phagostimulants, as pointed out by Cook (1976) and Aspirot and Laugé (1981). When testing a log series of sucrose concentrations, Cook found that the amount of substrate a locust consumes increases with increasing molarity up to 0.125 M and then decreases at 0.625 M. The amount of sucrose ingested, however, continues to increase. Meal size on the low-concentration discs may be regulated by stretch receptors in the crop and on the high-concentration discs by osmotic or chemosensory factors. These kinds of controls are likely to occur in other insects (Bernays and Simpson, 1982).

A. Amount Ingested: Direct Measures

If the test chemical is evenly applied, the amount of substrate ingested can be used to calculate the amount of chemical ingested. For substrates of appropriate consistency, e.g., discs or styropor lamellae, ingestion can be estimated from area loss measured by graph paper tracings, planimeter, ocular grid, electronic area meter, or an electronic digitizer attached to a computer (last method: Raffa and Berryman, 1982). This method requires a constant relationship between dry and fresh weight and that the insect feed completely through the substrate. For small insects photographic enlargements of the substrate can be measured in the above ways (Bristow et al., 1979). Visual estimate of area loss is best used only when frequently confirmed with an independent, though more time-consuming, measure (Jones et al., 1981).

Feeding on the above substrates and ones not measureable by area loss, e.g., diet, can be assessed by substrate dry weight loss. When comparing among insects, the amount eaten by an insect is divided by weight of the insect. For small insects feeding on diet, the only way to measure the amount ingested may be by radiolabeling a nutrient (see also Kogan, *this volume*).

B. Amount Ingested: Indirect Measures

Although fecal production is frequently used, there are several sources of error. The gut must be empty at the beginning of the test, but starved insects may in fact retain food in the gut (Waldbauer, 1964). Feeding rates, and hence fecal production, on less palatable or dilute food may exceed that on more palatable food due to controls on intake and rate of passage through the gut (Dethier, 1982).

The insect can be weighed before and after the test, controlling for weight loss due to respiration and excretion during the course of the test, although this can be tedious. Using crop weight (Bernays et al., 1976) may avoid these problems, but recent work suggests that food sometimes moves into or out of the crop during the course of a meal, particularly if the food is very wet (Bernays et al., unpublished observations). In some cases, inspection of the crop contents may be necessary to determine if feeding has occurred (Elkinton et al., 1981).

The duration of feeding can be used only if there is a constant relationship between amount ingested and rate of feeding, an assumption unlikely to be met when testing stimulants or deterrents against a control. For some insects, the existence of feeding marks and punctures has been used to determine if feeding has occurred (Kon et al., 1978). The position of the insect may be observed through time but this should not be the only measure of feeding, as it may reflect only arrestment and not feeding.

C. Observation

Observation of all activity through time may be more revealing of the effects of a chemical than a simple measurement of total substrate ingested. The amount of substrate lost during the course of a test may either reflect normal feeding with typical meal sizes and interfeeds, or many small feeds, taken with increasing frequency as hunger overrides the effect of a deterrent. Observations can either be direct or indirect. Indirect observations can be made with various forms of actographs (Ma, 1972; Blaney et al., 1973, Bernays, 1979) and feeding events inferred from the record of activity. Feeding may be monitored with more specialized devices that, for instance, record mandible movements or frequency and duration of meals (Kogan, 1973; Jones et al., 1979; Shimizu et al., 1980). The patterning of meals can be analyzed with statistical techniques discussed in Simpson (1981, 1982).

Direct observation, although more time-consuming, has the advantage of revealing flaws in cage design. Important behaviors associated with insect contact with a chemical would be missed without direct observation; e.g., colony disruption in sawflies (All and Benjamin, 1976) and associative learning in locusts (Blaney and Winstanley, 1982). Video recording or filming provides a permanent record but lighting requirements may be excessive.

4.2. Sucking Insects

A. *Amount Ingested: Direct Measures*

This approach has been used on artificial diets (Mittler, 1967), and weighing sachets before and after feeding by a group of 20+ aphids can give consistent results in times as short as 7 hr. The sachet weights need to be corrected for the weight loss of sachets on which no aphids have fed. Radioisotopes have also been used to measure food uptake from artificial diets (Erhardt, 1968).

B. *Amount Ingested: Indirect Measures*

With honeydew producers, the amount of honeydew excreted is accepted by most workers, without real evidence, as proportional to aphid feeding. Honeydew production is often measured as number of droplets (using bromocresol blue or ninhydrin as a stain, Auclair, 1958; Banks and Macaulay, 1964), but is more usefully measured by volume if it is to be correlated with ingestion. C.H.B. Honeyborne (unpublished observations) collected honeydew in a dish of light oil (specific gravity cá. 0.7) equilibrated with 10% sucrose solution to prevent significant exchange of water between honeydew and oil. The diameter of the dish was chosen to exceed the scatter of honeydew droplets and the dish was placed to give a short droplet falling distance to minimize evaporation. At the end of the collection period, the oil was centrifuged in a tube with 1-mm diameter capillary base, from which the volume of honeydew (which spins down out of the oil) can be recorded.

The longevity of adult aphids appears related to the intake of carbohydrates (Banks, 1965), and so can be used to distinguish large differences in ingestion rate on artificial diets to which secondary chemicals have been added (van Emden, 1978).

The growth rate of sucking insects cannot be used as an indirect measure of ingestion, unless losses due to excretion (honeydew) and body evaporation are measured simultaneously. Over a longer period, the weight of exuviae also needs to be recorded. This is a complex series of measurements, subject to cumulative errors; the appropriate techniques are discussed by Adams and van Emden (1972).

C. *Observation*

It is unfortunately impossible to assess visually when sucking insects are feeding, except under rather exceptional circumstances. The criterion sometimes suggested for aphids, that the antennae are at rest and held backwards along the body, merely means the insect is in a probing or feeding posture and not that liquid is being imbibed. T.C.R. White (un-published observations) was, however, able to time feeding bouts in *Myzus persicae* feeding on solutions through a very clear membrane ("Cling-film"™) and observing the insects under high magnification. Using only a single membrane, he found that pulsations at the front of the aphid's head were clearly visible. This seems a potentially useful technique, par-ticularly for testing the insect's response to secondary chemicals.

Otherwise the only approach to recording feeding activity through time is the use of an electronic method (originated by McLean and Kinsey, 1964). It can be applied to insects feeding on plants as well as on artificial diets. By coupling the aphid with gold wire as a link between an amplifier and a source of DC voltage applied to the substrate, reduction in resistance when the stylet canals are filled with liquid can be recorded continuously. Different patterns of record obtained have been interpreted as representing salivation, probing, and continuous feeding. The original apparatus has been modified by latter workers (Tjallingii, 1977; Kawabe and McLean, 1978).

V. Design and Analysis

5.1. Short- and Long-Term Tests

Since the behavioral effect of a chemical may or may not be independent of its nutritional value, these two properties must be experimentally sep-arated before designating a chemical as a feeding stimulant or deterrent. This is done by running short-term assays to avoid post ingestional effects on food choice or, alternatively, by bypassing the sensory system alto-gether. The latter method tests the effect of the chemical directly by de-livering it in capsules (Szentesi and Bernays, 1984) or by injection or can-nulation (Cottee and Mordue [Luntz], 1982). For long-term tests, consumption, growth, and digestive efficiency can all be measured sep-arately (chapter 6, *this volume*). Performing both short- and long-term tests provides the most information (Klocke and Chan, 1982; Lewis, 1984; Lewis and Bernays, 1985) as data on changes in behavior through lengthy exposure to a chemical may be useful.

5.2. No-Choice and Choice Tests

Test substances are typically compared with control or other test substrates in no-choice, binary, or multiple-choice designs. The number and ar-rangement of substrates varies greatly with species and cage design. A commonly used layout is based on Jermy et al. (1968) (Fig. 2).

With sucking insects, multi-choice arenas involving areas of Parafilm™ with artificial solutions have often been used (Cartier and Auclair, 1964;

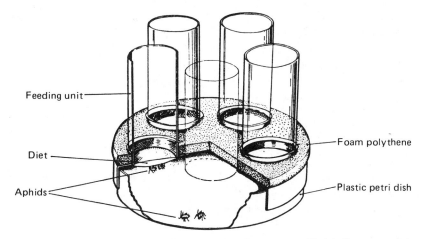

Figure 3. Chamber in which aphids can select between artificial diets containing various chemicals. (From Honyeborne CHB, 1969.)

Honeyborne, 1969; Fig. 3). For binary trials on plant leaves, techniques have been based on the principle of exposing equal areas of attached leaf to the insects. Either a single cage bridging two leaves or a common atrium leading to two open ports against the leaf surface (Kennedy and Booth, 1951) has been used. To determine the preferred substrate for feeding, systemic insecticide may be applied to the alternate choice in each of an even number of replicates (Matthieu and Verhoyen, 1980). Any behavioral interaction with the insecticide is thus allowed for in the analysis (van Emden, unpublished observations).

Results are frequently used to construct preference hierarchies. The term preference implies a choice made on the basis of a comparison. Its use in connection with no-choice tests should be avoided. Furthermore, in choice tests an insect may not actually make a comparison—an inadequate substrate may simply cause locomotion whereas a suitable one will lead to feeding. The insect will then appear to have made a comparison. That "preference" is the result of a differential movement (kinesis) rather than a taxis is inevitable with sucking insects from the very nature of their probing behavior, which is a precursor of any selection behavior.

Designs based on field observations of insect foraging in habitats varying in plant composition and distribution would be more desirable, although seldom used. For insects from natural rather than agricultural habitats, this might require varied ratios of test substrates offered simultaneously or sequentially. Data on latency to feeding and time between feeding bouts (Lewis and Bernays, 1985) could be used with amount eaten for assigning preference rules. For agriculturally important insects, a no-choice test might be most suitable, and is certainly needed in plant resistance studies and work on the biological control of weeds by insects.

Until such tests are designed, choice tests will predominate. These have revealed that the palatability of a chemical or plant is relative to the available alternatives (Meisner and Ascher, 1972; Jermy and Szentesi, 1978; Richardson and Whitaker, 1982), and to the gustatory background against which the chemicals are presented. Measures of palatability based on choice tests between two concentrations may differ from ones based on tests between each concentration and a control. For example, Cook (1976) found in a choice test between a high-sucrose disc and a control that locusts consumed less from the sucrose disc than in a test between a low-sucrose disc and a control, presumably because hemolymph osmotic factors lead to the termination of the meal in the first case and volumetric factors in the second. When given a binary test between the two sucrose discs, locusts consumed more of the high-sucrose disc.

5.3. Analysis

The ratio of the amount of test substrate consumed to the total test and control consumed is calculated and tested with appropriate parametric or nonparametric statistics. Binary choice tests (which are sometimes experimentally the most practicable) can be combined into a multi-choice hierarchy by seeking a pattern of consistency (Wearing, 1968). Nielson et al. (1979) and others discard results if the total amount ingested is above or below certain extreme values. The appropriate statistic will vary with the number of chemicals tested, normality of the data, and number of replicates. Cluster analysis can be an effective technique when testing more than one chemical (Wasserman, 1982; Chararas et al., 1982). Techniques developed in vertebrate chemoreception and preference studies may be useful as designs become more complex (Kroeze, 1979; Farentinos et al., 1981, Partridge, 1981; Partridge and Maclean, 1981).

One factor causing an increase in complexity of experiments is an awareness of the importance of combinations of deterrents and phagostimulants in determining host choice. Mixtures of sugars and salts, phagostimulants, deterrents, and plant extracts have been found to interact in synergistic or additive ways (Hsiao and Fraenkel, 1968; Ma, 1972; Adams and Bernays, 1978; Kon et al., 1978; Woodhead and Bernays, 1978; Nielson et al., 1979; Blaney, 1980).

Variability in thresholds, inducibility, and other behavioral responses are common to many bioassays (Jermy et al., 1968; Ma, 1972). Variability can occur within and between insects in samples of inbred and wild-type insects. If all sources of variability in testing procedure and pretreatment of insects are controlled, genetic variation may be suspected. Information on the genetic background of test insects would be useful for interpretation of the data.

Sample size determinations based on the variability of the data are the most desirable. Constraints on insect or chemical availability may be overriding. Still (1982) offers a number of suggestions for maintaining power with a reduced sample size, while cautioning that demonstrating nonpreference requires a greater sample size than does a demonstration of preference. Still's suggestions include increasing levels of alpha; this may be applicable to some bioassays. Chi-square analysis is very commonly used in this field of work. This test is particularly dependent on sample size; at the very least heterogeneity between replicates needs to be tested.

There is no standard terminology for effective doses of potential pest control agents. Probit analysis has been borrowed from pesticide studies to construct dose–response curves (Ascher and Nissim, 1965). Threshold units (Jermy et al., 1981) and the protective concentration (PC), the concentration necessary for a given percentage of leaf protection, have been used. Calculating PC values allows comparison of the effectiveness of a chemical among insects (Kubo and Klocke, 1982).Comparisons between species must be made with caution, as differences in feeding regulation may be great.

VI. Field and Laboratory

The majority of feeding bioassays with plant-derived chemicals has been performed under laboratory or, in some cases, greenhouse conditions, although field tests of metal-containing fungicides as antifeedants have been done for some time (Jermy and Matolcsy, 1967; Ascher, 1979). Field tests have at times confirmed results from laboratory tests for naturally occurring chemicals (All and Benjamin, 1976; Trial and Dimond, 1979; Metcalf et al., 1980). However, laboratory results cannot necessarily be extrapolated to field conditions for one or more of the following reasons: (1) alteration of the chemical in the field by environmental factors such as sunlight (Stokes and Redfern, 1982); (2) habituation of insects with no alternatives available; (3) use of artificial substrate in the laboratory (Jermy et al., 1982); (4) insect movement away from treated plants (Karieva, 1982); and (5) composition of the plant community (Risch, 1980). Many of the principles discussed in the preceding sections on design of laboratory bioassays apply for field tests, but field conditions will amplify differences among insects, particularly with respect to movement patterns and therefore require procedures adapted for the particular system.

VII. Conclusions

Although technological advances in substrate materials and chemical application may ease some of the problems discussed above, significant progress in bioassays can be made simply by improving behavioral observations and experimental design. Although variability in insect response is considered to be an inevitable problem, bioassays can be designed to uncover its source, thereby increasing our ability to predict evolutionary steps in the interaction between plant and insect. This task will be aided by neurophysiological techniques, with chemicals tested both on whole organisms and chemosensilla.

Acknowledgments. A.C.L. thanks E. Bernays, A. Szentesi, W. Blaney, R. Chapman, T. Jermy, K. Ascher, J. Meisner, C. Jones, D. Reed, D. Waller, and C. Lewis for papers and/or comments.

References

Adams CM, Bernays EA (1978) The effect of combinations of deterrents on the feeding behaviour of *Locusta migratoria*. Entomol Exp Appl **23**:101–109.

Adams JB, van Emden HF (1972) The biological properties of aphids and their host plant relationships. In: Aphid Technology. van Emden HF (ed), Academic Press, London, pp 47–104.

Albert PJ, Jerrett PA (1981) Feeding preferences of spruce budworm (*Choristoneura fumiferana* Clem) larvae to some host-plant chemicals. J Chem Ecol **7**:391–402.

Alfaro RI, Pierce HD, Borden JH, Oehlschlager AC (1980) Role of volatile and nonvolatile components of Sitka spruce bark as feeding stimulants for *Pissodes strobi* Peck (Coleoptera: Curculionidae). Can J Zool **58**:626–632.

Alfaro RI, Pierce HD, Borden JH, Oehlschlager AC (1981) Insect feeding and oviposition deterrents from western red cedar foliage. J Chem Ecol 7:39–48.

All JN, Benjamin DM (1976) Potential of antifeedants to control larval feeding of selected *Neodiprion* sawflies (Hymenoptera: Diprionidae) Can Entomol **108**:1137–1143.

Antonious AG, Saito T (1981) Mode of action of antifeeding compounds in the larvae of the Tobacco Cutworm, *Spodoptera litura* (F.) (Lepidoptera: Noctuidae) I. Antifeeding activities of chlordimeform and some plant diterpenes. Appl Entomol Zool **16**:328–334.

Ascher KRS (1979) Fifteen years (1963–1978) of organotin antifeedants—a chronological bibliography. Phytoparasitica 7:117–137.

Ascher KRS, Gurevitz E (1972) A further use of the styropor method: evaluating the response of the fruit bark beetle, *Scolytus (Ruguloscolytus) mediterraneus* Eggers, to extracts of its host plants. Z Pflkrankh Pflschutz **79**:216–222.

Ascher KRS, Gurevitz E, Renneh S, Nemny NE (1975) The penetration of females of the fruit bark beetle *Scolytus mediterraneus* Eggers into antifeedant-treated twigs in laboratory tests. Z Pflkrankh Pflschutz **82**:378–383.

Ascher KRS, Meisner J (1973) Evaluation of a method for assay of phagostimulants with *Spodoptera littoralis* larvae under various conditions. Entomol Exp Appl **16**:101–114.

Ascher KRS, Meisner J Flowers HM (1976) Effects of amino acids on the feeding behavior of the larva of the Egyptian cotton leafworm, *Spodoptera littoralis* Boisd. Phytoparasitica **4**:85–91.

Ascher KRS, Nemny NE (1979) Use of foamed polyurethane as a carrier for phagostimulant assays with *Spodoptera littoralis* larvae. Entomol Exp Appl **25**:346–349.

Ascher KRS, Nissim S (1965) Quantitative aspects of antifeeding: comparing "antifeedants" by assay with *P. litura*. Int Pest Control **7**:21–23.

Ascher KRS, Schmutterer, Glotter E, Kirson I (1981) Withanolides and related ergostane-type steriods as antifeedants for larvae of *Epilachna varivestis* (Coleoptera: Chrysomelidae). Phytoparasitica **9**:197–205.

Aspirot J, Laugé G (1981) Etude expérimentale de l'action phagostimulante du saccharose et de la sinigrine et mise en évidence de phénomenes de regulation chez le criquet pèlerin *Schistocerca gregaria* (Orthoptère: Acrididae). Reprod Nutr Dév **21**:695–704.

Auclair JL (1958) Honeydew excretion in the pea aphid, *Acyrthosiphon pisum* (Harr.) (Homoptera: Aphididae). J Insect Physiol **2**:330–337.

Banks, CJ (1965) Aphid nutrition and reproduction. Ann Rep Rothamsted Exp Stn 1964: pp 299–309.

Banks CJ, Macaulay EDM (1964) The feeding, growth and reproduction of *Aphis fabae* Scop on *Vicia faba* under experimental conditions. Ann Appl Biol **53**:229–242.

Barnes OL (1963) Food plant tests with the differential grasshopper. J Econ Entomol **56**:396–399.

Beck SD (1956) A bimodal response to dietary sugars by an insect. Biol Bull **110**:219–228.

Beck SD (1965) Resistance of plants to insects. Annu Rev Entomol **10**:202–232.

Bernays EA (1979) The use of Doppler actographs to measure locomotor activity in locust nymphs. Entomol Exp Appl **26**:136–141.

Bernays EA, Chapman RF, Horsey J, Leather EM (1974) The inhibitory effect of seedling grasses on feeding and survival of Acridids (Orthoptera). Bull Entomol Res **64**:413–420.

Bernays EA, Chapman RF, Leather EM, McCaffery AR (1977) The relationship of *Zonocerus variegatus* (L.) (Acridoidea: Pyrgomorphidae) with cassava. Bull Entomol Res **67**:391–404.

Bernays EA, Chapman RF, MacDonald J, Salter JER (1976) The degree of oligophagy in *Locusta migratoria* (L.) Ecol Entomol **1**:223–230.

Bernays EA, Simpson SJ (1982) Control of food intake. Adv Insect Physiol **16**:59–118.

Blaney WM (1980) Chemoreception and food selection by locusts. Olfact Taste **7**:127–130.

Blaney WM, Chapman RF, Wilson A (1973) The pattern of feeding of *Locusta migratoria* (L.) (Orthoptera: Acrididae). Acrida **2**:119–137.

Blaney WM, Winstanley C (1982) Food selection behavior in *Locusta migratoria* In: Proceedings of the 5th International Symposium on Insect–Plant Relationships, Wageningen, 1982. Visser JH, Minks AK (eds), Pudoc, Wageningen, pp 365–366.

Bright AM, Lea P, Kueh J, Woodcock C, Holloman D, Scott G (1982) Proline content does not influence pest and disease susceptibility of barley. Nature 295:592–593.

Bristow PR, Doss RP, Campbell RL (1979) A membrane filter bioassay for studying phagostimulatory materials in leaf extracts. Ann Entomol Soc Am 72:16–18.

Cartier JJ, Auclair JL (1964) Pea aphid behavior: colour preference on a chemical diet. Can Entomol, 96:1240–1243.

Chan BG, Waiss AC, Stanley WL, Goodban AE (1978) A rapid diet preparation method for antibiotic phytochemical bioassay. J Econ Entomol 71:366–368.

Chapman RF (1974) The chemical inhibition of feeding by phytophagous insects: a review. Bull Entomol Res 64:339–363.

Chararas C, Revolon C, Feinberg M, Ducauze C (1982) Preference of certain Scolytidae for different conifers: a statistical approach. J Chem Ecol 8:1093–1109.

Clark JV (1981) The glass microfibre disc method used to quantify feeding in the African armyworm, *Spodoptera exempta*. Entomol Exp Appl 30:195–197.

Cook AG (1976) A critical review of the methodology and interpretation of experiments designed to assay the phagostimulatory activity of chemicals to phytophagous insects. Symp Biol Hung 16:47–54.

Cook AG (1977) Nutrient chemicals as phagostimulants for *Locusta migratoria*. Ecol Entomol 2:113–121.

Cottee P, Mordue (Luntz) A (1982) An investigation into the physiological action of feeding deterrents. In: Proceedings of the 5th International Symposium on Insect–Plant Relationships, Wageningen, 1982. Visser JH, Minks AK (eds), Pudoc, Wageningen, pp 379–380.

Cull DC, van Emden HF (1977) The effect on *Aphis fabae* of diel changes in their food quality. Physiol Entomol 2:109–115.

Darwish YM, Matolcsy G (1981) Feeding inhibitory action of some heterocyclic quartenary ammonium salts on Colorado beetle adults (*Leptinotarsa decemlineata* Say (Col., Chrysomelidae). Z Ang Entomol 91:252–256.

Dethier V (1947) Chemical Insect Attractants and Repellents. Blakiston, Philadelphia.

Dethier VG (1976) The Hungry Fly. Harvard University Press, Cambridge.

Dethier VG (1980) Food-aversion learning in two polyphagous caterpillars, *Diacrisia virginica* and *Estigmene congrua*. Physiol Entomol 5:321–325.

Dethier VG (1982) Mechanism of host-plant recognition. Entomol Exp Appl 31:49–56.

Dethier VG, Barton Browne L, Smith CN (1960) The designation of chemicals in terms of the responses they elicit from insects. J Econ Entomol 53:134–136.

Doskotch RW, Cheng H-Y, Odell TM, Girard L (1980) Nerolidol: an antifeeding sesquiterpene alcohol for gypsy moth larvae from *Malalenca Leucadendron*. J Chem Ecol 6:845–851.

Elkinton JS, Wood DL, Browne LE (1981) Feeding and boring behavior of the bark beetle *Ips paraconfusus*, in extracts of Ponderosa pine phloem. J Chem Ecol 7:209–220.

Erhardt P (1968) Nachweisseiner durch symbiotische Mikroorganismen be wirkten Sterinsynthese in kunstlich ernährten Aphiden (Homoptera, Rhynchota, Insecta). Experientia 24:82.

Erikson JM, Feeny P (1974) Sinigrin: a chemical barrier to the black swallowtail butterfly, *Papilio polyxenes*. *Ecology* **55**:103–111.

Farentinos RC, Capretta PJ, Kepner RE, Littlefield VM (1981) Selective herbivory in tassel-eared squirrels: role of monoterpenes in Ponderosa pines chosen as feeding trees. *Science* **213**:1273–1275.

Freedman B, Nowak LJ, Kwolek WF, Berry EC, Guthrie WD (1979) A bioassay for plant-derived pest control agents using the European cornborer. *J Econ Entomol* **72**:541–545.

Hamilton MA (1930) Notes on the culturing of insects for virus work. *Ann Appl Biol* **17**:487–492.

Higgins RA, Pedigo LP (1979) Evaluation of guazatine triacetate as an antifeedant feeding deterrent for the green cloverworm on soybeans. *J Econ Entomol* **72**:680–686.

Honeyborne CHB (1969) An Investigation of the Responses of Aphids to Plants Treated with Growth Regulators. Ph.D. thesis, University of Reading.

Hsiao TH, Fraenkel G (1968) Isolation of phagostimulative substances from the host plant of the Colorado potato beetle. *Ann Entomol Soc Am*, **61**:476–484.

Isman MB, Duffey SS (1982) Toxicity of tomato phenolic compounds to the fruitworm, *Heliothis zea*. *Entomol Exp Appl* **31**:370–376.

Jermy T (1966) Feeding inhibitors and food preference in chewing phytophaogus insects. *Entomol Exp Appl* **9**:1–12.

Jermy T, Bernays EA, Szentesi A (1982) The effect of repeated exposure to feeding deterrents on their acceptability to phytophagous insects. In: Proceedings of the 5th International Symposium on Insect–Plant Relationships, Wageningen 1982. Visser JH, Minks AK (eds), Pudoc, Wageningen, pp 25–32.

Jermy T, Butt BA, McDonough, Dreyer DL, Rose AF (1981) Antifeedants for the Colorado potato beetle—I. Antifeeding consituents of some plants from the sagebrush community. *Insect Sci Appl* **1**:237–242.

Jermy T, Hanson F, Dethier V (1968) Induction of specific food preference in lepidopterous larvae. *Entomol Exp Appl* **11**:211–230.

Jermy T, Matolcsy G (1967) Antifeeding effect of some systemic compounds on chewing phytophagous insects. *Acta Phytopath Acad Sci Hung* **2**:219–224.

Jermy T, Szentesi A (1978) The role of inhibitory stimuli in the choice of oviposition site by phytophagous insects. *Entomol Exp Appl* **24**:458–471.

Jones CG (1979) An automatic feeding detector (AFD) for use in insect behavioral studies. *Entomol Exp Appl* **25**:112–115.

Jones CG, Hoggard MP, Blum MS (1981) Pattern and process in insect feeding behavior: a quantitative analysis of the Mexican bean beetle, *Epilachna varivestis*. *Entomol Exp Appl* **30**:254–264.

Kaethler F, Pree DJ, Bown AW (1982) HCN: a feeding deterrent in peach to the oblique-banded leafroller, *Choristoneura rosaceana* (Lepidoptera: Tortricidae). *Ann Entomol Soc Am* **75**:568–573.

Karieva P (1982) Exclusion experiments and the competitive release of insects feeding on collards. *Ecology* **63**:696–704.

Kawabie S, McLean DL (1978) Electronically recorded waveforms associated with salivation and ingestion behavior of the aster leafhopper, *Macrosteles fascifrons* Stal (Homoptera: Cicadellidae). *Appl Entomol Zool* **13**:143–148.

Kennedy JS, Booth CO (1951) Host alternation in *Aphis fabae* Scop. I. Feeding preferences and fecundity in relation to the age and kind of leaves. Ann Appl Biol **38**:25–64.

Klocke JA, Chan BG (1982) Effects of cotton condensed tannin on feeding and digestion in the cotton pest, *Heliothis zea*. J Insect Physiol **28**:911–915.

Kogan M (1973) Automatic recordings of masticatory motions of leaf-chewing insects. Ann Entomol Soc Am **66**:66–69.

Kon RT, Zabik MJ, Webster JA, Leavitt RA (1978) Cereal leaf beetle response to biochemicals from barley and pea seedlings. I. Crude extract, hydrophobic and hydrophilic fractions. J Chem Ecol **4**:511–522.

Kroeze HA (ed) (1979) Preference Behavior and Chemoreception, Proceedings of a Symposium, Horst, The Netherlands, May 1979. Information Retrieval Limited, London.

Kubo I., Klocke JA (1982) An insect growth inhibitor from *Trichilia roka* (Meliaceae). Experientia **38**:639–640.

Lenz M, Williams ER (1980) Influence of container, matrix volume and group size on survival and feeding activity in species of *Coptotermes* and *Nasutitermes* (Isoptera: Rhinotermitidae, Termitidae). Material Organismen **15**:25–46.

Lewis AC (1979) Feeding preference for diseased and wilted sunflower in the grasshopper *Melanoplus differentialis*. Entomol Exp Appl **26**:202–207.

Lewis AC (1982) Leaf wilting alters a plant species ranking by the grasshopper *Melanoplus differentialis*. Ecol Entomol **7**:391–395.

Lewis AC (1984) Plant quality and grasshopper feeding: effects of sunflower condition on preference and performance in *Melanoplus differentialis*. Ecology **65**:836–843.

Lewis AC, Bernays EA (1985) Feeding behavior: selection of both wet and dry food for optimal growth by *Schistocerca gregaria* nymphs. Entomol Exp Appl **37**:105–112.

Ma W-C (1972) Dynamics of feeding responses in *Pieris brassicae* Linn as a function of chemosensory input: a behavioral, ultrastructural and electrophysiological study. Mededelingen Landbouwhogeschool, Wageningen, Nederland, **72**:1–162.

Ma W-C, Kubo I (1977) Phagostimulants for *Spodoptera exempta*: identification of adenosine from *Zea mays*. Entomol Exp Appl **22**:107–112.

Marek J (1961) Uber das Einstich—und Saugverhalten der Zwiebellaus, *Myzus ascalonicus* Doncaster. Z Pflkrankh Pflschutz **68**:155–165.

Matthieu JL, Verhoyen M (1980) Facteurs influencant les aphides dans leur comportement de selection de l'hôte. IIde partie: efféts engendres par les insecticides et les huiles. Med Fac Landb Rijksuniv Gent **45**:501–511.

McLean DL, Kinsey MG (1964) A technique for electronically recording aphid feeding and salivation. Nature (Lond) **202**:1358–1359.

Meisner J, Ascher KRS (1972) Feeding stimulants for the larva of the Egyptian cotton leafworm, *Spodoptera littoralis* Boisd. Z Ang Entomol **4**:337–349.

Meisner J, Ascher KRS (1973) Attraction of *Spodoptera littoralis* larvae to colors. Nature **242**:332–334.

Meisner J, Ascher KRS, Aly R, Warthen JD (1981) Response of *Spodoptera lit-*

toralis (Boisd.) and *Earis insulana* (Boisd.) larvae to azadirachtin and salannin. Phytoparasitica **9**:27–32.

Meisner J, Ascher KRS, Lavie D (1974) Phagostimulants for the larva of the potato tuber moth, *Gnorimoschema operculella* Zell. Z Ang Entomol **77**:77–106.

Meisner J, Fleischer A, Eizick C (1982) Phagodeterrency induced by (–)-carvone in the larva of *Spodoptera littoralis* (Lepidoptera: Noctuidae). J Econ Entomol **75**:462–466.

Meisner J, Kehat M, Zur M, Eizick C (1978) Response of *Earis insulana* Boisd. larvae to neem (*Azadirachta indica* A. Juss) kernel extract. Phytoparasitica **6**:85–88.

Meisner J, Skatulla U (1975) Phagostimulation and phagodeterrency in the larva of the gypsy moth, *Porthetria dispar* L. Phytoparasitica **3**:19–26.

Metcalf RL, Metcalf RA, Rhodes AM (1980) Cucurbitacins as kairomones for diabroticite beetles. Proc Natl Acad Sci USA **77**:3769–3772.

Mittler TE (1967) Effect of amino acid and sugar concentrations on the food uptake of the aphid *Myzus persicae*. Entomol Exp Appl **10**:39–51.

Mittler TE, Dadd RH (1962) Artificial feeding and rearing of the aphid, *Myzus persicae* (Sulzer), on a completely defined synthetic diet. Nature (Lond) **195**:404.

Mittler TE, Koski P (1974) Meridic artificial diets for aphids. Entomol Exp Appl **17**:524–525.

Mordue (Luntz) AJ (1979) The role of the maxillary and labial palps in the feeding behavior of *Schistocerca gregaria*. Entomol Exp Appl **25**:270–288.

Müller HJ (1966) Uber die Ursachen der unterschiedichen Resistenz von *Vicia faba* L. gegenuber der Bohenblattlaus, *Aphis (Doralis) fabae* Scop. IX. Der Einfluss ökologischer Faktoren auf das Wachstum von *Aphis fabae* Scop. Entomol Exp Appl **9**:42–66.

Nielson JK, Larsen LM, Sorenson H (1979) Host plant selection of the horseradish flea beetle *Phyllotreta armoraciae:* Identification of two flavonyl glycosides stimulating feeding in combination with glucosinalates. Entomol Exp Appl **26**:40–48.

Ohigashi H, Wagner MR, Matsumura F, Benjamin DM (1981) Chemical basis of differential feeding behavior of the larch sawfly, *Pristiphora erichsonii* (Hartig). J Chem Ecol **7**:599–614.

Partridge L (1981) Increased preferences for familiar foods in small mammals. Anim Behav **29**:211–216.

Partridge L, Maclean R (1981) Effects of nutrition and peripheral stimuli on preferences for familiar foods in the bank vole. Anim Behav **29**:217–220.

Raffa KF, Berryman AA (1982) Gustatory cues in the orientation of *Dendroctonus ponderosae* (Coleoptera: Scolytidae) to host trees. Can Entomol **114**:97–104.

Raina AK, Benepal PS, Sheikh AQ (1980) Effects of excised and intact leaf methods, leaf size and plant age on Mexican bean beetle feeding. Entomol Exp Appl **27**:303–306.

Reed DK, Jacobson M, Warthen JD, Uebel EC, Tromley NJ, Jurd L, Freedman B (1981) Cucumber beetle antifeedants: laboratory screening of natural products. USDA Tech Bull No 1641, 13 pp.

Reed DK, Warthen JD, Uebel EC, Reed GL (1982) Effects of two triterpenoids from neem on feeding by cucumber beetles (Coleoptera: Chrysomelidae). J Econ Entomol 75:1109–1113.

Rembold H, Sharman GK, Czoppelt C, Schmutterer H (1980) Evidence of growth disruption without feeding inhibition by neem seed fractions. Z Pflkrankh Pflschutz 87:290–297.

Richardson B, Whittaker J (1982) The effect of varying the reference material on ranking of acceptability indices of plant species to a polyphagous herbivore, *Agriolimax reticulatus*. Oikos 39:237–240.

Risch S (1980) The population dynamics of several herbivorous beetles in a tropical agroecosystem: the effect of intercropping corn, beans and squash in Costa Rica. J Appl Ecol 17:593–612.

Rowell-Rahier M, Pasteels JM (1982) The significance of salicin for a *Salix*-feeder *Phratora (Phyllodecta) vitellinae*. In: Visser JH, Minks AK (eds) Proc 5th Int Symp Insect-Plant Relationships, Wageningen, 1982. Pudoc, Wageningen, pp. 73–79.

Rust MK, Reierson DA (1977) Using wood extracts to determine the feeding preferences of the western drywood termite, *Incisitermes minor* (Hagen). J Chem Ecol 3:391–399.

Schoonhover LM (1982) Biological aspects of antifeedants. Entomol Exp Appl 31:57–69.

Schoonhoven LM, Jermy T (1977) A behavioral and electrophysiological analysis of insect feeding deterrents. In: Crop Protection Agents—Their Biological Evaluation. MacFarlane NR (ed), Academic Press, New York, pp 133–146.

Shimizu T, Matsuzawa K, Yagi S, Robbins R (1980) A simple method for measuring the mandibular movements of the cabbage armyworm (*Mamestra brassicae* L.) Appl Entomol Zool 15:352–355.

Simpson SJ (1981) An oscillation underlying feeding and a number of other behaviors in fifth-instar *Locusta migratoria* nymphs. Physiol Entomol 6:315–324.

Simpson SJ (1982) Patterns in feeding: a behavioral analysis using *Locusta migratoria* nymphs. Physiol Entomol 7:325–336.

Smith CM (1978) Factors for consideration in designing short-term insect-host plant bioassays. ESA Bull 24:393–395.

Städler E, Hanson FE (1976) Influence of induction of host preference on chemoreception of *Manduca sexta:* behavioral and electrophysiological studies. Symp Biol Hung 16:267–273.

Städler E, Hanson FE (1978) Food discrimination and induction of preference for artificial diets in the tobacco hornworm, *Manduca sexta*. Physiol Entomol 3:121–133.

Still AW (1982) On the number of subjects used in animal behavior experiments. Anim Behav 30:873–880.

Stokes JB, Redfern RE (1982) Effect of sunlight on azadirachtin: antifeeding potency. J Environ Sci Health (A) Environ Sci Eng 17:57–66.

Szentesi A (1981) Antifeedant-treated potato plants as egg-laying traps for the Colorado beetle (*Leptinotarsa decemlineata* Say, Col, Chrysomelidae). Acta Phytopath Acad Sci Hung 16:203–209.

Szentesi A, Bernays EA (1984) A study of behavioral habituation to a feeding deterrent in nymphs of *Schistocerca gregaria*. Physiol Entomol 9:329–340.

Szentesi A, Greany PD, Chambers DL (1979) Oviposition behavior of laboratory-reared and wild Caribbean fruit flies (*Anastrepha suspensa;* Diptera: Tephritidae): I. Selected chemical influences. Entomol Exp Appl 26:227–238.

Tjallingii, WF (1977) Electronic recording of penetration behavior by aphids. Med Fac Landb Rijksuniv Gent 42:721–730.

Trial H, Dimond JB (1979) Emodin in buckthorn: a feeding deterrent to phytophagous insects. Can Entomol 111:207–212.

van Drongelen W (1979) Contact chemoreception of host plant specific chemicals in larvae of various *Ypononmeuta* species (Lepidoptera). J Comp Physiol 134:265–279.

van Emden HF (1977) Failure of the aphid, *Myzus persicae,* to compensate for poor diet during early growth. Physiol Entomol 2:53–58.

van Emden HF (1978) Insects and secondary substances—an alternative viewpoint with special reference to aphids. In: Phytochemical Aspects of Plant and Animal Co-evolution. Harborne JB (ed), Academic Press, London, pp 309–323.

van Emden HF, Bashford MA (1976) The effect of leaf exision on the performance of *Myzus persicae* and *Brevicoryne brassicae* in relation to the nutrient treatment of the plants. Physiol Entomol 1:67–71.

Waldbauer GP (1964) The consumption, digestion and utilization of solanaceous and non-solanaceous plants by larvae of the tobacco hornworm *Protoparae sexta*. Entomol Exp Appl 7:253–269.

Waller DA (1982) Leaf-cutting ants and live oaks: the role of leaf toughness in seasonal and intraspecific host choice. Entomol Exp Appl 32:146–150.

Wasserman SS (1982) Gypsy moth (*Lymantria dispar*): induced feeding preferences as a bioassay for phenetic similarity among host plants. pp 261–267 In: Proceedings of the 5th International Symposium on Insect-Plant Relationships, Wageningen, 1982. Visser JH, Minks AK (eds), Pudoc, Wageningen.

Wearing CH (1968) Host-Plant Relations of Aphids with Special Reference to Water Status. Ph.D. thesis, University of London.

Wensler RJD (1962) Mode of host selection by an aphid. Nature (Lond) 195:830–831.

Weston PA, Miller JR (1985) Influence of cage design on precision of tube-trap bioassay for attractants of the onion fly, *Delia antigua*. J Chem Ecol 11:435–439.

Whelan RJ (1982) Response of slugs to unacceptable food items. J Appl Ecol 19:79–87.

Woodhead S, Bernays EA (1977) Changes in release rates of cyanide in relation to palatability of *Sorghum* to insects. Nature 270:235–236.

Woodhead S, Bernays EA (1978) The chemical basis of resistance of *Sorghum bicolor* to attack by *Locusta migratoria*. Entomol Exp Appl 24:123–144.

Wyatt IJ (1969) Factors affecting aphid infestation of chrysanthemums. Ann Appl Biol 63:331–337.

Chapter 5

Postingestive Effects of Phytochemicals On Insects: On Paracelsus and Plant Products

May Berenbaum[1]

I. Introduction

The eating habits of herbivorous insects have long been of more than passing interest to humans, particularly since the domestication of plants some 8000 years ago, but the precise relationship between herbivorous insects and their host plants has persistently defied characterization. That plant chemicals, rather than nutritive substances, determine at least in part the peculiar likes and dislikes of insects, as they do for humans, is an idea of rather recent origin. The suggestion that the diversity of plant chemistry might be defensive in function was introduced when M. Leo Errera, in an address to the Royal Botanical Society of Belgium in 1886, remarked that, "many of the chemical compounds may serve the plant as means of defense against animals, and when we camphorize our furniture and poison our flower-beds, we are only imitating and reinventing what the plants practiced before the existence of man" (Abbott, 1887). The importance of insects as selective agents in determining patterns of plant chemistry in the "parallel evolution" (Brues, 1920) of insects and plants was not articulated until Fraenkel (1959) suggested that "reciprocal adaptive evolution" determined patterns of host-plant use by insects. Ehrlich and Raven (1964) coined the term "coevolution" to describe the stepwise process by which plants elaborate chemical defenses and insects evolve mechanisms of resistance or tolerance to those defenses.

Since the publication of these seminal papers, interest in plant chemicals

[1]Department of Entomology, University of Illinois, Urbana, Illinois 61801, U.S.A.

and their effects on insects has burgeoned. As phytochemical methodology increased in sophistication, the isolation and identification of individual plant chemical constituents has proliferated; at present, literally thousands of secondary substances have been catalogued from angiosperm plants. Measuring the behavioral or physiological effects of these chemicals on insects has been facilitated considerably by this phytochemical expertise, and methods of bioassay have proliferated concomitant with the increase in knowledge of plant secondary chemistry.

II. Preingestive Effects

Plant constituents other than nutritive substances have a variety of physiological and behavioral effects on insects. At the behavioral level, plant chemicals can affect insects by increasing, decreasing, or altogether suppressing ingestion. Chemicals that decrease or suppress the feeding response are known variously as rejectants, gustatory repellents, feeding deterrents, or antifeedants (Wright, 1963; Schoonhoven, 1982). Although technically these chemicals exert their effects prior to ingestion (gustation or olfaction being a sufficient force to halt the feeding response) there is a tacit correlation or association between nonpreference or antixenosis and unsuitability or antibiosis (Painter, 1951; Kogan and Ortman, 1978) over evolutionary time; an insect species can avoid a plant that adversely affects growth and development via the detection of proximate chemical cues associated invariably with disruption of metabolism following ingestion. These chemical cues may themselves be responsible for the disruption or they may simply co-occur with toxicants.

Thus, it is appropriate to investigate the preingestive effects of plant constituents on feeding behavior—deterrency is ostensibly associated with toxicity under ecological circumstances. Without doubt, detection of antifeedant activity is the most common approach to evaluating plant products with respect to insects. The methods of evaluating plant products for feeding deterrency are reviewed in this volume by Lewis and van Emden (Chapter 4, *this volume*). Interpretation of deterrency or repellency in an evolutionary context is subject to several caveats, however. Feeding deterrency without attendant toxicity is unlikely to continue to evoke an avoidance response in an insect over evolutionary time. An avoidance response is likely to remain adaptive to a herbivore only if the deterrent is itself toxic or is without exception associated with a toxin or poor survivorship. Otherwise, any herbivore that no longer displays an avoidance response to a distasteful but edible plant will be able to utilize a hitherto unavailable food resource. Thus, Bernays (1981) suggests that tannins deter feeding by insects not because they reduce digestibility but because they act as token stimuli to provide information to a herbivore as to the suitability of a particular plant. Since tannins tend to increase in concentration as leaf tissue toughness increases and protein and water levels decline, they are effective cues to indicate deteriorating food-plant quality. Similarly, azadirachtin, a feeding inhibitor, may be associated with nondeterrent growth-disrupting substances (Schmutterer and Rembold, 1980).

Demonstration of deterrency, then, is not tantamount to demonstration of toxicity, and any arguments advocating that a particular chemical evolved as a plant defensive compound must be couched in terms of evidence demonstrating toxicity *over and above* deterrency (e.g., Zalkow et al., 1979). It is difficult to conceive a scenario in which a repellent evolves and is maintained in a plant population without accompanying toxicity. Resistance to chemicals that are repellent and not toxic simply involves a loss of the behavioral avoidance response. There is ample evidence that insects are capable of such habituation. *Schistocerca gregaria* (Orthoptera: Acrididae) accommodates within 4 days to azadirachtin, initially an antifeedant (Gill, 1972, cited in Schoonhoven, 1982) and *Spodoptera litura* (Lepidoptera: Noctuidae) accommodates to shiromodial (a sesquiterpene) in a few hours (Munakata, 1977). This line of logic has regrettably received little attention from advocates of topical antifeedants for insect pest control (Jermy, 1971, Munakata, 1977).

III. Dosage: Acute versus Chronic Effects

Postingestive effects of plant constituents can be acute or chronic. In the case of acute toxicity, the chemical in question kills the organism outright due to involvement in intermediary metabolism or cellular functions. The incidence of mortality is generally an acceptable measure of potency expressed either as a percentage relative to a control or as an LC_{50} or LD_{50} value (lethal concentration or dose—the amount required per unit weight to kill 50% of the test population; Matsumura, 1976). Chronic toxicity is far more subtle and therefore more difficult to demonstrate unequivocally. Toxicologically, there are two levels at which chronic effects can be detected; first, at the systems level, by documentation of tissue and organ impairment or malfunction, and, second, at the organism level, by documentation of such secondary effects as reduced fecundity, egg viability, and the like (Matsumura, 1976). One useful measure in chronic toxicity studies is the ED_{50} (effective dose), the amount required to reduce growth to 50% of the control.

This apparent dichotomy in physiological effects has been formalized in the ecological literature as "qualitative" or "toxic" (= acute) and "quantitative" or "digestibility-reducing" (= chronic). Qualitative defenses are plant compounds of low molecular weight that are easily transported within the plant. Due to structural requirements for activity, they are associated with specific biochemical properties and are easily detoxified by adapted insects, as by the mixed-function oxidase enzyme complex. Typical toxins are alkaloids, glucosinolates (Feeny, 1976), and furanocoumarins (Berenbaum, 1983). Quantitative defenses, as exemplified by tannins and resins, are large amorphous compounds that exert their effects via nonspecific binding with plant proteins and digestive enzymes and concomitant reduction in digestive efficiencies. Due to their undefined structure, it is presumably difficult to counteradapt to or detoxify these defenses (Feeny, 1975, 1976; Rhoades and Cates, 1976; Fox, 1981).

Assumptions concerning the presumed mode of action of a phytochemical can influence the manner of testing for toxicity. Qualitative toxins are ostensibly active at very low concentrations, whereas quantitative toxins are increasingly effective as dosage increases, often up to very large amounts (Feeny, 1976; Rhoades and Cates, 1976; Fox, 1981). Dosage level has been a bête noire in all forms of toxicological testing, ranging from tests of allelochemics on insects through tests of food additives on humans. As Paracelsus maintained, dosage does determine effect (viz., "dose makes the poison" [Pachter, 1951]) and virtually any plant toxicant in sufficient quantity can disrupt insect growth and function. That this relationship between dose and effect is intuitively the case is illustrated by the considerable reluctance on the part of the public, scientific and otherwise, to accept maximum tolerable overdose testing, a "methodology made necessary by the difficulty of picking up a low level effect in a small group of animals" (Martin, 1980):—hence skepticism in the face of Canadian saccharin tests using doses "in excess of the amount that a consumer would receive from drinking 800 diet sodas daily" (Smith, 1980). Using high dosage presents the possibility not only of detecting low-level effects but also of creating high-dosage artifacts. In a more entomological frame, Klowden (1980), in a technical comment on a report by Beach (1979) purporting to show that injecting mosquitos with ecdysterone inhibits biting behavior, pointed out that the concentrations required to suppress the biting response were 18,000 times the physiological concentration in *Aedes aegypti*. Klowden (1980) offered the alternative explanation that the mosquitoes failed to respond to *any* stimuli because they were so "drugged."

In the study of plant–insect interactions it is certainly appropriate to select as maximum and minimum dosage levels the maximum and minimum concentrations reported in nature in the plants at the time during which insects are actively feeding (see Jones and Firn, 1978). These levels can be determined by consulting the literature or, preferably, by direct measurement. Methods of phytochemical analysis (Chapter 8, *this volume*) are manifold; precise techniques depend on the nature of the plant constituent to be analyzed. Harborne (1973) and Paech and Tracey (1954–1964) summarize several standard procedures. Any dosage above or below these levels is of academic interest probably but assuredly of limited ecological significance; a range of concentrations representing the continuum with which the insect is faced naturally is perhaps the happiest compromise. Qualitative toxins can and do have dosage-dependent effects *within the range of concentrations* that occur naturally in plants (e.g., Blau et al., 1978)—the difference between qualitative and quantitative effects is one of degree and not one necessarily of mechanism.

IV. Route of Administration

4.1. Injection

The route of administration influences the toxicity of plant constituents. Direct injection into the body cavity is a suitable means of administration if interest resides solely in the physiological properties of the plant chemical but it is inappropriate in an ecological or evolutionary context and can give rise to misconceptions about the defensive nature of secondary chemicals. Phytoecdysteroids provide a case in point. Nakanishi et al. (1966) discovered three compounds in *Podocarpus nakaii* (Podocarpaceae) that were structurally similar to ecdysone and ecdysterone (20-hydroxyecdysone), the insect molting hormones. Since that time, over three dozen ecdysone analogues, collectively called phytoecdysteroids, have been identified from 84 different plant families (Hikino et al., 1975), with ecdysterone the most common in occurrence. The isolation and identification of these components were facilitated by the use of bioassays for molting hormone activity. Two types of bioassay were used. In the *Calliphora* injection technique (Fraenkel, 1934), mature maggots (*Calliphora, Musca,* or *Sarcophaga*) are ligated behind the prothoracic gland, the site of ecdysone production. The plant extract or purified compound is then injected into the maggot behind the ligation, and the formation of puparial cuticle posterior to the ligation is an indication of ecdysone activity. In the *Chilo* dip test, abdomens of *Chilo suppressalis* larvae (Lepidoptera: Pyralidae) are isolated by ligation and dipped into a methanolic extract of plant material; again, pupal cuticle formation is indicative of molting hormone activity (Sato et al., 1968; Imai et al., 1969). Neither of these methods is suitable as a bioassay for ecological purposes because they circumvent the route by which insects encounter phytoecdysones in nature—that is, by ingestion—and thus circumvent the digestive and detoxication systems of the insect.

The effects of oral administration of phytoecdysones or plant extracts with molting hormone activity are still equivocal. Carlisle and Ellis (1968) fed pinnae of bracken fern *(Pteridium aquilinum),* reported to contain ecdysone and ecdysterone (Kaplanis et al., 1967), to nymphs of *Schistocerca gregaria*. Controls were fed wheat leaves. At maturity, the two groups of nymphs differed in size but the group fed bracken manifested no developmental aberrations. Direct injection of bracken extract, however, promoted early molting. Robbins et al. (1970) administered ecdysterone and ponasterone A in artificial diet to *Musca domestica* larvae and adults and observed a significant reduction in emergence and in ovarian development. Since house flies are not herbivorous, the ecological significance of these results is unclear.

Jones and Firn (1978) dismiss the notion that phytoecdysteroids are "antiherbivore" compounds in bracken fern in an ecological context. They introduced a series of bioassay insects into feeding chambers (petri dishes 4.8 cm in diameter lined with moistened filter paper) and provided them with host-plant leaf discs that had been dipped in methanol control or in a methanolic solution of ecdysteroids and then air-dried; as a measure of deterrency they measured the amount of disc eaten per hour up to 8 or 24 hr by photographically enlarging the discs onto graph paper. They demonstrated that, for the seven species studied, the dose required to affect feeding behavior was 9×10^3 to 2×10^6 times that occurring in frond tissue of *P. aquilinum* at the time during which the insects are active. They argue that measurements of ecdysterone concentration in foliage taken in October are irrelevant in that insects are not feeding on the fronds at that time; at the concentrations in the plant parts fed upon by phytophagous insects, phytoecdysteroids are not currently effective plant defenses. That ecdysterone is the most widespread of phytoecdysones may be less a reflection of selective pressure by insect herbivory and more likely of the technique used for screening. Plants lacking molting hormone activity in the *Calliphora* or *Chilo* bioassay are not screened further for phytoecdysteroids; they may well contain phytoecdysteroids that have no molting hormone activity (Jones and Firn, 1978).

Kubo et al. (1983) have developed a sensitive and appropriate bioassay technique for phytoecdysteroids based on ingestion via incorporation into an artificial diet (vide infra). The criterion used for assessing activity is molting cycle failure, manifested as increased development time, increased mortality, and retention of one or more "exuvial" head capsules (Fig. 4, Chapter 8, *this volume*). This latter effect is distinctive and dramatic. Included in their report is a photograph of a pink bollworm *(Pectinophora gossypiella)* after ingestion of a methanolic extract of *Ajuga remota* root; head capsules were retained through three molts and the insect, its mouthparts blocked, starved to death.

4.2. Contact or Volatile Toxicity

Contact or dermal toxicity is not widely used for purposes of bioassay, although there is an a priori reason to expect that natural products can function in this manner. For most herbivorous insects, host plants represent shelter as well as food (Southwood, 1972), so there is frequent bodily contact with plant tissues. This is particularly true for insects that roll leaves, mine leaves, or bore into stems or for very small insects that encounter directly glandular trichomes or exudates on the leaf surface. Contact toxicity bioassays have been developed extensively for the use of assaying plants for juvenile hormone activity or anti-juvenile hormone activity. Two types of bioassay are generally used. In the *Tenebrio* assay (Jacobson et al., 1975), 500 μg of extract dissolved in 1 μl of acetone is applied topically to the sternites of the last three abdominal segments of newly molted pupae of *Tenebrio molitor* (Coleoptera: Tenebrionidae) or nymphs of *Oncopeltus fasciatus* (Hemiptera: Lygaeidae). At the next molt, individuals are scored for the presence of immature characters. In *Tenebrio,* these include retention of gin traps, urogomphi, and pupal cuticle; in *Oncopeltus,* retention of nymphal coloration, reduced wings, or supernumerary nymphal stages (Bowers et al., 1976) are useful characters. In the second type of bioassay, 20 third or fourth instar nymphs of *O. fasciatus* are confined to a 9-cm petri dish coated with the plant extract. Molting to miniature adults is an indication of antiallatotrophic effects (Bowers et al., 1976); in identifying precocenes from *Ageratum,* Bowers and co-workers varied petri dish coverage from 0.9 to 3.9 μg/cm^2.

These techniques can be modified to suit both the test chemical and insect of interest. Dimock et al. (1982) investigated the contact toxicity of exudates from glandular trichomes in wild tomato (*Lycopersicon hirsutum* f. *glabratum*) toward the tomato fruitworm *Heliothis zea* (Lepidoptera: Noctuidae). These trichomes contain 2-tridecanone, a methyl ketone toxic on contact with *Manduca sexta* (tobacco hornworm) (Williams et al., 1980). For bioassay of tridecanone and analogues, they pipetted solutions of the test chemicals in a concentration of 10 mg/ml chloroform onto filter paper discs 5.5 cm in diameter placed in glass petri dishes. Solvent was evaporated off and 20 neonate *H. zea* larvae placed on the discs. The petri dishes were sealed with parafilm to keep caterpillars and vapors from escaping. The dishes were then placed in the dark for 6 hr after which time the larvae were moved to standard diet cups for an additional 18 hr. Those substances exhibiting activity were reassayed at a series of dosages from 42.1 to 420.9 μg/cm^2 in increments of 42.1 μg to calculate an approximate LD$_{50}$ range. Once the range was narrowed, an additional bioassay was conducted with 2.1 μg increments; LD$_{50}$ and 95% confidence limits were then calculated.

The possibility exists that, with a volatile compound such as 2-tridecanone, the observed toxicity was fumigant and not contact—the petri dish bioassay chamber may have become saturated with toxic vapors during the course of the experiment. This sort of bioassay is perhaps most appropriate (or least ambiguous) for chemicals with comparatively low volatility. One quick and effective way to distinguish contact from volatile toxicity was developed by Isman and Proksch (1985). They bioassayed the deterrent and insecticidal chromenes and benzofurans from *Encelia* spp. by coating the inner walls and bottom of 20-ml glass scintillation vials with a solution of the test compound in Me_2CO; the solvent was then allowed to evaporate and the insects (five neonate *Peridroma saucia* larvae) introduced into the vial for a 4-hr period, after which time they were given a 2-g plug of artificial diet and checked again after 24 hr. Death under these conditions could have resulted either from direct contact with a treated wall or bottom surface or from fumigant action of volatile residues. To rule out the latter possibility, they also applied 1 μmol of encecalin, one of the test compounds, to the cap of the vial only; since this treatment did not cause significant larval mortality, as compared to comparable amounts of encecalin applied to wall and bottom surfaces, they suggested that contact toxicity was the primary mode of action for these compounds.

The assessment of contact toxicity is also ecologically justifiable with aquatic insects. Dhillon and co-workers (1982a,b) examined the effects of chemicals from algal species and from the hydrophyte *Myriophyllum spicatum (haloragacehae)* on midge and mosquito larvae, since these insects normally encounter water-soluble components of aquatic plants. Mosquitoes were placed in petri dishes containing 100 ml tap water to which was added 70 mg mosquito food (lab chow and yeast) every other day. Each glass dish (11 cm diameter) contained 20 first or fourth instar larvae for tests. Midges were placed 10 per paper cup (7 cm diameter) in 100 ml water with 2 g sand in the bottom. Graded doses of plant extract were added to the water and mortality assessed either after 24 hr or after development was completed.

Fumigant toxicity of plant products can be bioassayed as well. Volatile substances such as terpenoid essential oils are logical candidates for such testing. Eisner (1964) simply pointed a fine capillary tube filled with nepetalactone, a cyclopentanoid monoterpene responsible for the cat-attracting properties of catnip *(Nepeta cataria)*, at representatives of 18 insect families and observed the behavioral response. More quantitative is the method used by Marcus and Lichtenstein (1979) to identify the toxic component of the highly fragrant anise plant *(Pimpinella anisum)* (Umbelliferae). Macerated anise tips (3 g) were placed in moistened filter paper in jars with or without cylindrical screen inserts. Fifty *Drosophila melanogaster* (Diptera: Drosophilidae) were introduced into the jar and mortality over a 24-hr period recorded. The presence of the screen barrier allowed for distinguishing between fumigant and contact toxicity. Mortality was compared to that in jars with wet filter paper only. In jars without screens, 50% mortality was effected in 1.2 hr and in jars with screens after 14.4 hr. Aliquots of anise extracts were subsequently pipetted into bioassay jars and anise and estragole, monoterpenoid essential oils, were identified as the toxic components. Volatile toxicity is certainly an important ecological possibility in the Umbelliferae. A disproportionate number of umbellifer associates live in leaf rolls (Berenbaum, 1978, 1981c); these leaf rolls may well function as "gas chambers" in which the volatile components of umbelliferous essential oil are concentrated and made more lethal.

4.3. Oral Administration

Probably the most common means of assessing either acute or chronic effects of plant constituents is via oral adminstration. Since in the vast majority of cases, plant allelochemics are tested against foliage-feeding herbivores, the approach is not unreasonable—most herbivores encounter allelochemics via ingestion. There are four primary means of administering a test chemical orally.

A. Incorporation of the Chemical into an Artificial Diet

This technique is applicable to many taxa displaying various feeding modes (e.g., folivorous Leidoptera, Mabry et al., 1977; root-feeding Coleoptera, Sutherland et al., 1980; phloem-feeding Homoptera, Schoonhoven and Derksen-Koppers, 1976). In recent years, artificial diet has proved to be the method of choice, at least in terms of frequency of usage. The benefits of artificial diet are legion. Artificial diet formulation allows for precise control of nutritional factors. Plant material is difficult to standardize; individual plants may vary in protein content and composition, mineral content, and water level. These factors can all be standardized with the use of artificial diet, either completely defined or semi-synthetic (incorporating a homogeneous admixture of plant material). In addition, the use of artificial medium precludes the introduction of pathogenic organisms that are often present in or on plants in nature. Insects can be reared, and bioassays conducted, year-round, and artificial diet facilitates the rearing of large numbers of synchronous individuals (Singh, 1977).

Although there are advantages, there are serious drawbacks to the use of artificial medium. Artificial medium is not, strictly speaking, inert; since it contains nutritive substances, the possibility for synergistic interactions between nutrients and allelochemics exists. These have been documented by Reese (1979); the amount of protein present in artificial medium significantly affects the toxicity of various plant-derived phenolics. Synergism or antagonism between test chemicals and substances present in crushed leaves or leaf-powder in semidefined diets could well obscure test results. Plants in many families, for example, contain methylenedioxphenyl compounds (Newman, 1962) with known inhibitory effects on insect mixed-function oxidase activity (Lichtenstein and Casida, 1963). Incorporation of such plant material into an artificial diet to assess toxicity of co-occurring toxins might interfere with insect detoxification. By the same token, trace amounts of insecticide residue on plant tissues could serve as a low-grade stress-inducing agent and render an insect more susceptible to the allelochemic in question (Meisner et al., 1977c), though in fairness it must be noted that artificial diet components are not free from contamination; Ahmad and Forgash (1978) detected chlorinated hydrocarbons, including DDT, in the parts per billion level in agar, casein, and wheat germ in their artificial diet. Problems with plant variability can be obviated with the adoption of a completely defined diet, but this approach has its limits in that not all insects obligingly eat completely defined diet, or, for that matter, eat any artificial diet at all.

Other hazards of artificial diet in bioassays include uneven incorporation of toxicant throughout the medium, resulting in high variability of response, and artifact formation due to the conditions of preparation, including heating and oxidation. Chan et al. (1978a) developed a method to ensure homogeneous incorporation of phytochemicals into artificial diets. They recommend dissolving the purified chemical or plant extract in a minimum volume of solvent; this mixture is then added to alphacel, polyamide, or casein as a carrier. Remaining solvent can be removed by applying a stream of nitrogen followed by applying a vacuum to a desiccator with a water aspirator. Preferred solvents include ethanol and ethyl acetate, both natural products of metabolism; in the event that the test chemicals are water-soluble, no pretreatment with solvent is necessary (see also Rehr et al., 1973). This mixture is combined in sequence with vitamin solution and nutrient powder, heated to 60°C, and finally mixed with the gelling agent via vigorous stirring. This treatment, in lieu of mixing in a blender and heating to dissolve the gelling agent, may also reduce the risk of artifacts produced by drastic preparation conditions. Using a variation of this technique, in which the secondary substance is pulverized with the dry diet components for 5 min, Campbell and Duffey (1980) checked the dispersion of water-soluble (salicylic acid) and water-insoluble (digitoxin) supplements by spectroscopy and radiolabel, respectively, and found a coefficient of variation of 1%, suggesting that this particular technique provides excellent homogeneity.

There are several other methods to incorporate plant extract or secondary chemicals into artificial diet. Both dry material and solutions can be mixed with the complete diet before it solidifies; the diet can be poured into a tared beaker into which a preweighed amount of phytochemical has been added. Diet and phytochemical can then be mixed vigorously with a sterile toothpick or spatula. It is often helpful to incorporate a vegetable dye along with the phytochemical to ensure homogeneous mixing; the dye in this case should be added to the control diet as well.

Gillette et al. (1978) incorporated molt inhibitors (TH6308 and difluron) into artificial diet by reliquefying in a double boiler diced diet combined with insecticide; the mixture was then allowed to solidify. One difficulty with this technique is that several typical diet ingredients (vitamins, ascorbic acid, antibiotic) are extremely heat-labile and may be altered depending on the experimental conditions.

Freedman et al. (1979) evaluated several methods of incorporation, including: (1) adding 1 ml ethanolic extract to diet surface followed by evaporation; (2) thorough mixing of ethanolic extract and vacuum drying overnight; (3) evaporating ethanolic extract in diet cup and mixing diet with residue; and (4) incorporating 1 ml aqueous extract directly with the aqueous components of the diet. They found that, with the first method, caterpillars simply bored beneath the diet surface and avoided contact with the additive, and, with the second method, vacuum drying left toxic amounts of ethanol remaining in the diet. The method of choice, evaporating the extract in the diet cup and subsequently adding diet, is suitable provided that the secondary chemical is not likely to evaporate with the ethanol.

An additional drawback to using artificial diets for bioassay is that artificial diet alone can impair normal development. Many growth abnormalities are associated with everyday rearing on artificial medium, including supernumerary instars, color aberrations (Dahlman, 1969; Clark, 1971), greatly enlarged size at maturity, and such pathologies as "pupal syndrome." According to Bracken (1982), pupal syndrome, the failure of the pupa to contract properly after ecdysis, is absent in insects raised on leaves and results rather from a deficiency in dietary fats in artificial medium. The precise characteristics of pupal syndrome—unsclerotization of the membranous area to the fourth abdominal sternite ventral to the wing pads and adult emergence with crumpled wings (Fig. 1)—is reported with alarming frequency in studies of allelochemics administered in artificial diet. Pupae of *Spodoptera* (= *Prodenia*) *eridania* (Lepidoptera: Noctuidae) raised on artificial diet containing ground *Mucuna* seeds display precisely these symptoms (Rehr et al., 1973). Rehr et al. interpreted these results to mean that L-dopa, a constituent of the seeds, interferes with tyrosinase, an enzyme involved in cuticle formation. While Rehr et al. (1973) did not obtain deformities on their control diet, these exact symptoms can be obtained in *Spodoptera eridania* raised on bean-leaf diet lacking allelochemics (Berenbaum, unpublished observations). Similar pupal disfiguration was observed by Dahlman and Rosenthal (1976) in *Manduca sexta* (Lepidoptera: Sphingidae) fed artificial diet containing L-canavanine. Whenever possible, artificial diet tests should be supplemented with observations in the field or on intact plants in the laboratory. Dahlman (1980) conducted a field test of canavanine by spraying a solution onto tobacco plants in field plots; although he did observe slower growth rates in *Manduca sexta,* he did not report the presence of pupal deformities similar to those observed in laboratory diet studies.

Figure 1. Pupal deformities of *Heliothis zea* from routine maintenance on artificial diet. Deformities include larval head capsule retention (top row), incomplete sclerotization (second row), and size reduction (third row). Bottom row are normal pupae (Pupae provided by G. Waldbauer and R. Cohen).

B. Application of the Chemical onto Leaf Surfaces (Bernays 1978, Berenbaum and Feeny 1981, Wagner et al. 1983).

This technique is not particularly suitable for chemicals that degrade in the presence of oxygen or light, nor is it appropriate for the bioassay of highly volatile substances

C. Incorporation of the Chemical into the Vascular Tissues of Excised Plants via Stem-Feeding (viz., Erickson and Feeny 1973, Blau et al. 1978).

Erickson and Feeny (1973) cultured stems of celery (*Apium graveolens:* Umbelliferae) in groups of 10 for 18 hr in 10-ml solutions of sinigrin at various concentrations. Uptake by the stems was measured quantitatively by hydrolysis of sinigrin to allylisothiocyanate and conversion to allyl-thiourea, which was then quantified spectrophotometrically after visualizing spots on thin-layer plates with Grote's reagent (see Erickson and Feeny, 1973 for details). Uptake could also be measured volumetrically; since the concentration of sinigrin in the culture solutions did not change over the 18-hr period, uptake could be estimated by measuring the loss in volume of the culture solutions.

Administering toxins in or on plant tissue has certain advantages over artificial diet. If natural host plants are used, the appropriate nutrient balance to maintain normal growth is present, although, as previously mentioned, it is difficult to ensure uniformity among different plant individuals. Individual variability among plants is definitely one source of error in these studies. Plant water status (Miles et al., 1982), age (Marian and Pandian, 1980; Tamura, 1981), and overall health (Hare and Dodd, 1978) can affect insect performance and none of these variables is easily controlled, even in a greenhouse. Despite such variability, intact plant material has the distinct advantage that insects generally accept it readily. Artificial diet is frequently rejected outright by many insects; in other cases, insects will reject artificial diet after it ages (Berenbaum, 1981b), necessitating frequent replacement and attendant wastage of both diet and allelochemic. In that preparation of test material is oftentimes laborious, time-consuming, and expensive, the economy that accompanies feeding on natural hosts is no small consideration.

Nevertheless, there are problems associated with the use of plant material. Designing a bioassay to test the effects of furanocoumarins on the adapted specialist *Papilio polyxenes* (Lepidoptera: Papilionidae) presented such a problem. *P. polyxenes* (black swallowtail) is restricted to members of the family Umbelliferae, and many of its natural hosts, including *Pastinaca, Conium, Angelica,* and *Heracleum,* already contain furanocoumarins. Wild carrot, *Daucus carota,* however, proved suitable as a component for an artificial diet to which could be added the test chemicals, since the plant is one of few apioid umbellifers lacking furanocoumarins (Ivie et al., 1982) that at the same time is an acceptable host for *P. polyxenes* (Berenbaum, 1981b). Finding a test plant that lacks the phytochemical to be tested and that is acceptable to the bioassay insect is particularly difficult for oligophagous insects.

Compatibility between the test chemical and the plant material to be used in testing is particularly critical in studies in which the chemical is incorporated into the plant via stem-feeding. Many insecticidal chemicals are phytotoxic as well, and uptake by the test plant might well kill it before the bioassay can even begin. Blau et al. (1978) administered sinigrin to *Spodoptera eridania* via culturing stems of *Phaseolus lunatus* in a solution of sinigrin. This method proved unsuitable for testing the toxicity of furanocoumarins to *S. eridania;* stems of *P. lunatus* placed in a solution of xanthotoxin, a furanocoumarin, died within 4 hr (Berenbaum, unpublished observations.) Xanthotoxin, is however, taken up by cut stems of *Heracleum lanatum,* an umbellifer containing furanocoumarins, with no adverse consequences, despite the fact that the phloem does not normally contain furanocoumarins (Camm et al., 1976).

Another drawback with the use of plant material is that chemical and physiological changes occur in plants when they are tampered with—and rarely do space and labor considerations allow rearing bioassay insects on intact plants. The nature of the physiological and chemical changes induced is a function of the type of damage. Leaf tissue has been excised, torn, cut, and punched in the name of bioassay, and each type of damage is associated with changes in the nutritional quality and possible chemical defenses of the leaves (e.g., McCaffery, 1982). One serious consideration— the loss in turgor pressure and water content resulting from tearing tissue— can be ameliorated via water supplementation in Aqua-pik containers (Scriber, 1977), available at most florists, and by running tests under 100% humidity conditions, unless such conditions are conducive to the establishment of pathogenic organisms.

Homogeneous distribution of test chemical is as much a problem with intact tissues as it is in artificial diet. This problem is obviated if chemicals are translocated by cut stems (Erickson and Feeny, 1973). If leaf discs are used, application to the leaf surface is preferably by micropipette rather than by dipping or immersion; generally, the flow of material can be controlled to increase the probability of even dispersion. The carrier solvent is a major factor. Acetone is especially suitable in that it disperses and evaporates quickly. A surfactant can be added (e.g., 0.2% Lab-Brite + 0.5% tannic acid in water; Bernays et al., 1980), in which case it is advisable to incorporate the surfactant to the control treatment as well.

As for practical considerations, the amount of test chemical to be added should be estimated on a wet weight basis and recalculated to determine dry-weight concentration. A subset of the leaves to be used can be lyophilized or oven-dried to obtain an estimate of water content. After leaves are weighed, an appropriate amount of solvent containing the test chemical can be pipetted onto the surface of the leaf. The phytochemical should be dissolved in a minimum of solvent or else the investigator risks dissolving lipophilic components of the leaf tissue. Incremental application can be used in the event that large amounts must be added; each application should be allowed to dry thoroughly before the next is made. A gentle stream of air or nitrogen expedites drying.

D. Use of Near Isogenic Lines or Related Genotypes Differing in Chemical Composition.

This technique has received limited attention, due to the infrequency with which known single mutations occur (e.g., DePonti, 1980). Scriber (1978) successfully used cyanogenic and acyanogenic strains of *Lotus corniculatus* (Leguminosae) to test toxicity of cyanogenic glycosides on *Spodoptera eridania* (Lepidoptera: Noctuidae), and glanded and glandless strains of cotton have been used to assess the effects of gossypol on various insects (Singh and Weaver, 1972; Meisner et al., 1977a).

4.4. Methods and Interpretations

The differences between the leaf tissue and artificial diet approaches are nowhere better illustrated than in the debate over digestibility reduction and antinutritive effects. Feeny (1976) and Rhoades and Cates (1976), among others, proposed that certain plant secondary substances exert their effects by reducing digestibility—that is, by complexing either with leaf protein or with gut enzymes and thereby interfering with nitrogen and biomass utilization. Most frequently cited in support of this proposed mechanism is Feeny's (1970) study of winter moth, *Operophtera brumata* (Lepidoptera: Geometridae) on oak, *Quercus robur*. Feeny found that larvae feeding on artificial diet to which extracted oak leaf tannins had been added developed more slowly and weighed less at pupation than did caterpillars growing on diet without tannins. Moreover, he demonstrated that, at physiological pH levels, oak tannins complex with leaf proteins (Feeny, 1968), suggesting that one possible reason for reduced growth was reduced availability of dietary protein. Chan et al. (1978a,b) also observed retarded growth rates and reduced size at pupation in *Heliothis virescens* raised on artificial diet with added cotton square (flower) tannin *(Gossypium hirsutum)*. In contrast, Fox and McCauley (1977) found that tannin content of *Eucalyptus* leaves had no effect on the growth rate of the eucalyptus associate *Paropsis atomaria* (Coleoptera: Chrysomelidae) and Bernays (1978) found that, for the grasshopper *Schistocerca gregaria*, mortality, growth rates, dry matter digestibility, and growth efficiencies were not affected by the addition of tannic acid or condensed tannin (derived from quebracho) to wheat leaves (Bernays et al., 1980; 1981). Moreover, Berenbaum (1983) compared the effects of ingesting tannins derived from *Liriodendron tulipifera*, the tulip-poplar, on two species of swallowtails—*Papilio polyxenes*, a specialist on Umbelliferae, and *P. glaucus*, a general feeder on tanniniferous trees—and found no effects on digestion, consumption rate, growth rate, or nitrogen utilization.

It is instructive that in all cases purporting to demonstrate digestibility reduction artificial diet was used and in all cases failing to demonstrate digestibility reduction leaf material was used. The mode of administration may well affect the outcome. There are two possible explanations for apparent digestibility reduction in artificial diet. First, tannins are generally localized in special plant organs or cell structures. Foliar application may be a better simulation of the physical separation of tannins and proteins in intact plants; in artificial diet, tannins and proteins are free to form intractable complexes. Secondly, neither Feeny (1970) nor Chan et al. (1978b) took into account differences in consumption rate among treatments. Slow growth rate and reduced pupal weight can both result from reduced intake; tannins might act simply as feeding deterrents. On re-examination, Klocke and Chan (1982) demonstrated that, for third instar *Heliothis zea* caterpillars grown for 5 days on artificial diet containing condensed tannin, approximate digestibility and efficiency of conversion of digested food (see Kogan, *this volume*) were not depressed while consumption rate and final larval weight were reduced. Furthermore, Reese at al. (1982) reported a negative correlation between condensed tannin concentration in the diet and the amount ingested by *H. zea* larvae and at the same time observed no reduction in assimilation or efficiency of conversion of assimilated diet.

The apparent depressive effect of tannins on consumption rate may also be due to direct toxicity. Bernays (1978), Bernays et al. (1981), and Berenbaum (1983) suggest that tannins disrupt epithelial cell integrity and Klocke and Chan (1982) demonstrate interference by tannins with gut protease activity.

If no other lesson emerges from the tannin imbroglio, the importance of distinguishing between toxicity and deterrency cannot be overemphasized. This point was dramatically made by Blau et al. (1978), who introduced an ingenious and straightforward methodological approach to separating the confounding effects of toxicity and deterrency on weight loss in caterpillars. Basically, a "calibration curve" is obtained in which consumption rate is regulated by varying the amount of food available and the subsequent effects on larval growth can be plotted. Food is then offered with added toxicant and consumption and growth rate recorded. Regression lines for all treatments are then calculated through a common origin (the rate at which caterpillars starve irrespective of treatment) and compared by standard statistical methodology for linear models (see Fig. 2). For any given consumption rate, then, the corresponding growth rate can be compared. If a compound is simply a feeding deterrent, then plots of growth rate against consumption rate will not differ from the calibration curve; if a compound is a toxicant, growth will be correspondingly less for a given consumption rate and the regressions will have slopes that are significantly different. This method has proved useful in detecting toxic effects that fall short of outright mortality (Berenbaum and Feeny, 1981: Rausher, 1981; Miller and Feeny, 1983; Usher and Feeny, 1983).

One possible solution to the toxicity/deterrency problem is the use of microencapsulation techniques to incorporate deterrents into artificial medium (M. Martin, personal communication.) By administering allelochemicals in microcapsules of cellulose acetate phthalate or similar materials (Merkle and Speiser, 1973; Bala and Vasudevan, 1981), chemo-

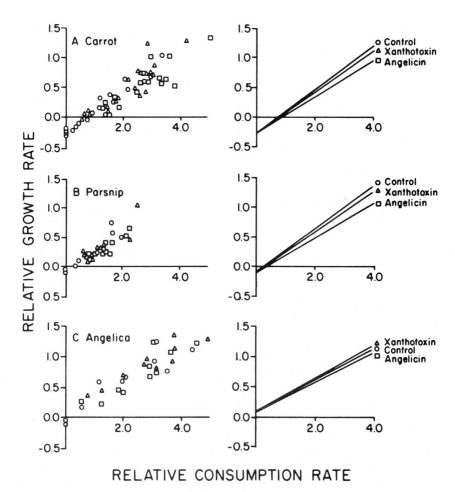

Figure 2. Analysis of relative growth rate versus relative consumption rate for *Papilio polyxenes* larvae raised on different host plants with and without test chemicals (common-intercept regression model). The data for each host plant treatment are on the left; the corresponding calculated regression lines are on the right. A *t* test was used for comparing steps. Several of the starved larvae raised on angelica consumed filter paper and thus did not lose weight after starvation; the position of the intercept (statistically not different from zero) does not affect the analysis. Experimental data are as follows: From Berenbaum and Feeny, Science 212:927–929. Copyright 1981 by the AAAS.

receptors in the oral region are bypassed and, presumably, behavioral effects on ingestion can be eliminated; once ingested, the cellulose acetate phthalate microcapsules dissolve under the alkaline conditions of caterpillar midgut to release the allelochemical contained therein. Microencapsulation may also prove to eliminate another potential source of confusion regarding tannin bioassay. Containing potentially protein-binding material in stable microcapsules could conceivably prevent the formation of artifactual protein-tannin complexing during diet preparation.

Several other methods for circumventing the complications of deterrency in bioassays of toxicity deserve at least mention in passing. Waldbauer and Fraenkel (1961) discovered that the amputation of the maxillae of *Manduca sexta* greatly increased the range of suitable host plants. Maxillae were amputated by pinching with a pair of straight watchmaker's forceps ground to a fine point and sterilized in 95% ethyl alcohol. Larvae were initially anesthetized by chilling or by CO_2 (details in Waldbauer and Fraenkel, 1961). Unfortunately, maxillectomy does not eliminate discrimination altogether; maxillectomized larvae still rejected over a dozen species of plants (Waldbauer and Fraenkel, 1961). Similar methods that have been explored include electric cauterization of the sensory sensillae (Berenbaum, unpublished observations) and administering narcotic drugs, a procedure that appears to eliminate many aversive responses (G. Fraenkel, personal communication, 1983).

V. Designing a Bioassay

5.1. Selecting a Species

First and foremost in designing an effective bioassay for assessing the physiological effects of allelochemicals on insects is selecting an appropriate bioassay species. If a particular plant constituent is of pharmacological or toxicological interest only, then the choice of species is not nearly as restrictive as when a plant constituent is of ecological or evolutionary interest. In the former case, the bioassay species is generally a pest, the potential target of a chemical control program (e.g., Freedman, et al., 1982). In the latter case the preferred species is one that is likely to encounter the plant or plant constituent during the course of its existence. Too often, bioassay insects are selected on the basis of criteria external to the system under study, e.g., on the basis of ease of rearing and/or availability. Thus, components from the seeds from a tropical tree have been assayed against a north temperate foliage feeder (Rehr et al., 1973); oligophagous crucifer feeders used to evaluate extracts from ferns (Jones and Firn, 1979); and the African desert locust, a grass-feeder, used to evaluate tannins extracted from a South American tree (Bernays et al., 1980). These tests can document the physiological properties of plant chemicals but do not necessarily reflect the evolutionary or ecological implications of these properties.

Criteria to be considered in the selection of a bioassay species include:

1. The growth form of the source plant. *Heliothis zea* (Lepidoptera: Noctuidae), one of the more abused of bioassay species, certainly has a

very broad host range, encompassing over 100 species of plants—but all but a handful of these plants are herbaceous. In nature, despite its voracity, *H. zea* is unlikely to encounter tree foliage and it is unlikely to be adapted to the lower leaf water and nitrogen content or increased toughness of tree leaves (Scriber and Slansky, 1981). Demonstration of toxicity of an allelochemical derived from a tree is not equivalent to demonstrating a protective function of that chemical if the bioassay organism is not already predisposed or preadapted to that type of food.

2. Taxonomic range of the host plant. Oligophagous insects deserve special consideration in the context of bioassays in that they are behaviorally or otherwise restricted to a particular group of plants. Again, demonstration of a toxic or deterrent effect of a plant chemical from outside the host range does not identify that chemical as a defensive chemical in nature. That black swallowtails *(Papilio polyxenes)* cannot tolerate sinigrin does not *necessarily* mean that sinigrin is a barrier to colonization to crucifers (Erickson and Feeny, 1973), at least in ecological time. *P. polyxenes* does not feed on plants outside the Umbelliferae, apparently because such plants lack token stimuli (Dethier, 1941), and it is difficult to imagine how it would encounter glucosinolates unless an umbelliferous plant acquired them.

3. Feeding mode of the bioassay insect. Most insects are behaviorally or physiologically restricted to a particular mode of feeding or to particular plant structures. In that seed and leaf protein differ substantively in composition, incorporating seed powder in place of leaf powder in a bioassay involving folivorous insects (e.g., Rehr et al., 1973) might introduce confounding effects on simple digestibility. Arnault (1979), for example, discovered that immature larvae of the oligophagous species *Acrolepiosis assectella*, a specialist on leaves of leek *(Allium porrecta)*, die when leek-flower powder is incorporated into artificial diet in place of leaf powder, and larvae of *Acanthoscelides obtectus* (Coleoptera: Bruchidae), which feed on seeds of *Phaseolus vulgaris,* die when seed coat flour is incorporated in artificial diet in place of cotyledon or embryo flour (Stamopoulos and Huignard, 1980). Similarly, insects that restrict their feeding to particular plant parts may never encounter certain plant chemicals in nature. Camm et al. (1976) tested the effects of furanocoumarins in *Heracleum lanatum* on the aphid *Cavariella aegopodii,* despite the fact that furanocoumarins are not known to occur in the phloem, the vascular tissue to which *C. aegopodii* is restricted. Again it is difficult to assess the ecological significance of a physiological phenomenon that is unlikely to occur.

It can certainly be argued that the degree of improbability of encounter is irrelevant—that perhaps host-plant switches are the result of such "improbabilities" during the course of evolution. There is certainly evidence that unlikely scenarios do occur; witness the abundance of records of oviposition "mistakes" by butterflies, ovipositing on host plants outside the normal range (Straatman, 1962; Chew, 1977; Berenbaum, 1981a). However, host shifts entail more than a simple genetic shift in host preference. Such a shift must be accompanied by simultaneous genetic shifts in host-plant utilization efficiencies and survival as well (Bush, 1975; Feeny et al., 1983), reducing the evolutionary significance of "mistakes" still further.

4. Source of insects for bioassay. There is a frying pan-fire dilemma in deciding between lab-raised or wild-collected insects for bioassay. Bringing insects in from the field introduces problems with disease and parasitism—diseased and parasitized larvae can demonstrate markedly different feeding behavior and digestive efficiencies (e.g., Mohamed, 1982; Huebner and Chiang, 1982). On the other hand, lab-raised insects, particularly those in culture for long periods of time, suffer from genetic and behavioral change due to inbreeding. Berlocher and Friedman (1981) demonstrated a loss of genetic variation in a laboratory culture of *Phormia regina* (Diptera: Calliphoridae) after only a single generation; some laboratory cultures have been maintained for over 300 generations (see also Dickerson et al., 1979). Behavior may change as a result of laboratory rearing; raising some parasitoids in culture for many generations has resulted in insects that flee at the sight of their natural host (G. Waldbauer, personal communication, 1983). Presumably, maintaining herbivores on artificial diet for many generations can bring about changes in preference and behavior with respect to normal host plants.

5.2. Number of Individuals

The number of individuals to use in a bioassay depends in large part on the magnitude of the effect being measured. Since there is no way to anticipate *a priori* what effects are in store, a pilot study is advisable if time and materials permit, if for no other reasons than to establish appropriate dosage levels. Irrespective of the number, maintaining individuals in separate containers is strongly recommended for several reasons: it allows for greater accuracy in observing effects on growth and development and facilitates gravimetric measurements; it reduces the risk of horizontal disease transmission; and it prevents cannibalism, a serious problem in many noctuid caterpillars used for bioassay purposes (e.g., *Heliothis zea;* Gould et al., 1980). Grouping insects may, however, be desirable for species that aggregate in nature, since rearing individuals in isolation may reduce survivorship (e.g., *Porthetria dispar*—personal observation).

5.3. Pretest Conditions: Handling the Insects

Pretest conditions can profoundly affect the outcome of a bioassay. It is almost superfluous to note that test microclimate conditions should resemble those under which the bioassay animals are reared since such factors as temperature and humidity can influence duration of feeding bouts and consumption rates. Conditioning is another complicating factor. Many insects develop a relative preference for a particular host plant or food material on which they have fed previously (Jermy et al., 1968; Greenblatt et al., 1978; Wilson and Starks, 1981). Behavioral induction may well be a general phenomenon, at least among Lepidoptera (Grabstein and Scriber, 1982a,b). In terms of preingestive effects, conditioning can influence feeding rate and preference; arctiids (*Diacrisia virginica* and *Estigmene congrua*) following massive regurgitation on consuming foliage of *Petunia hybrida* display a reduced preference for *P. hybrida* in subsequent feeding trials (Dethier, 1980). Physiologically, conditioning can affect the ability of insects to detoxify plant constituents (Brattsten, 1979). Mixed-function oxidases show a graded response to concentrations of allelochemics, and brief contact with a particular secondary substance can stimulate a demonstrable increase in mixed-function oxidase activity; larvae with induced mixed-function oxidase levels showed decreased sensitivity to nicotine with respect to uninduced control larvae (Brattsten et al., 1977). Prior exposure can also improve nitrogen utilization efficiencies (Scriber, 1982).

One way to avoid problems with behavioral induction is to use neonate larvae for testing. This approach is appropriate for testing of both acute and chronic toxins in that, in most cases, neonates of necessity must encounter plant toxicants as soon as they begin to feed in nature. Use of neonate larvae has the additional advantage of increased sensitivity to toxicants. Neonate and early instar *Heliothis zea,* for example, suffer from rutin toxicosis whereas later stage larvae do not (Isman and Duffey, 1982a,b) and Reese et al. (1982) demonstrated a dramatic difference in sensitivity to maysin between neonates and day-old larvae (see also Shaver and Parrott, 1970). The drawbacks to using neonates include difficulties in obtaining accurate estimates of consumption rate or growth by gravimetric means (Scriber and Slansky, 1981); neonate larvae from even the largest species of insects require more than a conventional laboratory balance to estimate body weight gain and fecal production accurately. A second problem is that, if chronic effects are to be measured, larvae must be maintained for the duration of their developmental period; this necessitates large quantities of plant material and/or artificial medium and of the chemical under study. Finally, neonates are extremely sensitive to the effects of handling; particular care must be taken if test insects are to be moved.

Photographic estimates of insect weight may eliminate some of these problems. Smiley and Wisdom (1982) have developed such a technique. Larvae of *Heliconius cydno* (Lepidoptera: Nymphalidae) were photographed either in the focal plane of a metric ruler attached in front of a

Nikon RE camera, or with a 55-mm macro lens with magnification calibrations marked on the lens barrel, in which the lens was placed on the desired setting and the camera moved to focus. The developed film was projected onto a sheet of paper with a photographic enlarger and the insect silhouette traced onto the paper. Caterpillars were divided into regions of constant diameter and volume for each section calculated as $(D_i/2)^2 \cdot L_i$, where D_i is the diameter of section i and L_i the length down the center. Summing the sections gives an estimate of total insect volume. Volume alone is a sufficient measure where relative estimates are sought (e.g., relative growth rates); to estimate absolute volume, the relative measures can be divided by the cube of the magnification factor, determined by the enlargement of the ruler in the photograph. This technique is two to four times more accurate than are measures based on head capsule width or body length and is particularly appropriate when insects cannot be moved or disturbed, as with early instars or in the field.

A second approach to the problem of behavioral induction is to raise the test insects to the desired instar strictly on artificial medium, to minimize the risk of conditioning to plant constituents. This again has limitations, most conspicuous of which is the fact that not all insects have been or can be raised on synthetic medium successfully. In addition, insects have been known to alter their feeding behavior after rearing on artificial medium (Schoonhoven, 1967).

5.4. Endpoint Considerations

Determination of the endpoint of the bioassay may or may not be obvious. In the case of a highly toxic substance, complete mortality is a possibility. In this event, a series of dosages should be administered in order to calculate an LD_{50} or LC_{50}. LD_{50} calculation has been facilitated in recent years by the increasing availability of computer programs (e.g., Robertson et al., 1981). If acute effects are not observed within a 24-hr period, a test can be conducted for the duration of an instar, the developmental period to pupation, or the developmental period through eclosion. In that chronic toxicants often have secondary effects, measurement of fecundity, egg viability, and survivorship of offspring are logical criteria for examination. Parameters that have been used to assess toxicity of allelochemics include:

1. Mortality: during a 24-hr period, during instar, through pupation and eclosion.
2. Rate of growth: relative growth rate during a 24-hr period, number of days to pupation, number of days to eclosion.
3. Amount of growth: larval weight, pupal weight, adult weight at eclosion. An important consideration in using amount of growth as an objective criterion is to standardize the timing of data acquisition. Pupae, for example, should be weighed at a set time after ecdysis, generally after

sclerotization is complete and before weight loss has commenced. Length, although correlated with weight (Schoener, 1982), is an unreliable estimate, particularly in soft-bodied immature forms such as caterpillars, grubs, and maggots, due to individual variability and effects of posture.

4. Fecundity: number of eggs per female, hatchability, or viability of eggs (e.g., Palumbo and Dahlman, 1978).
5. Change in digestive efficiencies or nitrogen utilization (reviewed by Kogan, Chapter 8 *this volume*).
6. Incidence of anatomical abnormalities: supernumerary instars, "pupal syndrome," discoloration, molting-cycle failure.
7. Incidence of behavioral abnormalities: ataxia, uncoordination, paralysis, regurgitation (Dethier, 1980), inability to spin silk (Sharma and Agarwal, 1982).
8. Increased susceptibility to disease (Stubblebine and Langenheim, 1977).
9. Histological changes (Bernays, 1978; Bernays et al., 1981).

5.5. Statistical Analysis

The method of statistical analysis of a bioassay depends on the specific bioassay and the parameters measured. In general, where changes in continuous variables are compared (as in 2–5 above), tests comparing means are appropriate. Where there are two treatments, t tests or the nonparametric analogue Wilcoxon two-sample test can be used; for more than two treatments, single classification ANOVA (parametric) or Kruskal-Wallis (nonparametric) can be used. These tests establish only that there is a significant added variance component; to distinguish precisely which means differ, an a posteriori test such as Student-Newman-Keuls or Duncan's new multiple range test can be done. When the effects of two or more factors are considered simultaneously, multiple ANOVA techniques can be employed.

Many bioassay parameters are frequency measures (e.g., 1, 6–9 above) and as such can be compared with tests of independence, including the G test, chi-square, or Fisher's exact test. Which test should be used depends on sample size and calculated expected values.

This is by no means a definitive discussion; these statistical methods represent a tiny fraction of the possible methods for analysis. Precise procedures and criteria for selecting tests can be found in many standard texts (e.g., Steele and Torrie, 1980; Sokal and Rohlf, 1981).

VI. Applications: Future Prospects

The bioassay procedures outlined here are flexible and can be adapted to test for physiological activities of plant substances irrespective of their mode of action. Thus, these methods have been applied to evaluate the effects of ingestion of antimetabolites such as ecdysone (Jones and Firn, 1978) and proteinase inhibitors (Gatehouse et al., 1979) as well as chronic and acute toxicants. Up to this point in time, the bulk of bioassays of plant constituents have been conducted in order to isolate and identify potential insecticides for commercial application (e.g., Willomitzer and Tomanek, 1981). Only recently has research been motivated by interest in coevolutionary interactions between insects and their plant hosts. This change in orientation and objective has necessitated changes in methodology. First, selection of a species for bioassay is no longer based on convenience or economic necessity (i.e., potential targets for insecticide control). Bioassay organisms must be an integral component of the evolutionary regimen under which the plant in question elaborated its chemical defenses—if not an actual member of the ecosystem, then at least an ecological analogue of such members. Secondly, dosage determination is different; only effects produced by concentrations present in the plant at the time during which it is susceptible to attack are germane to ecological or coevolutionary arguments. Third, a critical consideration virtually ignored in the past is the importance of interactions among secondary chemicals. Janzen (1973) was among the first to recognize the implications of the chemical complexity of plants:

> I have been speaking as though a bite of a plant contained only one defensive compound. More than likely, it contains many (and the possible synergistic effects are immense).

Jermy (1982, cited in Schoonhoven 1982) pointed out that there is no a priori reason to expect that single chemical bioassays will be instructive when multicomponent defensive systems are operative in nature; he and his co-workers (Jermy et al., 1981) in fact demonstrated the importance of synergism vis-à-vis feeding behavior in the Colorado potato beetle *(Leptinotarsa decemlineata)*. Adams and Bernays (1978) also demonstrated synergistic effects of plant chemicals on feeding. In addition, third-trophic level effects of secondary chemicals (Price et al., 1980) are now receiving attention. Studies on the toxicity of plant constituents on parasitoids such as those conducted by Campbell and Duffey (1979, 1980) on the effects of tomatine, a constituent of the hostplant of *Heliothis zea,* on its parasitoid *Hyposoter exiguae* deserve emulation, as does the work by Jones et al. (1981) on the volatile toxicity of 2-furaldehyde from baldcypress toward the gut symbionts of *Bombyx mori* (Lepidoptera: Bombycidae) (see also Barbosa and Saunders, 1985).

These and other considerations will no doubt receive increasing attention in future research. In the field, a multiplicity of ecological factors can modify an insect's response to a plant chemical (competitive ability, dispersal potential, and the like) any of which can alter the physiological response of an insect and none of which is easily modelled under laboratory conditions. Not just biological factors can temper toxicity; physical factors that are not a priori suspect might well affect toxicity in the field. Light is one such factor. Many chemicals display increased toxicity in the presence of ultraviolet light (e.g., furanocoumarins, Berenbaum, 1978), and wavelength specifications for insect growth chambers are rarely a consideration when designing a bioassay. In the field, many more dimensions are likely to be identified in the response of insects to plant chemicals and by rights ecologists ought to keep one foot in the field in order to keep their findings in perspective.

New techniques in gene manipulation have potential applications in the bioassay of plant products for behavioral and physiological effects on insects. Genes can now be removed from the nucleus of one species, cloned, vectored into the cells of another plant species, and, in some cases, expressed during development and transmitted to subsequent generations (Foard et al., 1983; Murdock, 1985). The possibility exists, then, that toxic gene products, such as proteinaceous lectins, lipoxygenases, and enzyme inhibitors, can be bioassayed by incorporating them into an appropriate test plant; this technique also would facilitate testing an allelochemical against a variety of chemical/genetic backgrounds in order to detect synergistic interactions. Even the more traditional genetic techniques, such as utilization of near-isogenic or isogenic lines (viz., Phillips, 1968), have not really been exploited to full advantage with respect to evaluating the defensive potential of allelochemicals.

Bioassay of plant constituents is a powerful tool in developing both theory and practice of coevolution. *Why* a plant species evolved a particular suite of secondary chemicals is an evolutionarily moot point, difficult if not impossible to prove; yet, in contemporary time, an insect feeding on a plant must encounter the secondary chemicals in the plant and must respond. Measuring *how* different insects—oligophagous or polyphagous—respond to those chemicals can yield insights into the nature of possible selective forces exerted by insects in the course of the "coevolutionary arms race" (Whittaker and Feeny, 1971) and aid in reconstructing the processes of adaptation and counteradaptation between plant and insect.

Acknowledgments. I thank my associates at the Department of Entomology, University of Illinois Urbana-Champaign S. Berlocher, R. Cohen, J. Nitao, W.S. Sheppard, and G. Waldbauer for improving the manuscript. I would also like to thank P. Feeny, Section of Ecology and Systematics, Cornell University, for setting a good example and continuing to render assistance, whenever needed.

References

Abbott H (1887) Comparative chemistry of higher and lower plants. Am Nat **21**:800–810.

Adams CM, Bernays EA (1978) The effect of combinations of deterrents on the feeding behaviour of *Locusta migratoria*. Entomol Exp Appl **23**:101–109.

Ahmad S, Forgash AJ (1978) Gypsy moth mixed-function oxidases: gut enzyme levels increased by rearing on a wheat germ diet. Ann Entomol Soc. Am **71**:449–452.

Arnason T, Swain T, Wat C-K, Graham EA, Partington S, Towers GHN (1981) Mosquito larva activity of polyacetylenes from species in the Asteraceae. Biochem Syst Ecol **9**:63–68.

Arnault C (1979) Influence de substances de la plante-hote sur le developpement larvaire d'*Acrolepiopsis assectella* (Lepidoptera, Acrolepiidae) en alimentation artificielle. Entomol Exp Appl **25**:64–74.

Bala K, Vasudevan P (1981) Polymeric microcapsules for drug deliver. J Macromol Sci Chem **A16**:819–827.

Barbosa P, Saunders J (1985) Plant allelochemicals: linkages between herbivores and their natural enemies. Recent Adv Physochem **19**:107–137.

Beach R (1979) Mosquitoes: biting behavior inhibited by ecdysone. Science **205**:829–831.

Berenbaum M (1978) Toxicity of a furanocoumarin to armyworms: a case of biosynthetic escape from insect herbivores. Science **201**:532–534.

Berenbaum M (1981a) Note on an oviposition "mistake" by *Papilio glaucus* (Papilionidae). J Lepidopt Soc **35**:75.

Berenbaum M (1981b) Effects of linear furanocoumarins on an adapted specialist insect *(Papilio polyxenes)*. Ecol Entomol **6**:345–351.

Berenbaum M (1981c) Furanocoumarin distribution and insect herbivory in the Umbelliferae: plant chemistry and community structure. Ecology **62**:1254–1266.

Berenbaum M (1983) Coumarins and caterpillars: a case for coevolution. Evolution **37**:163–179.

Berenbaum M (1983) Effects of tannin ingestion on two species of papilionid caterpillars. Entomol Exp Appl **34**:245–250.

Berenbaum M, Feeny P (1981) Toxicity of angular furanocoumarins to swallowtails: escalation in the coevolutionary arms race. Science **212**:927–929.

Berlocher SH, Friedman S (1981) Loss of genetic variation in laboratory colonies of *Phormia regina*. Entomol Exp Appl **30**:205–208.

Bernays EA (1978) Tannins: an alternative viewpoint. Entomol Exp Appl **24**:44–53.

Bernays EA (1981) Plant tannins and insect herbivores: an appraisal. Ecol Entomol **6**:353–360.

Bernays EA, Chamberlain DJ (1980) A study of tolerance of ingested tannin in *Schistocerca gregaria*. J Insect Physiol **26**:415–420.

Bernays EA, Chamberlain DJ, Leather EM (1981) Tolerance of acridids to ingested condensed tannin. J Chem Ecol **17**:247–256.

Bernays DA, Chamberlain DJ, McCarthy P (1980) The differential effects of ingested tannic acid on different species of Acridoidea. Entomol Exp Appl **28**:158–166.

Blau PA, Feeny P, Contardo L, Robson DS (1978) Allylglucosinolate and herbivorous caterpillars: a contrast in toxicity and tolerance. Science **200**:1296–1298.

Bowers WS, Ohta T, Cleere JS, Marsella PA (1976) Discovery of insect antijuvenile hormones in plants. Science **1933**:542–547.

Bracken GK (1982) The bertha armyworm, *Mamestra configurata* (Lepidoptera: Noctuidae). Effects of dietary linolenic acid on pupal syndrome, wing syndrome, survival, and pupal fat composition. Can Entomol **114**:567–573.

Brattsten L (1979) Biochemical defense mechanisms in herbivores against plant allelochemics. In: Herbivores: Their Interaction with Secondary Plant Metabolites. Rosenthal GA, Janzen DH (eds), Academic Press, New York pp 199–270.

Brattsten L, Wilkinson DF, Eisner T (1977) Herbivore-plant interactions: mixed-function oxidases and secondary plant substances. Science **196**:1349–1352.

Brues CT (1920) The selection of food-plants by insects, with special references to lepidopterous larvae. Am Nat **54**:313–332.

Bush GL (1975) Sympatric speciation in phytophagous parasitic insects. In: Evolutionary Strategies of Parasitic Insects and Mites. Price PW (ed), Plenum, New York.

Camm EL, Wat C-K, Towers GHN (1976) An assessment of the roles of furanocoumarins in *Heracleum lanatum*. Can J Bot **54**:2562–2566.

Campbell BC, Duffey S (1979) Tomatine and parasitic wasps: potential incompatibility of plant antibiosis with biological control. Science **205**:700–702.

Campbell BC, Duffey S (1980) Alleviation of α-tomatine-induced toxicity to the parasitoid *Hyposoter exiguae,* by phytosterols in the diet of the host, *Heliothis zea*. J Chem Ecol 7:927–946.

Carlisle DB, Ellis PE (1968) Bracken and locust ecdysones: their effects on molting in the desert locust. Science **159**:1272–1474.

Chan BG, Waiss AC, Binder RG, Elliger CA (1978b) Inhibition of lepidopterous larval growth by cotton constituents. Entomol Exp Appl **24**:94–100.

Chan BG, Waiss AD, Stanley WL, Goodban AE (1978a) A rapid diet preparation method for antibiotic phytochemical bioassay. J Econ Entomol **71**:366–367.

Chew F (1977) Coevolution of pierid butterflies and their cruciferous foodplants. II. The distribution of eggs on potential foodplants. Evolution **31**:568–579.

Clark RM (1971) Pigmentation of *Hyalophora cecropia* larvae fed artificial diets containing carotenoid additives. J Insect Physiol **17**:1593–1598.

Dahlman DL (1969) Cuticular pigments of tobacco hornworm *(Manduca sexta)* larvae: effects of diet and genetic differences. J Insect Physiol **15**:807–814.

Dahlman DL (1977) Effect of L-canavanine on the consumption and utilization of artificial diet by the tobacco hornworm, *Manduca sexta*. Entomol Exp Appl **22**:123–131.

Dahlman DL (1980) Field tests of L-canavanine for control of tobacco hornworm *(Manduca sexta)*. J Econ Entomol **73**:279–281.

Dahlman DL, Rosenthal GA (1976) Further studies of the effect of L-canavanine on the tobacco hornworm, *Manduca sexta*. J Insect Physiol **22**:265–271.

DePonti OMB (1980) Resistance of *Cucumis sativus* L., to *Tetranychus urticae* Koch. 6. Comparison of near isogenic bitter and non-bitter varieties for resistance. Euphytica **29**:261–265.

Dethier VG (1941) Chemical factors determining the choice of food plants by *Papilio* larvae. Am Nat **75**:61–73.

Dethier VG (1980) Food-aversion learning in two polyphagous caterpillars, *Diacrisia virginica* and *Estigmene congrua*. Physiol Entomol **5**:321–325.

Dhillon MS, Mulla MS, Hwang Y-S (1982a) Allelochemics produced by the hydrophyte *Myriophyllum spicatum* affecting mosquitoes and midges. J Chem Ecol **8**:517–526.

Dhillon MS, Mulla MS, Hwang Y-S (1982b) Biocidal activity of algal toxins against immature mosquitoes. J Chem Ecol **8**:557–566.

Dickerson WA, Hoffman JD, King EG, Leppla NC, Odell TM (1979) Arthropod species in culture in the United States and other countries. USDA, Hyattsville.

Dimock MB, Kennedy GG, Williams, WG (1982) Toxicity studies of analogs of 2-tridecanone, a naturally occuring toxicant from a wild tomato. J Chem Ecol **8**:837–842.

Ehrlich PR, Raven PH (1964) Butterflies and plants: a study in coevolution. Evolution **18**:586–608.

Eisner T (1964) Catnip: its raison d'etre. Science **146**:1318–1320.

Erickson JM, Feeny P (1973) Sinigrin: a chemical barrier to the black swallowtail butterfly, *Papilio polyxenes*. Ecology **55**:103–111.

Feeny PP (1969) Inhibitory effect of oak leaf tannins on the hydrolysis of proteins by trypsin. Phytochemistry **7**:2119–2126.

Feeny P (1970) Seasonal changes in oak leaf tannins and nutrients as a cause of spring feeding by winter moth caterpillars. Ecology **51**:565–581.

Feeny P (1975) Biochemical coevolution between plants and their insect herbivores. In: Coevolution of Animals and Plants. Gilbert LE, Raven PH (eds), University of Texas Press, Austin.

Feeny P (1976) Plant apparency and chemical defense. Recent Adv Phytochem **10**:1–40.

Feeny P, Rosenberry L, Carter M (1983) Chemical aspects of oviposition behavior in butterflies. In: Herbivorous Insects: Host-Seeking Behavior and Mechanisms. Ahmad S (ed), Academic Press, New York.

Foard DE, Murdock L, Dunn P (1983) Engineering of crop plants with resistance to herbivores and pathogens: an approach using primary gene products. In: Plant Molecular Biology. Goldberg R (ed), Alan R. Liss, New York, pp 224–233.

Fox, LR (1981) Defense and dynamics in herbivore-plant systems. Am Zool **24**:853–864.

Fox LR, McCauley BJ (1977) Insect grazing on *Eucalyptus* in response to variation in leaf tannins and nitrogen. Oecologia (Berl) **29**:145–162.

Fraenkel GS (1934) Pupation of flies initiated by a hormone. Nature **133**:834.

Fraenkel GS (1959) The raison d'etre of secondary plant substances. Science **129**:1466–1470.

Freedman B, Nowak LJ, Kwolek WF, Berry DC, Guthrie WD (1979) A bioassay for plant-derived pest control agents using the European corn borer. J Econ Entomol **72**:541–545.

Freedman B, Reed DK, Powell RG, Madrigal RV, Smith CR (1982) Biological activities of *Trewia nudiflora* extracts against certain economically important insect pests. J Chem Ecol **8**:409–418.

Gatehouse AMR, Gatehouse JA, Dobie P, Kilminister AM, Boulter D (1979) Biochemical basis of insect resistance in *Vigna unguiculata*. J Sci Food Agric 30:948–958.

Gillette ML, Robertson JL, Lyon RL (1978) Bioassays of TH6038 and difluron applied to western spruce budworm and Douglas-fir tussock moth. J Econ Entomol 71:319–323.

Gould F, Holtzman G, Rabb RL, Smith M (1980) Genetic variation in predatory and cannibalistic tendencies of *Heliothis virescens* strains. Ann Entomol Soc Am 73:243–250.

Grabstein EM, Scriber JM (1982a) The relationship between restriction of host plant consumption, and post-ingestive utilization of biomass and nitrogen in *Hyalophora cecropia*. Entomol Exp Appl 31:202–210.

Grabstein EM, Scriber JM (1982b) Host-plant utilization by *Hyalphora cecropia* as affected by prior feeding experience. Entomol Exp Appl 32:262–268.

Greenblatt JA, Calvert WH, Barbosa P (1978) Larval feeding preferences and inducibility in the fall webworm, *Hyphantria cunea*. Ann Entomol Soc Am 71:605–606.

Harborne JB (1973) Phytochemical Methods. Chapman and Hall, London.

Hare JD, Dodd JA (1978) Changes in food quality of an insect's marginal host species associated with a plant virus. J NY Entomol Soc 86:292.

Hikino H, Ohizujmi Y, Takekmoto T (1975) Detoxication mechanism of *Bombyx mori* against exogenous phytoecdysone ecdysterone. J Insect Physiol 21:1953–1963.

Huebner LB, Chiang HC (1982) Effects of parasitism by *Lixophaga diatraea* (Diptera: Tachinidae) on food consumption and utilization of European corn borer larvae (Lepidoptera: Pyralidae). Environ Entomol 11:1053–1057.

Imai S, Toyosato T, Sakai M, Sat Y, Sugioka S, Murata E, Cota M (1969) Screening results of plants for phytoecdysones. Chem Pharmacol Bull 17:335–339.

Isman MB (1977) Dietary influence of cardenolides on larval growth and development of the milkweed bug *Oncopeltus fasciatus*. J Insect Physiol 23:1183–1187.

Isman MB, Duffey SS (1982a) Toxicity of tomato phenolic compounds to the fruitworm *Heliothis zea*. Entomol Exp Appl 31:370–376.

Isman MB, Duffey SS (1982b) Phenolic compounds in foliage of commercial tomato cutivars as growth inhibitors to the fruitworm, *Heliothis zea*. J Am Horticult Sci 107:167–170.

Isman M, Proksch P (1985) Deterrent and insecticidal chromenes from *Encelia* (Asteraceae). Phytochemistry 24:1949–1951.

Ivie GW, Beier RC, Holt DL (1982) Analysis of the garden carrot (*Daucus carota* L.) for linear furocoumarins (psoralens) at the sub parts per million level. J Agric Food Chem 30:413–415.

Jacobson M, Redfern R, Mills G (1975) Naturally occurring insect growth regulators. II. Screening of insect and plant extracts as insect juvenile hormone mimics. Lloydia 38:455–472.

Janzen DH (1973) Community structure of secondary compounds in nature. Pure Appl Chem 34:529–538.

Jermy T (1971) Biological background and outlook of the antifeedants approach to insect control. Acta Phytopath Acad Sci Hung 6:253–260.

Jermy T, Butt BA, McDonough L, Dreyer D, Rose AF (1981) Antifeedants for the Colorado potato beetle. I. Antifeeding constituents of some plants from the sagebrush community. Insect Sci Appl I:237–242.

Jermy T, Hanson FE, Dethier VG (1968) Induction of specific food preference in lepidopterous larvae. Entomol Exp Appl 11:211–230.

Jones CG, Aldrich JR, Blum MS (1981) Baldcypress allelochemics and the inhibition of silkworm enteric microorganisms. J Chem Ecol 7:103–114.

Jones SB, Burnett WC, Coile NC, Mabry TJ, Betkouski MR (1979) Sesquiterpene lactones of *Vernonia*—influence of glaucolide-A on the growth rate and survival of lepidopterous larvae. Oecologia (Berl) 39:71–77.

Jones CG, Firn RD (1978) The role of phytoecdysteroids in bracken fern, *Pteridium aquilinum* (L.) Kuhn as a defense against phytophagous insect attack. J Chem Ecol 4:117–138.

Jones CG, Firn RD (1979) Resistance of *Pteridium aquilinum* to attack by nonadapted phytophagous insects. Biochem Syst Ecol 7:96–101.

Kaplanis JN, Thompson MJ, Robbins WE, Bryce BM (1967) Insect hormones: alpha ecdysone and 20-hydroxyecdysone in bracken fern. Science 157:1436–1438.

Klocke JA, Chan BG (1982) Effects of cotton condensed tannin on feeding and digestion in the cotton pest, *Heliothis zea*. J Insect Physiol 28:911–915.

Klowden MJ (1980) Large doses of ecdysterone may inhibit mosquito behavior nonspecifically. Science 208:1062–1063.

Kogan M, Ortman EF (1978) Antixenosis—a new term proposed to define Painter's "nonpreference" modality of resistance. Bull Entomol Soc Am 24:175–176.

Kubo I, Klocke JA, Asano S (1983) Effects of ingested phytoecdysteroids on the growth and development of two lepidopterous larvae. J Insect Physiol 29:307–316.

Lichtenstein EP, Casida JE (1963) Myristicin, an insecticide and synergist occurring naturally in the edible parts of parsnip. J Agric Food Chem 11:410–415.

Mabry TJ, Gill JE, Burnett WC, Jones SB (1977) Antifeedant sesquiterpene lactones in the Compositae. In: Host Plant Resistance to Pests. Hedin PA (ed), ACS Symposium Series No. 62.

Marcus C, Lichtenstein EP (1979) Biologically active components of anise: toxicity and interaction with insecticides in insects. J Agric Food Chem 27:1217–1223.

Marian MP, Pandian TJ (1980) Effects of feeding senescent leaf of *Calotropis gigantea* on food utilization in the monarch butterfly *Danaus chrysippus*. Entomon 5:257–264.

Martin JG (1980) Saccharin controversy. Science 208:1086.

Matsumura F (1976) Toxicology of Insecticides. Plenum, New York.

McCaffery AR (1982) A difference in the acceptability of excised and growing cassava leaves to *Zonocerus variegatus*. Entomol Exp Appl 32:111–115.

McFarlane JE, Distler JHW (1982) The effect of rutin on growth, fecundity and food utilization in *Acheta domesticus* (L.) J Insect Physiol 28:85–88.

McKey D (1979) The distribution of secondary compounds within plants. In: Herbivores their Interaction with Secondary Plant Metabolites. Rosenthal GA, Janzen DH (eds), Academic Press, New York.

Meisner J, Ascher KRS, Sur M, Kabonci I (1977c) Synergistic and antagonistic effects of gossypol for phosfolan in *Spodoptera littoralis* larvae on cotton leaves. J Econ Entomol 70:717–719.

Meisner J, Navon A, Zur M, Ascher KRS (1977b) The response of *Spodoptera littoralis* larvae to gossypol incorporated in an artificial diet. Environ Entomol 6:243–244.

Meisner J, Zur M, Kabonci E, Ascher KRS (1977a) The influence of the gossypol content of leaves of different cotton strains on the development of *Spodoptera littoralis* larvae. J Econ Entomol 70:714–716.

Merkle H, Speiser P (1973) Preparation and in vitro evaluation of cellulose acetate phthalate coacervate microcapsules. J Pharm Sci 62:1444–1448.

Miles PW, Aspinall D, Correll AT (1982) The performance of two chewing insects on water-stressed food plants in relation to changes in their chemical compositions. Aust J Zool 30:347–251.

Miller JS, Feeny P (1983) Effects of benzylisoquinoline alkaloids on the larvae of polyphagous Lepidoptera. Oecologia (Berl) 58:332–339.

Mohamed AKA (1982) Effect of *Nomuraea rileyi* on consumption and utilization of food by *Heliothis zea* larvae. J Ga Entomol Soc 17:356–362.

Munakata K (1977) Insect feeding deterrents in plants. In: Chemical Control of Insect Behavior. Shorey HH, McKelvey JJ (eds), Wiley, New York, pp 93–102.

Murdock LL (1985) How to make the world safe for elephant-eye corn, amber waves of grain and the lowly bean. Eleventh Annual Illinois Crop Protections Workshop, March 5–7.

Nakanishi K, Koreeda M, Sasaki S, Chang ML, Hsu HY (1966) Insect hormones I. The structure of ponasterone A, an insect-moulting hormone from the leaves of *Podocarpus nakaii* Hay. Chem Commun 1966:181–222.

Newman AA (1962) The occurrence, genesis and chemistry of the phenolic methylene dioxy ring in nature. Chem Prod 25:161–166.

Pachter M (1951) Magic in Science. Schuman, New York, p 86.

Paech K, Tracey MV (1954–1964) (eds) Moderne Methoden der Pflanzenanalyse. 7 Vol. Springer-Verlag, Berlin.

Painter RH (1951) Insect resistance in crop plants. Macmillan, New York.

Palumbo RE, Dahlman DL (1978) Reduction of *Manduca sexta* fecundity and fertility by l-canavanine. J Econ Entomol 71:674–676.

Phillips RL (1968) Cyanogenesis in *Lotus* species. Crop Sci 8:123–124.

Picman AK, Elliott RH, Towers GHN (1978) Insect feeding deterrent property of alantolactone. Biochem Syst Ecol 6:333–335.

Price PW, Bouton CE, Gross P, McPheron BA, Thompson JN, Weis AE (1980) Interactions among three trophic levels: influence of plants on interactions between insect herbivores and natural enemies. Annu Rev Ecol Syst 11:41–65.

Rausher MD (1981) Host plant selection by *Battus philenor* butterflies: the roles of predation, nutrition and plant chemistry. Ecol Monogr 51:1–20.

Reese JC (1979) Interactions of allelochemicals with nutrients in herbivore food. In: Herbivores: their Interaction with Secondary Plant Metablites. Rosenthal GA, Janzen DH (eds), Academic Press, New York, pp 309–330.

Reese JC, Chan BC, Waiss AC (1982) Effects of cotton condensed tannin, maysin (corn) and pinitol (soybeans) on *Heliothis zea* growth and development. J Chem Ecol 8:1429–1436.

Rehr SS, Janzen DH, Feeny PP (1973) L-Dopa in legume seeds: a chemical barrier to insect attack. Science 181:81–82.

Rembold H, Sharm GK, Czoppelt CH, Schmutterer H (1982) Azadirachtin: a potent insect growth regulator of plant origin. Angew Entomol 93:12–17.

Rhoades DR, Cates RG (1976) Toward a general theory of plant antiherbivore chemistry. Recent Adv Phytochem 10:168–213.

Robbins WE, Kaplanis JN, Thompson MJ, Shortino TJ, Joyner SC (1970) Ecdysones and synthetic analogs: molting hormone activity and inhibitive effects on insect growth metamorphosis and reproduction. Steroids 16:105–125.

Robertson JL, Russell RM, Savin NE (1981) POLO2: A computer program for multiple probit or logit analysis. Bull Entomol Soc Am 27:210–211.

Sato Y, Sakai M, Imai S, Fujioka T (1968) Ecdysone activity of plant-originated moulting hormones applied on the body surface of lepidopterous larvae. Appl Entomol Zool 3:49–51.

Schmutterer H, Rembold H (1980) Zur Wirkung einiger Einfraktionen aus Samen vor *Azadirachta indica* auf Frassaktivitat und Metamorphose von *Epilachna varivestris* (Col. Coccinellidae). Angew Entomol 89:179–189.

Schoener TW (1982) Length-weight regressions in tropical and temperate forest understory insects. Ann Entomol Soc Am 73:106–109.

Schoonhoven LM (1967) Loss of host plant specificity by *Manduca sexta* after rearing on an artificial diet. Entomol Exp Appl 10:270–272.

Schoonhoven LM (1982) Biological aspects of antifeedants. Entomol Exp Appl 31:57–69.

Schoonhoven LM, Derksen-Koppers I (1976) Effects of some allelochemics on food uptake and survival of a polyphagous aphid, *Myzus persicae*. Entomol Exp Appl 19:52–56.

Schoonhoven LM, Meerman J (1978) Metabolic cost of changes in diet and neutralization of allelochemics. Entomol Exp Appl 24:489–493.

Scriber JM (1977) Limiting effects of low leaf-water content on the nitrogen utilization, energy budget, and larval growth of *Hyalophora cecropia* (Lepidoptera: Saturniidae). Oecologia (Berl) 28:269–287.

Scriber JM (1978) Cyanogenic glycosides in *Lotus corniculatus* Their effect upon growth, energy budget and nitrogen utilization of the southern armyworm, *Spodoptera eridania*. Oecologia (Berl) 34:143–155.

Scriber JM (1978) Cyanogenic glycosides in *Lotus corniculatus*. Oecologia (Berl) 34:143–155.

Scriber JM (1982) The behavior and nutritional physiology of southern armyworm larvae as a function of plant species consumed in earlier instars. Entomol Exp Appl 31:359–369.

Scriber JM, Slansky F (1981) The nutritional ecology of immature insects. Annu Rev Entomol 26:183–211.

Sharma HC, Agarwal RA (1982) Effect of some antibiotic compounds in *Gossypium* on the post-embryonic development of spotted bollworm *(Earias vittella)*. Entomol Exp Appl 31:225–228.

Shaver TN, Parrott WL (1970) Relationship of larval age to toxicity of gossypol to bollworms, tobacco budworms and pink bollworms. J Econ Entomol 63:1802–1803.

Singh ID, Weaver JB (1972) Growth and infestation of boll weevils on normal-glanded, glandless, and high-gossypol strains of cotton. J Econ Entomol 65:821–824.

Singh P (1977) Artificial Diets for Insects, Mites, and Spiders. Plenum, New York.

Smiley JT, Wisdom CS (1982) Photographic estimation of weight of insect larvae. Ann Entomol Soc Am **75**:616–619.

Smith RJ (1978) NAS saccharin report sweetens FDA position, but not by much. Science **202**:852–853.

Smith RJ (1980) Latest saccharin tests kill FDA proposal. Science **208**:154–156.

Sokal RR, Rohlf FJ (1981) Biometry. Freeman, San Francisco.

Southwood TRE (1972) The insect/plant relationship—an evolutionary perspective. Royal Entomol Soc Symp **6**:3–30.

Stamopoulos D, Huignard J (1980) L'influence des divers parties de la harine de haricot *(Phaseolus vulgaris)* sur le developpement des larves d'*Acanthoscelides obtectus* (Coleoptere: Bruchidae). Entomol Exp Appl **28**:38–46.

Steel RGD, Torrie JH (1980) Principles and Procedures of Statistics. A Biometric Approach. McGraw-Hill, New York.

Straatman R (1962) Notes on certain lepidoptera ovipositing on plants which are toxic to their larvae. J Lepidopt Soc **16**:99–103.

Stubblebine SH, Langenheim JH (1977) Effects of *Hymenaea courbaril* leaf resin on the generalist herbivore *Spodoptera exigua* (beet armyworm). J Chem Ecol **3**:633–647.

Sutherland ORW, Russell BG, Biggs, DA, Lane BA (1980) Insect feeding deterrent activity of phytoalexin isoflavonoids. Biochem Syst Ecol **8**:73–75.

Tamura M (1981) Influence of the growth stages of the host plant leaves on the assimilation rate of *Pryeria sinica* Moore (Lepidoptera: Zygaenidae). Jpn J Appl Entomol Zool **25**:121–122.

Towers OHN (1982) Photosensitizers from plants and their photodynamic action. Prog Phytochem **6**:183–202.

Usher BF, Feeny P (1983) Atypical secondary compounds in the family Cruciferae: tests for toxicity to *Pieris rapae,* an adapted crucifer-feeding insect. Entomol Exp Appl **34**:257–265.

Wagner MR, Benjamin DM, Clancy KM, Schuh BA (1983) Influence of diterpene resin acids on feeding and growth of larch sawfly, *Pristiphora erichsonii* (Hartig). J Chem Ecol **9**:119–127.

Waldbauer GP, Fraenkel G (1961) Feeding on normally rejected plants by maxillectomized larvae of the tobacco hornworm, *Protoparce sexta* (Lepidoptera, Sphingidae). Ann Entomol Soc Am **54**:477–485.

Williams WG, Kennedy GG, Yamamota RT, Thacker JD, Bordner J (1980) 2-Tridecanone: a naturally occurring insecticide from the wild tomato *Lycopersicon hirsutum* f. *glabratum*. Science **207**:888–889.

Willomitzer J, Tomanek J (1981) Larvicidal efficiency of some inorganic compounds and plant extracts against the house fly. Acta Vet Brno **50**:105–112.

Wilson RL, Starks KJ (1981) Effect of culture-host preconditioning on greenbug (Homoptera, Aphididae) response to different plant species. Southwest Entomol **6**:229–232.

Whittaker RH, Feeny PP (1971) Allelochemics: chemical interactions between species. Science **171**:757–770.

Wright DP (1963) Antifeeding compounds for insect control. Adv Chem Ser **31**:56–63.

Zalkow LH, Gordon MM, Lanir N (1979) Antifeedants from rayless goldenrod and oil of pennyroyal: toxic effects for the fall armyworm. J Econ Entomol **72**:812–815.

Chapter 6

Bioassays for Measuring Quality of Insect Food

Marcos Kogan[1]

I. Introduction

Measurement of food quality requires evaluation of the effects of a diet on physiological processes that maximize progeny production and survival, the key parameters of fitness (Williams, 1966). Slansky (1982) suggested that the amount, rate, and quality of food consumed by larvae affected growth rate, developmental time, final body weight, movement, and survival. Amount, rate, and quality of food for adults influence fecundity, longevity, movement, and competitive ability. Larval food quality may additionally affect pupal and adult phenotypic characteristics. Obvious effects of inadequate larval diets are pupal distortions and wing malformations in the imago. For example, such effects characterize the toxicity of L-canavanine to *Manduca sexta* L., the tobacco hornworm (Rosenthal and Dahlman, 1975).

   Since most of the criteria to measure food quality and postingestive effects of diets are based on detection of symptoms, insect ecologists and nutritionists have used growth analysis and utilization indices in an attempt to establish a preliminary link between effects and their probable causal mechanisms. In measuring food quality it is essential to be able to distinguish between the preingestive behavioral responses of insects to diet components and the postingestive physiological effects. M. Berenbaum (Chapter 5, *this volume*) has stressed this need and covered the procedures to bioassay postingestive effects of phytochemicals on insects. This chapter

[1]Illinois Natural History Survey and University of Illinois at Urbana-Champaign, Urbana, Illinois 61801, U.S.A.

concentrates on measurement of food intake and utilization as a basis for defining diet quality. However, I also call attention to information on mass production of insects for use in biological and genetic control methods. The criteria for assessing quality control of the mass-produced insects are applicable also in insect dietetics and nutritional ecology.

II. Basic Insect Nutritional Requirements

Except for a general requirement for a source of a steroid nucleus (Svoboda et al., 1975), insects have the same basic nutritional requirements of all higher animals, with very few exceptions. Fraenkel (1953) suggested that an insect's optimal diet should (a) provide the required nutrients in adequate proportions and availability; (b) favor a high conversion rate (adequate digestibility and assimilation); (c) lack antibiotic allomones; (d) have adequate feeding excitants to guarantee continuous feeding and ingestion, e.g., sucrose as a biting factor and fiber for swallowing (Hamamura et al., 1962); and (e) have a physical structure compatible with the insect's feeding mechanism and ingestion capabilities.

The most comprehensive review to date of insect nutritional requirements was compiled by Dadd (1977). Nutritional requirements of insects are qualitatively and quantitatively specific. There is a defined ratio of protein to carbohydrate that must be maintained for optimal growth, survival, and reproduction. These proportions vary according to the nature of the diet of the insect. In a recent study, Waldbauer et al. (1984) showed that corn earworm, *Heliothis zea* (Boddie), larvae achieved best utilization and conversion of digested food on a diet containing an 80:20 protein/carbohydrate proportion. When offered a choice of media containing either casein or sucrose the larvae selectively ingested enough of each medium to approach the optimal 80:20 ratio. The ability to self-select an optimal diet composition may have far-reaching ecological implications in a variable environment (see Whitham, 1983, for instance).

III. Measurement of Food Intake, Utilization, and Growth

The fields of insect dietetics and nutritional ecology have made great strides in recent years based mainly on the extensive use of techniques of intake, utilization, and growth analyses (Gordon, 1968; Waldbauer, 1968; Petrusewicz and Macfadyen, 1970; Klein and Kogan, 1974; Slansky and Scriber, 1982; Scriber, 1984). This section covers experimental methods in analyses of food intake and utilization, as well as computations of the most useful nutritional indices, statistical procedures, and restrictions in the adoption of these indices.

3.1. Experimental Procedures

The same basic precautionary steps recommended for selection of number of individuals, pretest conditions, and standardization of both test animals and test diets, discussed by M. Berenbaum (see Chapter 5, *this volume*) should guide experimental design and protocols in nutritional analyses. Nutritional indices are extremely sensitive to the variance of parameters used in their computation (Schmidt and Reese, 1986). Extreme care in reducing extraneous sources of variation is, therefore, essential.

A complete nutritional analysis experiment provides sets of data obtained during a defined time interval t (t may correspond to an instar, a complete stage of growth [larval], or a predetermined number of hours or days). Symbols used here are basically those proposed by Waldbauer (1968), slightly modified by Klein and Kogan (1974) and Scriber and Slansky (1981) and now reassessed to avoid duplication and incongruences. The data sets necessary for quantitative nutritional analyses are:

1. Amount of food consumed (F_t) during t;
2. Amount of total body weight change (B_t) during t;
3. Amount of total excretion (E_t) during t (including cast skins [exuviae], secretions, cocoons, and feces);
4. Volume of CO_2 respired (desirable but seldom included in nutritional analyses).

The basic method of data acquisition is gravimetric. Measurements are usually obtained in the form of fresh weights although most analyses are based on dry weights because dry weight data are usually less variable. Aliquots must be used to compute conversion factors of fresh weights into dry weights, if these cannot be obtained directly.

A. Measurement of Food Consumed (F_t)

The amount of food consumed by an insect in time interval t is obtained by measuring the food given (F_i = initial amount of food) and the food remaining (F_r = food left) after the feeding period t. The amount of food consumed is then:

$$F_t = F_i - F_r$$

Measurements of F_i are necessarily made on fresh weights. Conversion to dry weights is made by obtaining the fresh and dry weights of an aliquot as similar to the test food as possible. When leaves are used, it is common practice to split the leaf along the mid-rib, and to use one half-leaf in the feeding test and the other half as the aliquot. The dry weight/fresh weight ratio of the aliquot multiplied by the fresh weight of the test food provides the dry weight of the food given. What seems to be a straightforward experimental procedure is, perhaps, one of the main sources of error in quantitative nutritional analyses (Schmidt and Reese, 1986). Some of the problems stem from the heterogeneity of the plant material and others are inherent in the experimental design. Plant materials are highly hygroscopic and dried leaves absorb moisture as shown for dried soybean leaves held at 40 and 70% relative humidity environments (Fig. 1).

In addition to their use in the computation of dry weight/fresh weight ratios, aliquots are used to adjust F_t for errors due to natural weight loss (or gain) of the food. The aliquot is kept under exactly the same conditions as the experimental part of the food. If a is the initial weight of the aliquot and b the final weight, then the two following ratios are computed as correction factors: $\alpha = (a-b)/a$ and $\beta = (a-b)/b$. The two ratios are then applied in the formula to obtain the corrected weight of food eaten (Waldbauer, 1968):

$$F_t = [1 - (\alpha/2)] [F_i - (F_r + \beta F_r)]$$

To reduce the error introduced by variation in food weight, the food remaining (F_r) should be just above the amount needed to allow individuals to feed ad libitum (all consumption experiments must provide food ad libitum) unless effects of starvation are being tested. However, it is often difficult to determine a priori the size of F_i. Excessive residue (large F_r) is a major source of error. When measuring food consumption by young larvae the amount of foliage consumed is so small that plant parts kept under moist conditions may grow more than the insect is capable of con-

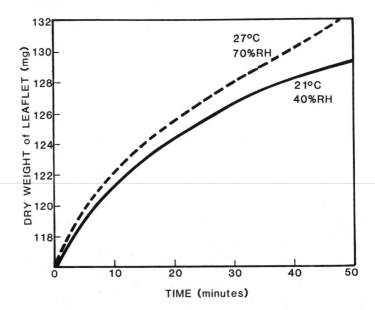

Figure 1. Absorption of moisture by dry soybean leaves over a 50-min period of exposure to two sets of temperature and relative humidity.

suming. The result is negative consumption ($F_r > F_i$). Growth of excised plant parts during the course of the experiment is therefore a concern. Excised soybean leaflets kept on moist filter paper in closed petri dishes grew 5.8% in area and 23% in weight in 72 hr (personal observations). One solution in tests with young larvae is to use cohorts instead of individuals, provided that the species is not cannibalistic and that aggregation does not interfere with normal feeding behavior. A general recommendation is to keep F_i as small as possible, closely monitor feeding rates, and replace the food as often as possible. Logistics of such experiments become extremely cumbersome and may lead to risky compromises with number of replicates.

It is essential to maintain the quality of the diet throughout the entire feeding period. One major difficulty is to preserve the degree of hydration of the plant material. Water content of the diet is an important factor influencing consumption (Scriber, 1979). Experiments with excised plant parts are usually done in closed petri dishes lined with moist filter paper. In our laboratory all tests are conducted with a layer of plaster of Paris that is soaked and covered with filter paper. The plaster of Paris helps retain humidity within a wide range of temperatures.

Ideally, one should be able to measure consumption on the standing plant. The only method that permits this uses determination of leaf area consumed. In our laboratory we have used a clip-cage for this purpose (Fig. 2). Insects confined in the cage are allowed to feed until the entire area is consumed. By determining the specific leaf weight (mean leaf weight per unit area) of aliquots it is possible to convert leaf area to leaf weight. The introduction of additional sources of error by this method is, however, inevitable. We tested conversions of measurements of food consumption with the soybean looper, *Pseudoplusia includens* (Walker), by area and by weight. Fifth and sixth instars ate an average of 830 mg of soybean foliage, corresponding to about 92 cm^2. When the leaf area was converted to leaf weight by a regression procedure, the value obtained was 918 mg, an error of about 10% (Kogan and Cope, 1974). The heterogeneity of the plant material introduces difficulties that often result in absurd consumption values. Researchers must be aware of this and take every possible precaution to minimize these sources of error.

B. Measurement of Weight Gain (B)

Weights of insects (AW) are taken at time t_n and time t_{n+1} starting at $n = 1$, the onset of the experiment. Weight gain (B) is given by:

$$B = AW_{t_{n+1}} - AW_{t_n}$$

Data are usually recorded as fresh weights and then converted to dry weights by ratios obtained with a comparable cohort of the insect pop-

Figure 2. (A) Clip cage used to measure food consumption based on leaf area and specific leaf weight. (B) Soybean loopers on soybean leaflet showing circular areas eaten.

ulation maintained under identical experimental conditions. Weights of very young larvae are often below the limits of precision of most common balances (except highly sensitive microbalances with precision on the order of 10^{-3} mg). Initial weights of these larvae are obtained by measuring cohorts (we use cohorts of 100 larvae for newly emerged noctuids). The use of parallel cohorts is a source of error but it provides the only non-destructive gravimetric procedure to obtain dry weight values for the individuals in the test cohort. Of course the terminal dry weight is obtained by direct measurement of the experimental individuals, but those in the parallel cohort are also measured for an additional control of the sensitivity of the dry weight conversion procedure.

A major source of error in weight gain measurements is the degree of satiation of animals at the time they are weighed. There may be considerable differences in weight between larvae with full and larvae with empty guts (Waldbauer, 1968). Larvae just prior to a molt usually stop feeding and gradually lose weight until they resume feeding again after ecdysis and sclerotization of the new cuticle. Measurements made with larvae at these stages may prevent errors due to the variable amounts of food in the gut.

When measuring large numbers of experimental animals it is important to avoid dehydration of live specimens. Figure 3 shows the weight loss of soybean looper larvae kept on dry and on wet filter paper over a 2-hr period. On the other hand, when measuring dry weights, it is necessary to avoid prolonged exposure to humid environments as the dried cuticle is hygroscopic and will absorb water, increasing the apparent weight of the animals (see Fig. 3). For greater precision of measurements it is desirable to obtain three successive readings and take the arithmetic mean.

C. Measurements of Egesta and Excreta (E)

With insects that excrete discrete fecal pellets, collection of excreta is not very dificult. It is, however, important to avoid contact of feces with the feeding substrate, as humidity of the substrate may remove soluble components of the feces, thus masking true values as well as changing the chemical nature of the diet. In our laboratory we keep lepidopterous larvae feeding on food positioned in a way that causes fecal pellets to collect in the bottom of the container. When artificial media are used, diet cups are kept upside down and feces are collected on the cardboard lid.

There are difficulties in collecting excreta of insects with watery or with highly concentrated feces. The Mexican bean beetle usually excretes droplets of yellowish feces that adhere to the foliage on which they feed. The uric acid technique (Bhattacharya and Waldbauer, 1970) addresses the problem of measuring excreta when mixed with food. This technique

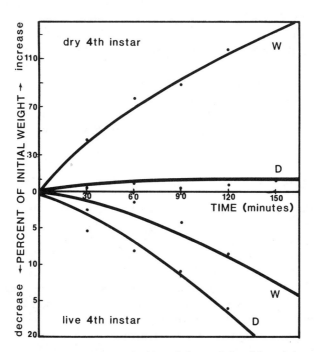

Figure 3. Change in fresh (lower graph) and dry weight of fourth instar soybean loopers in dry and in moist (saturated) chambers over a period of 150 min.

uses measurements of uric acid content in the mixture; determination of
E is based on previously determined uric acid concentration in the excreta
(see below for detailed description of procedure).

In work with mammals, distinction is made between *egesta,* the part
of consumption eliminated in feces or regurgitated, and *excreta,* products
of catabolism eliminated in urine or through skin (Petrusewicz and Macfadyen, 1970). However, in insects excreta and egesta are mixed in the
feces, a fact that influences computations of digestibility indices, particularly for insects feeding on high-nitrogen diets (Waldbauer, 1968).

3.2. Quantitative Nutritional Indices

In quantitative nutritional studies it is necessary to measure the amount
of food consumed that is digested, assimilated, converted into biomass,
and metabolized (Scriber and Slansky, 1981). Much of the current ecological literature in the United States uses symbols and concepts proposed
by Waldbauer (1968) recently modified by Scriber and Slansky (1981).
However, European researchers prefer the system proposed by Petrusewicz (1967), and by Petrusewicz and Macfadyen (1970); and others, yet,
use the conceptual framework proposed by Gordon (1968, 1972). Although
Petrusewicz and Macfadyen's (1970) methods are directed principally to
investigations of energy flow in communities and they apply better to populations rather than to individuals, there is still considerable convergence
of concepts. It would be extremely useful to strive for uniform terminology
and notation. The multiplicity of notations adds undue confusion to a literature that tends to be cluttered with cumbersome indices, some of which
contain biological meaning that is difficult to interpret. Table 1 attempts
to show the homologies among the most common indices and notations
proposed by various authors. The main nutritional indices are discussed
in detail.

A. Mean Animal Weight

An important parameter used in measurements of rates of food consumption and growth is the mean animal weight ($\overline{B}$ = mean biomass) (see
symbols used by various authors in Table 1). The normal growth of immature insects usually follows a sigmoid curve (Fig. 4). When measurements are taken during the phase of accelerated growth (e.g., instars 3–
5 in many lepidopterous larvae), growth approaches a straight line (segment
$\overline{AB}$ in Fig. 4) and the mean animal weight may be defined by the area of
the trapezoid $ABcd/T_{\overline{cd}}$, or more easily by the area of the rectangle $cdc'd'/$
$2T_{\overline{cd}}$; $T_{\overline{cd}}$, is the number of time units (usually days) contained in $\overline{cd}$.

There is, however, more interest in the computation of $\overline{B}$ for the entire
developmental stage, in which case it is necessary to integrate the area

Table 1. Parameters and indices most commonly used in quantitative nutrition and nutritional ecology: Comparison of symbols and concepts proposed by various authors

Parameter or index	Petrusewicz, 1967; Petrusewicz and Macfadyen, 1970	Gordon, 1968, 1972	Waldbauer, 1964, 1968	Kleskouski, 1970 in Stepien and Rodriguez, 1972	Klein and Kogan, 1974[a]	Scriber and Slansky, 1981	This chapter
Animal weight[b]	W	$W_{i,f}$	—	—	AW	—	AW
Weight gain	ΔW	ΔW	G	P	WG	B	B
Food weight[c]	—	F	WL	—	FO, FN	—	F_o, F_f
Food consumed	C	ΔF	F	C	FC	I	F
Mean animal weight[d]	q	W_e	A	—	MAW	B	$\bar{B}$
Excreta[e] and egesta[e]	FU	S	wt, feces	FU	FES	F	E
Respiration	R	ΔO	—	R	—	—	R
Consumption index or relative consumption rate:	—	F	CI	—	CI	RCR	RCR
computation[f]	—	$\Delta F/W_i T$	F/TA	—	$I/MAW \cdot ND$	$I/\bar{B}T$	$F/\bar{B}T$
Relative growth rate:	—	G	GR	—	GR	RGR	GR
computation	—	$\Delta W/W_i T$	G/TA	—	—	—	$B/\bar{B}T$
Mean relative growth rate:	—	—	—	—	$\overline{RGR}$	—	RGR
computation	—	—	—	—	$(\ln MAXAW - \ln AW)/T$	$B/\bar{B}T$	$(\ln B_n - \ln B_i)/T$
Efficiency of conversion of ingested food:	—	E	ECI	K_1	ECI	ECI	ECI
computation	P + R + U	$G/(\Delta F)$	wt gain/F	P/C	WG/FC	—	B/F
Approximate digestibility:	D	—	AD	U'	AD	AD	AD
computation	—	I/F	(F − wt feces)/F	A/C	(FC − FES)/FC	(I − F)/I	$(F_i − E)/F$
Efficiency of conversion of digested food:	—	—	ECD	K_2	ECD	ECD	ECD
computation	—	G/I	wt gain/(F − wt feces)	P/A	WG/(FC − FES)	B/(I − F)	$B/(F_i − E)$

[a] Symbols in Klein and Kogan (1974) followed by F or D denote fresh or dry weights.

[b] In Gordon (1968) subscripts i and f indicate initial and final weights, respectively.

[c] In Waldbauer (1968) W = weight of food given, L = weight of uneaten food. In Klein and Kogan (1974) FO = weight of food at start of period (food given), FN = weight of food at end of period (uneaten food).

[d] Computational procedures are presented in the text.

[e] FU = feces + urine in Petrusewicz and Macfadyen (1970), and Kleskouski (1970, in Stepien and Rodriguez, 1972).

[f] In Klein and Kogan (1974) parameters identified as code for computer program, so T (time) = ND (number of days).

underneath the sigmoid curve. Waldbauer (1964) proposed the use of weighted averages of daily weights, as follows: If B_1 is the initial weight and B_n the last, and t_1, t_2. . . .t_{n-1}, t_n the time interval in days between weighings, then the mean animal weight is given by:

$$\bar{B} = [0.5B_1 + (B_2 + B_3 + ...B_{n-1}) + 0.5\,(t_n - t_{n-1})\,B_n]/(t_n - t_1)$$

and $t_n - t_1$ = the total number of days between the last and the first weighings. (The method described here is slightly simplified from the original, which had an additional factor affecting the penultimate weight.) This method requires the measurement of a series of intermediate weights.

Gordon (1968) computed a mean animal weight ($W_e = \bar{B}$) by solving the integral of the growth curve such that:

$$W_e = \Delta W/[d \cdot \ln (W_f/W_i)],$$

where d is the dry weight/fresh weight ratio of an aliquot of the animal population, and W_f and W_i the final and initial weights of the animals, respectively. Klein and Kogan (1974) computed $\bar{B}$ by integrating:

$$\bar{B} = \int_{p=1}^{NP} B_p dp/T,$$

where p = growth period (usually an instar), NP = number of periods (instars), and T = duration of all periods, except the last one.

B. Growth Rate

The expression of growth as the weight increment per unit of extant body weight is useful in comparative studies involving animals of various sizes. There are several methods to express this relationship but the relative merits of each method have not been critically analyzed. Waldbauer (1968) and Gordon (1968) proposed computation of the relative growth rate as:

$$RGR = B\,/\,\bar{B}T,$$

where B is the weight gain during time interval T. Kogan and Cope (1974) proposed a mean relative growth rate as:

$$\overline{RGR} = (\ln B_n - \ln B_1)/T$$

that was adapted from methods of plant growth analysis (Radford, 1967). The value of $\overline{RGR}$ for the soybean looper feeding on soybean foliage was about 40% higher than the corresponding value for RGR, computed by the Waldbauer (1968) method. Grabstein and Scriber (1982) defined RGR as $RCR \times AD \times ECD$ where RCR = relative consumption rate, AD = approximate digestibility, and ECD = efficiency of conversion of digested

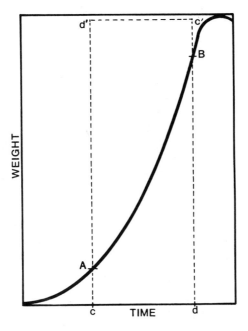

Figure 4. Generalized growth (weight gain) curve used to explain computation of mean animal weight $(\bar{B})$.

food. The units for *RGR* are milligrams weight gain per milligram body weight per day.

C. Consumption Rate

The rate of food consumed per unit of time per unit of animal body weight is given by:

$$RCR = F/\bar{B}T,$$

where F is ingested food during time interval T and $\bar{B}$ is the mean animal weight (mean biomass gained).

D. Conversion of Ingested Food

The food ingested is converted into animal biomass (B) depending on the ability of the animal to digest and assimilate the food to convert it into body matter. The expression of the efficiency of conversion of ingested food is:

$$ECI = B/F$$

usually expressed as percentage ($\times 100$). *ECI* represents the ratio of *RGR*/*RCR* or the rate of growth over the consumption rate. This relationship has been used extensively to separate the effect of reduced efficiency of conversion from the effect of feeding inhibition on growth rates (Blau et al., 1978, see also M. Berenbaum, Chapter 5, *this volume*). The relationships among *ECI, RGR,* and *RCR* have been discussed in detail by Scriber and Slansky (1981). Scriber (1984) demonstrated the effect of leaf water content and leaf nitrogen content on *ECI, RGR,* and *RCR.* The intelligent use of nutritional analysis in those studies provided an excellent tool for the assessment of food quality.

E. Digestion of Ingested Food

Indices used to measure digestibility of the food are: approximate digestibility (*AD*), and the efficiency of conversion of digested food (*ECD*). These indices are computed as:

$$AD = (F - E)/F \text{ and } ECD = B/(F - E)$$

both expressed as percentages ($\times 100$), where $B =$ biomass gained, $F =$ ingested food, and $E =$ excreta.

IV. Indirect Methods of Measurement of Food Intake

The experimental work discussed to this point uses gravimetry to measure all parameters. Despite the care and improvements in experimental techniques it is apparent that there are difficulties in measuring food intake and utilization with accuracy. The fresh weight/dry weight ratios used in most studies are likely to introduce errors even in the most carefully performed experiments (Schmidt and Reese, 1985). It is even more difficult to measure intake and utilization using artificial media and in situations in which the insect lives within the feeding substrate, such as stored product insects, leaf miners, stem and fruit borers, and dung feeders. Alternative methods have been proposed to solve these specific trophic situations. These methods can be grouped into five main categories: (1) colorimetric, (2) isotopic, (3) trace element, (4) enzymatic, and (5) immunological. In addition, in ecological research biomass turnover is often expressed in terms of energy. Calorimetry is then the method of choice. All methods, however, require some gravimetric measurements; thus each of these methods is a combination of gravimetry with some other indirect indicator of consumption. Description of the methods included below are based mainly on Parra and Kogan (1981) and the comparative analyses of the methods by Kogan and Parra (1981).

4.1. Colorimetric Methods

These methods use a nontoxic and nondegradable marker. The ideal marker should allow thorough incorporation into the medium, quantitative ingestion by the insect with no undesirable consequences, and it should have no nutritional value. Some markers are absorbed and incorporated into fat body tissues, while others pass through the gut and are quantitatively recovered in the feces. The following dyes have been used: chromic oxide (McGinnis and Kasting, 1964a), Calco Oil Red N-1700 (Daum et al., 1969), Solvent Red 26 and Soluble Blue 58® (Brewer, 1982), amaranth = Acid red 27 (Hori and Endo, 1977; Kuramochi and Nishijima, 1980.) The techniques employing Calco Oil Red (COR) and chromic oxide will be described in detail as they illustrate basic methodologies of general application in colorimetric methods.

A. COR Method

There are hundreds of dyestuffs potentially useful in quantitative nutrition, but only a few of them have been tested with insects. Calco Oil Red N-1700® or Solvent Red 26 (obtained from Keystone Aniline and Chemical Company, Chicago, IL, U.S.A.) is incorporated into artificial media at a rate of 1 g/liter (Hendricks and Graham, 1970). The dye is predissolved in an oil component of the medium. Parra and Kogan (1981), using a modified Henneberry and Kishaba (1966) medium, dissolved COR in wheat germ oil. The analytical procedure follows Daum et al. (1969). Larvae, prepupae, and pupae are removed from the medium for measurement and are killed in acetone to remove dye residues that contaminate the integument. The dye, incorporated into the animal fat body and in the feces, is extracted with spectrophotometric grade acetone in a tissue grinder. The extract is filtered and the filter paper is extracted in reflux in a Soxhlet extractor and the filtrates are combined. COR concentration in the filtrate is measured spectrophotometrically at 510 nm. Standard curves are prepared for each dye's absorbance maximum.

Use of dye concentrations only will provide a measurement of digestibility (*AD*) by the formula:

$$AD = (M_E - M_F)/M_E$$

where M_E = concentration of dye in excreta, and M_F = concentration of dye in ingested food. If the weight of the excreta (*E*) or the weight of the food (*F*) is known then it is possible to calculate the unknown parameter (Waldbauer, 1968):

$$F = (M_F/M_E)E, \text{ and } E = (M_F/M_E)F$$

Examples of uses of the method are found in Jones et al. (1975), Parra and Kogan (1981), and Brewer (1982).

B. Chromic Oxide Method

This method was proposed by McGinnis and Kasting (1964a). Chromic oxide (CR_2O_3) is incorporated into the medium at 4% concentration. Chromic oxide is more soluble at higher pH so it is more easily incorporated by mixing it with the KOH fraction used to adjust the medium pH. In the original procedure, CR_2O_3 in the feces and animals is oxidized to $Cr_2O_7^{-2}$ and the dichromate ion is measured colorimetrically with diphenylcarbazide. Samples of 10–25 mg are weighed and placed into 100-ml Kjeldahl flasks. Ten milliliters of a digestion mixture containing 10 g $Na_2MoO_4 \cdot 2H_2O$, 150 ml H_2SO_4 (conc.), and 200 ml of 70% $HClO_4$ in 150 ml distilled water is added to each sample. The mixture is heated for 30 min, in flasks held on a Kjeldahl digestion rack (we used a Kontes Products, NJ, U.S.A., rotary rack; Fig. 5). A NaOH trap is connected to the Kjeldahl rack exhaust to reduce the risk of spread of acid fumes, and digestion is carried out in an explosion-proof fume hood. After 15 min the digests turn from green to orange. Digestion proceeds for another 15 min and the mixture is cooled at room temperature. McGinnis and Kasting

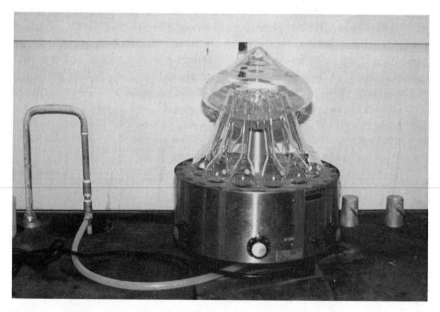

Figure 5. Rotary Kjeldahl digestion apparatus (Kontes®), used in the chromic oxide method.

(1964a) recommend adding 500 ml of distilled water to the digest, followed by 4.5 ml of 0.25 N sulfuric acid, and 0.5 ml of diphenylcarbazide. After 3 min, absorbance is measured at 540 nm against a blank consisting of 9.5 ml of 0.25 N sulfuric acid and 0.5 ml of the diphenylcarbazide reagent. Samples for the standard curve should contain 150–1200 μg of Cr_2O_3.

Parra and Kogan (1981) use atomic absorption (AA) spectroscopy with an AA emission spectrophotometer (IL 353, Instrumentation Laboratory, using a 5 cm N_2O burner head). Readings are given in g/ml and three subsamples are measured for each of the samples in the experiment. If AA is used, the methodology is analogous to that using trace elements; however, Cr is not incorporated into living tissue and it does not replace mineral nutrients.

Computations of parameters and indices use the following formulas:

Food consumed: $F = (E \times \%$ Cr in feces$) + (B \times \%$ Cr in animal$)$

in which F is computed indirectly and the other parameters are previously measured. With $F, B,$ and E known, it is then possible to calculate ECI and ECD.

Digestibility: $AD = 1 - (\%$ Cr in medium$/\%$ Cr in feces$) \times 100$

To illustrate use of the method under diverse conditions see Holter (1974), and Parra and Kogan (1981).

4.2. Isotope Method

Several isotopes have been used as markers in nutritional studies. For example: ^{22}Na with *Sitophilus granarius* L. on wheat flour (Buscarlet et al., 1974), with *Locusta migratoria* L. by injection (Buscarlet, 1974); ^{134}Cs with the spider *Pardosa lapidicina* (Nabholz and Crossley, 1978); ^{137}Cs with various arthropods, e.g., the grasshopper, *Trimeroptropis saxatilis* (Duke and Crossley, 1975); ^{134}Cs and ^{85}Sr as double markers, acting as metabolic analogs of K and Ca, respectively, to study ingestion rates of several invertebrates (Gist and Crossley, 1975); ^{51}Cr and ^{14}C as a non-assimilated and an assimilated tracer in food and feces, the ratio of which allows computation of carbon assimilated from the food (Cammen, 1977); ^{32}P, as $H_3{}^{32}PO_4$, to demonstrate plant feeding by predaceous phytoseiid mites (Porres et al., 1975); [^{3}H]inulin, to measure aphid feeding rates on artificial diets (Wright et al., 1985). One of the most commonly used tracers is ^{14}C, generally used as [^{14}C]glucose, ^{14}C-sucrose or ^{14}C-cellulose (McGinnis and Kasting, 1969). The following procedure adapted from Kasting and McGinnis (1965, 1966) was used by Parra and Kogan (1981) to investigate consumption of an artificial medium by the soybean looper.

[^{14}C]Glucose is dissolved in acetone and the solution, with an activity of 2.1×10^6 cpm/ml, is added to the artificial medium (AM). The medium is mixed in a Waring blender with 4.5 ml of the radioactive glucose solution added to about 800 ml of AM. The medium is poured into 29.5-ml plastic diet cups, 10 ml/cup.

Determinations are made for the entire larval stage through peak development of sixth instars or through pupation generally following the procedure described by Kasting and McGinnis (1965). Separate batches are kept for each set of determinations and two larvae/cup are used in all experiments. Extra [14C]glucose medium cups are seeded under the same experimental conditions. When insects molt to the second instar, the experimental cups are checked. If one or both larvae in the experimental cups die they are replaced with equivalent second instars from the reserve cups. For collection of CO_2, cups are kept under vacuum bell jars (Figs. 6 and 7). A hole is made in the lid of each cup and a glass tube is inserted through the hole. Air passing through a water bath is drawn into the cups by means of a single-stage vacuum pump set on a timer activated for 2 min every 15 min. Air flow is visually monitored by checking the air bubbles in the water flasks. The air removed from the rearing cups passes through two flasks of $CaCl_2$ pellets, to remove most of the water. The CO_2 expired within the cups is collected in two gas traps each containing 75 ml of Carbo-sorb® 11 (Packard Inst. Co., Downers Grove, IL, U.S.A.). The gas traps consist of two 125-ml gas washing bottles containing glass tubing that terminate in fritted glass cylinders of extra-coarse porosity. The two bottles are mounted in Dewar flasks packed with crushed ice. After passing through the gas traps the air circulates through a vapor trap

Figure 6. Vacuum bell jar used in [14C] glucose intake measurements with soybean loopers on artificial media. (From Parra and Kogan 1981.)

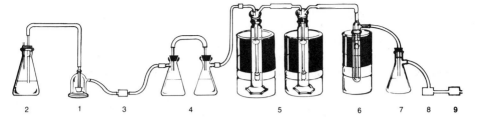

Figure 7. Scheme of the equipment used to measure intake and utilization of an artificial medium containing [^{14}C] glucose, by soybean looper larvae: (1) vacuum bell jar with cup containing the artificial medium and the larvae; (2) air inlet passing through a water bath to equalize air moisture; (3) aquarium valve; (4) dehumidification flasks, containing $CaCl_2$ pellets; (5) gas traps containing Carbo-sorb II; (6) cold trap for Carbo-sorb vapors; (7) water flask; (8) vacuum pump; (9) timer. (From Parra and Kogan 1981.)

and flask of water to remove NH_3 vapors (Fig. 7). Carbo-sorb is changed every 48 hr and activity is checked by adding 0.5 ml of Carbo-sorb directly to a scintillation vial containing 10 ml Permafluor® (Packard Inst. Co.). Activity is measured for 10 min on a liquid scintillation spectrometer.

Experimental animals at the proper stage of growth are removed from the cups and killed by quickly dipping in acetone; the procedure causes no weight loss. The pupal cocoon is weighed and analyzed with the feces. After drying the larvae, feces and AM are analyzed for total radioactivity. The entire animal is homogenized in about 7 ml acetone in a 15-ml glass tissue grinder. The homogenate is centrifuged for 10 min at 4830 rpm. The supernatant is decanted and the residual pellet is re-extracted with about 15 ml acetone and centrifuged for an additional 10 min. The second supernatant is combined with the first. Triplicate aliquots of this extract are taken and added to scintillation vials containing 10 ml Aquasol® (New England Nuclear, Boston, MA, U.S.A.). Activity is measured on the liquid scintillation counter.

Acetone-unextractable radioactivity in the pellet is removed from the dried pellet which is weighed in the centrifuge tube. The pellet is moistened with one or two drops of water and 2 ml of Protosol® (New England Nuclear) is added to each tube. Pellet and Protosol are mixed with a wooden stirrer. The tube is covered with Parafilm® and the pellet is allowed to dissolve for 48 hr. The tubes are placed on a shaker water bath at 48°C for 48 hr. The dissolved pellets are brought to 10 ml volume with Aquasol. Duplicate 0.5-ml aliquots are taken from each tube and placed in scintillation vials containing 10 ml Aquasol. To suppress chemoluminescence, 0.1 ml of glacial acetic acid is added to each vial after which the aliquots in the vials are refrigerated for 24 hr before reading on the liquid scintillation counter.

The same procedure used for the animals is used for the excreta and the medium. The entire fecal sample is analyzed. Data are converted to decompositions per minute (dpm) and the analyses are based on the results of n replications. When using ^{14}C as the label either as [U-^{14}C]sucrose, glucose, or acetate, computation of food consumed is given by:

$$F = \frac{\mu Ci \text{ larvae } + \mu Ci \text{ feces}}{\mu Ci/mg \text{ medium}}$$

Buscarlet (1974) provides the following derivation of food intake based on ^{22}Na turnover:

If sodium concentration (Q) in the insect is assumed constant, at any moment, the rate of Na loss dNa_l/dt is about equal to the rate of Na gain dNa_g/dt or

$$dNa_l/dt = -dNa_g/dt \tag{1}$$

In addition, if the insect is marked with ^{22}Na at time t and quantity q, the specific radioactivity—dq/dNa_l—of Na eliminated is equal to the specific radioactivity q/Q of Na in the insect, or:

$$dq/dNa_l = q/Q \tag{2}$$

Thus, the rate of decrease of radioactivity in the insect is given by:

$$dq/dt = \frac{dNa_l}{dt}\frac{q}{Q} = -\frac{dNa_g}{dt}\frac{q}{Q} \tag{3}$$

If dNa_g/dt in diet is assumed constant, then:

$$dq/dt = -Kq, \tag{4}$$

K being the coefficient of Na turnover defined as:

$$K = dNa_g/Qdt \tag{5}$$

Integration of Eq. (4) gives:

$$\ln q/q_0 = -Kt \tag{6}$$

where q_0 = initial quantity of ^{22}Na in medium.

If Na concentration in the medium is variable the expression dNa_g/dt is not constant and one cannot integrate Eq. (4) with time. Food consumed is then derived from Eqs. (1) and (2), restated as:

$$dq/dNa_g = -q/Q \tag{7}$$

which, integrated with variable Na_g, represents the cumulative uptake of Na in time t,

$$\log q/q_o = -\frac{1}{Q}Na_l \tag{8}$$

The food uptake (assimilation) dNa_g/dt, can then be calculated from the food consumed dF/dt and the amount of [Na] it contains:

$$dNa_g/dt = dF/dt \cdot [Na] \qquad (9)$$

with Eq. (9) taking the form:

$$\log q/q_o = 1/Q \int_o^t dF/dt \cdot [Na] \cdot dt \qquad (10)$$

which provides a means of computing food intake (F) of labelled insects using a food with a constant [Na] level as:

$$F = \int_0^t (dF/dt) \cdot dt = -\log(q/q_o) \cdot Q \cdot (1/[Na]) \qquad (11)$$

where Q = amount of Na in insect, and [Na] is determined from the food supply. The irradiation done with the use of ^{32}Na must be carefully determined as food consumption measured by ^{32}Na accumulation and respiration decreased in *Tribolium confusum* J. de V., with increased irradiation (Buscarlet, 1983).

The computation of AD of food using the twin tracers ^{51}Cr and ^{14}C is given by Cammen (1977) as:

$$AD = 100 \times 1 - [(dpm\ ^{51}Cr:dpm\ ^{14}C\ food)/(dpm\ ^{51}Cr:dpm\ ^{14}C\ feces)]$$

4.3. Enzymatic Methods

Certain metabolites occurring either in the food or in the insect can be used as natural markers. One such use is the uric acid method described below.

To overcome the difficulties in measuring food consumption when food and feces are inseparable, Bhattacharya and Waldbauer (1969a,b, 1970) developed a method that uses uric acid as a marker. Uric acid, although absent in the food, is present in the feces as a product of protein catabolism.

Use of the uric acid method requires separation of a sample of the feces from the mixture of feces and leftover food. The concentration of uric acid in the mixture is estimated in an aliquot of the feces and in a sample of the mixture. The amount of feces in the mixture is calculated by the equation:

$$\text{mg feces in mixture} = \frac{\text{mg uric acid in mixture}}{\text{mg uric acid/mg feces}} + \text{weight feces sample}$$

and the weight of food consumed (F) is calculated by:

F = amount of food introduced $-$ (amount left $-$ weight feces in mixture)

The enzymatic method for uric acid determination described by Bhattacharya and Waldbauer (1969a) was based mainly on Liddle et al. (1959). Uric acid is extracted from a weighed feces sample with 3–3.5 ml of a 0.6% aqueous lithium carbonate solution in a tissue grinder. After centrifugation the supernatant is brought to volume with 0.1 M (pH 9.4) glycine buffer. Ten successive extractions were performed by Bhattacharya and Waldbauer (1969a,b). Standard curve for uric acid is determined at 292 nm. Uric acid concentration is determined by measuring initial absorbance in 1-ml samples of the diluted extracts plus 2.0 ml glycine buffer. Subsequently 10μg of urease is added to the cuvette and absorbance is measured again after oxidation of uric acid to allantoin is completed. Absorbance is measured again (allantoin absorbs far less at the same wavelength) and amount (in μg) of uric acid in the 1-ml sample is given by:

$$\mu\text{g uric acid in cuvette} = \frac{\Delta \text{ absorbance} \times \text{volume sample in cuvette}}{0.074}$$

(0.074 = absorbance of 1 mg purified uric acid in 1 ml glycine buffer). Figure 8 illustrates the correlation between the determination of weight of feces by direct measurement and by the indirect uric acid method (from Bhattacharya and Waldbauer, 1970). The uric acid method has been used by Bhattacharya and Waldbauer (1970) with *Tribolium confusum* and various wheat and yeast diets; Chou et al. (1973) with *Argyrotaenia velutinana* and *Heliothis virescens,* on chemically defined media; and Cohen and Patana (1984) with *Heliothis zea* and artificial media or green beans.

4.4. Qualitative Indirect Methods

In addition to the above quantitative methods several qualititative methods have been used to ascertain the origin of an insect's diet. The trace element methods fall into this category. Rubidium and cesium are rare elements used to mark insects in various ecological studies, particularly in research on dispersal (Berry et al., 1972; Stimman, 1974; Shepard and Waddill, 1976; van Steenwyk et al., 1978; Alverson et al., 1980; Moss and van Steenwyk, 1982). These elements are readily absorbed by plant tissue and transferred quantitatively to insects upon feeding. Detection of these elements by atomic absorption spectroscopy is very efficient at trace amounts and it is conceivable that they can be useful in quantitative nutrition. If concentrations of the trace element in the food, in the animal, and in the excreta are known, the same equations described for the chromic oxide method can be used. Another qualitative method with increasing ecological applications is serology. This method was used by Lund and Turpin (1977) to determine consumption of black cutworm, *Agrotis ipsilon* (Hufnagel),

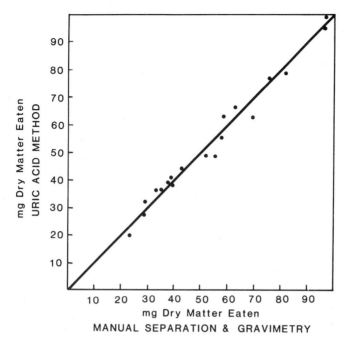

Figure 8. Correlation between values of consumption of whole wheat flour by last instar *Tenebrio molitor* larvae computed by the uric acid method and by manual separation of uneaten food and feces. (Redrawn from Bhattacharya and Waldbauer 1969b.)

larvae by carabids of the genus *Pterostichus*. Guidelines for the selection of appropriate immunological techniques in identifying animal diets are presented by Calver (1984).

4.5. Calorimetry

Ecological studies on trophic interactions at the community level may require expression of results in energy units, or caloric equivalents. In these studies interest lies in determining energy flow and the construction of energy budgets (Southwood [1978] and references therein provide a preliminary key to the literature in the area; a concise discussion of concepts is found in Krebs [1972]). Krebs (1972) presents the partitioning of foodstuffs and energy for an animal as a series of dichotomies (Fig. 9). The interrelation among these various levels of energy is expressed by Petrusewicz and Macfadyen (1970) in the equations:

$$C = D + F = A + U + F = R + P + (F + U)$$

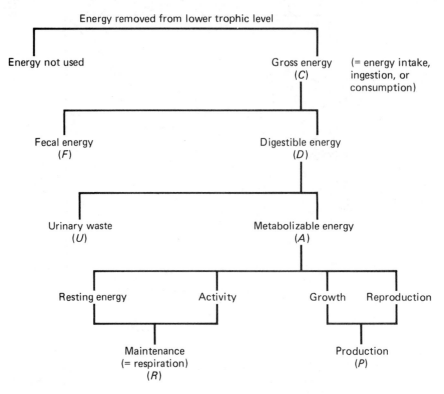

Figure 9. The dichotomic partitioning of foodstuffs and energy for an animal. (Based on Krebs 1972.)

A. Measurement of Caloric Values

The heat of combustion of larvae, feces, and food is determined by means of a bomb calorimeter. The heat of combustion is the energy liberated as heat when organic matter is totally oxidized to CO_2 and H_2O. There are several types of calorimeters, some of which are available commercially. In our laboratory we use a Parr 1200 oxygen bomb calorimeter provided with a 2601 temperature controller (Parr Instrument Co., Moline, IL, U.S.A.). The equipment requires considerable calibration but measurements can be made in small samples of food, dried larvae, pupae, and feces.

The parameters measured by calorimetry are akin to those used in quantitative nutrition. The most common parameters, using Waldbauer's (1968) system are:

Coefficient of metabolizable energy: $CME = [C - (FW)]/C$, or

$$CME = \frac{(\text{gross energy in food eaten}) - (\text{gross energy in feces})}{\text{gross energy in food eaten}}$$

Efficiency of storage of ingested energy: $ESI = P/C$, or

$$ESI = \frac{\text{gross energy stored in body}}{\text{gross energy in food eaten}}$$

Efficiency of storage of metabolizable energy: $ESM = P/[C - (F + U)]$, or

$$ESM = \frac{\text{gross energy in body}}{(\text{gross energy in food eaten} - \text{gross energy in feces})}$$

Precautions and details on calorimetric measurements can be found in Petrusewicz and Macfadyen (1970), Southwood (1978), and other basic ecology texts.

V. Interpretation of Data and Statistical Analyses

Interpretation of experimental results in quantitative nutrition is not easy. The relationships among the various nutritional indices provide some indication of processes, but ignorance of allelochemic (non-nutritional) effects of the food may lead the researcher to incorrect conclusions. Acute effects of inadequate food have clearly detectable symptoms. Chronic effects, however, are manifested through subtle variations in *ECI* or *ECD* (e.g., Reese, 1978, 1979). Digestibility reducing factors may result in higher consumption rates but lower growth rates. Homeostatic mechanisms seem to effect this compensatory interaction, but the precise pattern of compensation is incompletely understood. Recent expansion of research in nutritional ecology, however, is helping to bring greater understanding of the ecological meaning of these nutritional indices (e.g., Reese and Beck, 1978; Scriber and Slansky, 1981; Slansky, 1982; Scriber 1977, 1984).

Statistical procedures to define food quality based on nutritional analyses use regression of growth rate on consumption rates (e.g., Blau et al., 1978). Multivariate analysis of a combination of nutritional indices, behavioral responses, and mortality is used to classify the quality of foods within the host range of a phytophagous insect (Kogan, 1972). The use of cluster analysis in this case may not have been the most appropriate procedure because the method is not expected to establish hierarchical relationships among hosts. However, this method of ordination was useful in helping interpret the mass of data generated in those experiments. Figure 10 shows a dendrogram of food plants used in the Mexican bean beetle experiment.

5.1. Experimental Errors: The Influence of Methodology

Some of the main sources of variation in experimental procedures to measure food intake and utilization have been already discussed. In sum-

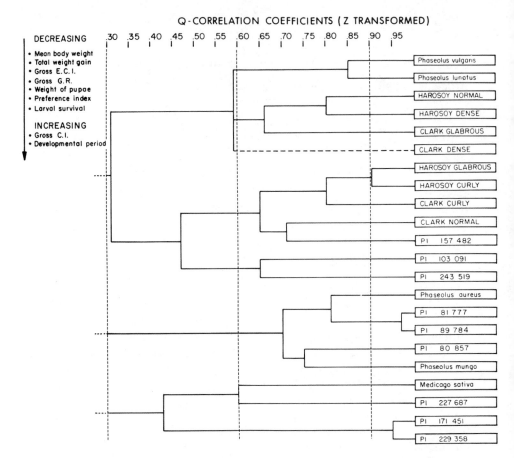

Figure 10. Dendrogram of relationships of five species of legumes and 17 varieties and lines of soybeans (named varieties and plant introductions) tested as food for Mexican bean beetle larvae. Cluster analysis performed on nutritional indices, fitness parameters, and behavioral analyses. (From Kogan 1972.)

mary they are: (a) the inherent variability among individual insects in a population, (b) fluctuations of moisture content in the food, (c) variability in feeding behavior caused by minor or major chemical components of the food, (d) differences in utilization of foods after ingestion, (e) errors due to instrumentation, and (f) errors in the manipulation of samples (Kogan and Parra, 1981). The influence of these various factors was investigated using an artificial medium, soybean looper larvae, and five experimental methods: (1) gravimetry, (2) Calco oil red method, (3) chromic oxide method, (4) isotope method, and (5) calorimetry (Parra and Kogan, 1981; Kogan and Parra, 1981).

Experiments conducted to test precision of those methods used gravimetry as the standard for comparison. Each medium was tested using the specific method (e.g., colorimetry with Calco oil red) and gravimetry (e.g., gravimetric measurements with larvae on a Calco oil red-containing medium). Precision, or the degree of dispersion about the mean, was determined for ECI and ECD using the criterion: $(1 - SD/\bar{x}) \times 100$.

Accuracy, or closeness to the true value, in this case, results obtained by gravimetry, was defined by the expression:

$$[(ECI_{(AM + COR\ grav)}) - (ECI_{(AM + COR\ color)})]/[ECI_{(AM + COR\ grav)}] \times 100$$

(or ECI for larvae on a medium with COR measured gravimetrically minus ECI with same medium measured colorimetrically, divided by ECI measured gravimetrically).

The same was done with all other media and results are summarized in Table 2. Gravimetry with the standard artificial medium (with no dye or isotope additives) was 85.7% precise, an acceptable degree of precision for such experimental data. The best indirect method was the chromic oxide procedure using atomic absorption to measure Cr. It yielded an 80.0% and 82.1% level of precision in measurements of ECI and ECD, respectively.

Effect of diet quality was measured by comparison with results obtained with the standard artificial medium. It was apparent that larvae ate more of the medium without additives than of any other medium. The Calco oil red (COR) was detrimental to the larvae, as larvae reared on a medium containing 0.1% COR were about 24% lighter than those reared on the standard artificial medium. The other media had little effect on the larvae. Food consumption, however, was substantially reduced in all indirect methods. It was apparent that larvae ate less of these media but utilized them more efficiently as weight gain was not greatly affected (except by COR).

ECI, ECD, and AD were consistently accurate when measured by the

Table 2. Precision[a] and accuracy [b] of measurements of ECI and ECD for larval stage on artificial media measured by four methods

Diet	Method of measurement	Precision (%)		Accuracy[b]	
		ECI	ECD	ECI	ECD
AM	Gravimetric	85.7	85.7	—	—
AM + COR	Colorimetric	33.3	34.7	−6.2	−28.9
AM + Cr₂O₃	Atomic absorption	80.0	82.1	−3.8	−8.5
AM + [¹⁴C]glucose	Radioisotopic	60.0	(19.4)	−44.4	−57.5

From Parra and Kogan 1981.
[a]Precision = $(1 - SD/\bar{x})$ 100.
[b]Accuracy I = deviation from value obtained by gravimetry but using same medium; e.g., for Calco oil red accuracy I = ([AM + COR grav) − (AM + COR color)]/[(AM + COR grav)]) × 100.

Cr_2O_3 method. All other methods departed substantially from their gravimetric estimates. Notice the very low *ECD* values measured by the radioisotopic method (Fig. 11), which can be explained by the fact that the expired CO_2 was explicitly used in these computations. Variations in utilization measured by the calorimetric method may suggest that efficiencies in energy conversion were not directly comparable to efficiencies of nutrient utilization.

The time required to process samples by the indirect methods varied from 6 (radioisotopic) to 18 times (Cr_2O_3) more than the time required by standard gravimetry. All indirect methods required the use of an analytical balance in addition to more expensive equipment used in specific determinations (spectrophotometers, scintillation counter, etc.). We gained little, if any, precision in our indirect measurements, and accuracy was also reduced by most indirect procedures. It seems therefore, improbable that the use of nongravimetric procedures is justified whenever gravimetry can be used. Under certain experimental conditions, where food consumed

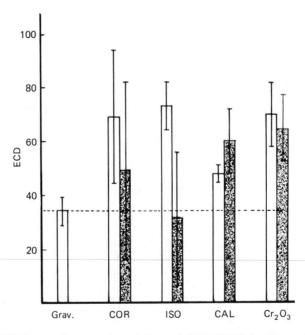

Figure 11. Efficiency of conversion of digested food (ECD) for the soybean looper feeding on an artificial medium measured by gravimetric (GRAV.), colorimetric (COR), isotopic (ISO), calorimetric (CAL) and chromic oxide methods. Horizontal lines indicate one standard deviation (n = 20). Open bar *ECI* based on gravimetric measurements and stipled bar measurement by indirect methods using the same diet. (From Kogan and Parra 1981.)

cannot be directly measured, alternative methods may be justified. Our results indicate that use of COR should be made judiciously as, at least for the soybean looper, there was a detrimental effect. Whenever caloric equivalents are needed, one must resort to a bomb calorimeter. But, instead of making caloric measurements for all specimens, it may be sufficient to obtain dry weights of all parameters and compute caloric equivalents of aliquots.

The value of data on intake and utilization in measuring food quality is unquestionable. There are, however, no shortcuts. Experiments involving gravimetric measurements of insect growth and food consumption are painstakingly tedious, but all the alternatives we tested were less precise, some rather inaccurate, and most were exceedingly more expensive.

VI. Quality Control of Mass-Produced Insects

Insects are mass-produced for use in basic and applied research, product development, and practical control applications. For these uses it is essential that the laboratory colonies preserve the genetic composition as well as the phenotypic characteristics of the wild populations that they are supposed to represent.

Insects in culture suffer the risk of genetic alterations by inbreeding and inadvertent selection (Berlocher, 1984). Populations of *Epilachna varivestis* Mulsant, the Mexican bean beetle, collected from various regions on either soybean or common bean, experienced a reduction in electrophoretic variability after just one generation in culture (Kogan and Berlocher, unpublished data). Peters and Barbosa (1977) studied the influence of density on size, fecundity, and developmental rates of insect populations in culture and reported many physiological and behavioral changes as a consequence of crowding. The competitive ability of field-released mass-produced insects in genetic, autocidal, and biological control programs is essential for success of the project. Consequently, researchers have developed a detailed protocol to monitor quality of these mass-produced insects. Chambers (1977) defined quality, in this context, as the degree of excellence in some traits or skills relative to a reference. The traits and skills refer to the vitality, aggressiveness, and behavior of the mass-produced population; the standards of reference are similar traits of a wild population. Although much of the variability of laboratory-produced insects is due to genetic factors affected by the manipulative techniques, the methodology developed to monitor quality in mass-produced insects is equally applicable to investigate food quality.

The massive program to produce sterile Mediterranean fruit fly, *Ceratitis capitata* (Wied.), to control its spread from Central America into Mexico and southern United States, has led scientists at the Insect Attractants,

Behavior, and Basic Biology Research Laboratory, U.S. Department of Agriculture, at Gainesville, FL, U.S.A. to develop a detailed protocol to measure and monitor the quality of mass-produced flies (Boller et al., 1981; Webb et al., 1981; Brewer, 1983; Chambers et al., 1983).

The criteria used in monitoring fruit fly quality may be divided into two sets: (1) criteria related to life-table parameters such as survival, fecundity, fertility, developmental rates, emergence rates, and sex ratio; and (2) behavioral or physiological responses associated with key behavioral mechanisms such as reproductive behavior, acoustical behavior, visual sensitivity, respiration rates, and motor activity. Because life-table parameters represent fundamental information obtained in most biological studies there is no need to elaborate on methodologies for determination and analysis of those parameters. However, some of the procedures for measuring other physiological and behavioral activities deserve brief mention. It must be stressed again that any attempt to investigate food quality using activity parameters requires rigorous control over the nature of the test population. As in nutritional analyses, insects used in all tests should be homogeneous as to sex, age, physiological state (starvation, dehydration, circadian rhythm, etc.). Possible sources of inducible response should be eliminated as the food previously eaten by insects often influences acceptance and feeding rates on diets subsequently offered to the same animal (e.g., Stadler and Hanson, 1978; Barbosa et al., 1979; Grabstein and Scriber, 1982).

6.1. Tests for Activity

Devices known as actographs are used to monitor motor activity. Several types have been developed based on diverse principles of operation. One of these devices uses change in equilibrium of a cage in which the animal is contained. The cage delicately rests on razor-thin edges and any movement within the cage tilts the assemblage; the direction of tilt is electronically detected and recorded (Viaud and Le Cain, 1975). This device was originally used to measure activity of mice but it could conceivably be adapted to use with caterpillars, grasshoppers, or other insects with adequate weight to trigger the mechanism. Video recording of animals as they pass a determined scanning line is used by van Lenteren et al. (1976). Leppla and Spangler (1971) use an isolated metal screen cage touching a vibration-sensitive transducer mounted in a metal canister. As the cage is struck by an insect the vibration is detected, amplified, and recorded. Commercial units are currently available for detecting activity of crawling insects (example is the crawling insect activity monitor, Columbus Instruments, Columbus, OH, U.S.A.). Some of these systems use infrared detectors, photocells, or displacement transducers to detect activity. Miller (1979) provides a comprehensive review of actographs used in neurophysiological research.

The most widely used system to detect flight activity is the flight mill, an example of which is provided by Michel et al. (1977). In these types of devices the insect is tethered to the extreme end of a blade balanced on a vertical axis held with negligible friction. The blade rotates about its axis under propulsion of the insect in flight. Number of rotations and amount of rest time are recorded as each revolution of the blade interrupts a light-photodiode connection that triggers an event counter. A much simpler system is used by Boller et al. (1981). In this system pupae of the test animal are placed in a petri dish at the bottom of a cardboard tube 20 cm in height. The tube is fitted into the petri dish and it is internally treated in a way that crawling insects are prevented from moving up the wall. To leave the enclosure, insects are forced to fly. The test is based on the number of individuals that remain in the unit. This test also provided information on pupal mortality and emergence rates of adults. Tests on the effect of several diets on the flight ability of the southern green stink bug, *Nezara viridula* (L.) use specimens tethered with a nylon thread suspended from a fluorescent tube (Kester and Smith, 1984). A mild air stream is generated by a fan suitably located and flight activity is observed and timed. There are significant differences in performance among individuals reared on various diets. Sustained flight ability of tethered large milk weed bugs, *Oncopeltus fasciatus* (Dallas), requires continuous feeding throughout the adult life (Slansky, 1980).

6.2. Acoustical Analysis

Insect sounds are associated with specific behaviors such as mating, aggression, recognition, and alarm or distress. Webb et al. (1981) use the recording of sounds produced by the fly *Dacus dorsalis* (Hendel) from Australia and from Hawaii. There are marked differences between these two populations. Sound production is recommended as a criterion for quality control of fruit flies as it is apparent that the quality of the food eaten by the animal affects the sounds it produces. Webb et al. (1981) conclude that "the signature analysis technique of measuring and monitoring differences in insect behavior is new and a considerable amount of work is needed before it can be applied to quality control and mass rearing." They feel, however, that there is a potential for use of sound in measuring behavioral differences among species, different activities among individuals of the same species, and changes that occur during colonization.

6.3. Visual Analysis

Visual analysis by electroretinograms may be useful in testing food quality. Agee and Park (1975) reared nine strains of the Caribbean fruit fly, *An-*

astrepha suspensa, in the laboratory, on various diets. Those flies on the more marginal diets had a visual sensitivity much lower than that of flies reared on the natural or on the best laboratory diet.

6.4. Monitoring Respiration

Webb et al. (1981) suggested that respirometry may be used to advantage in monitoring quality of insects reared in the laboratory. Many systems to collect and measure CO_2 produced by insects are currently available (see Petrusewicz and Macfadyen, 1970).

A rigorous mass rearing program should include a detailed protocol to monitor quality of the insects produced. Per se, none of the techniques described above is sufficient, but in combination with nutritional analyses and growth parameters they provide adequate means of quality control. Because insect behavior is so variable, the optimal set of assays must be chosen to take full advantage of the most characteristic and easily monitored behaviors. This set of bioassays in turn provides many additional options to evaluate the effect of food on insects.

VII. Concluding Remarks

Although insects' nutritional requirements are fairly uniform, their dietary patterns are incredibly diverse. The realization that food quality depends not only on its nutritional content but also on other chemical and physical components of the diet with nonnutritional value has had major impact on insect dietetics (Beck, 1972; Beck and Reese, 1976; Reese, 1979). When placed in its broader ecological context, insect dietetics becomes nutritional ecology, a field rapidly expanding since the 1970s (Scriber and Slansky, 1981; Slansky, 1982).

Progress in nutritional ecology is somewhat limited by the disinclination of researchers to perform the tedious weight gain and food consumption experiments. These, however, are essential for a full understanding of insect trophic interactions. The use of highly sensitive top-loading balances coupled with automatic data entry to microcomputers may so increase the efficiency of data collecting that researchers will no longer shun this type of research. Then detailed nutritional analyses will become an essential and required component of any ecological work involving trophic interactions.

References

Agee HR, Park ML (1975) Use of electroretinogram to measure the quality of vision of the fruit fly. Environ Let **10**:171–176.

Alverson DR, All JN, Bush PD (1980) Rubidium as a marker and simulated in-oculum for the black-faced leafhopper, *Graminella nigrifons,* the primary vector of maize chlorotic virus of corn. Environ Entomol **9**:29–31.

Barbosa P, Greenblatt J, Withers W, Cranshaw J, Harrington EA (1979) Host-plant preferences and their induction in larvae of the gypsy moth, *Lymantria dispar.* Entomol Exp Appl **26**:180–188.

Beck SD (1972) Nutrition, adaptation and environment. In: Insect and Mite Nutrition: Significance and Implications in Ecology and Pest Management. Rodriguez JG (ed), North-Holland Publishing, Amsterdam, pp 1–6.

Beck SD, Reese JC (1976) Insect-plant interactions: nutrition and metabolism. Recent Adv Phytochem **10**:41–92.

Berlocher SH (1984) Insect molecular systematics. Annu Rev Entomol **29**:403–433.

Berry WL, Stimman MW, Wolf WW (1972) Marking of native phytophagous insects with rubidium: a proposed technique. Ann Entomol Soc Am **65**:236–238.

Bhattacharya AK, Waldbauer GP (1969a) Quantitative determination of uric acid in insect feces by lithium carbonate extraction and the enzymatic-spectrophotometric method. Ann Entomol Soc Am **62**:925–927.

Bhattacharya AK, Waldbauer GP (1969b) Faecal uric acid as an indicator in the determination of food utilization. J Insect Physiol **15**:1129–1135.

Bhattacharya AK, Waldbauer GP (1970) Use of the faecal uric acid method in measuring the utilization of food by *Tribolium confusum.* J Insect Physiol **16**:1983–1990.

Blau PA, Feeny P, Cotardo L, Robson DS (1978) Allylglucosinolate and herbivorous caterpillars: a contrast in toxicity and tolerance. Science **200**:1296–1298.

Boller EF, Katsoyannos BI, Remund U, Chambers DL (1981) Measuring, monitoring and improving the quality of mass-reared Mediterranean fruit flies, *Ceratitis capitata* Wied. 1. The RAPID quality control system for early warning. Z Ang Entomol **92**:67–83.

Brewer FD (1982) Development and food utilization of tobacco budworm hybrids fed artificial diet containing oil soluble dyes. J Ga Entomol Soc **17**:248–254.

Brewer FD (1983) Evaluation of selected parameters as quality control criteria for mass producing a tobacco budworm (Lepidoptera: Noctuidae). Ann Entomol Soc **76**:339–342.

Buscarlet LA (1974) The use of ²²Na for determining the food intake of the migratory locust. Oikos **25**:204–208.

Buscarlet LA (1983) Effects of irradiation on respiration and on food consumption measured through ²²Na in *Tribolium confusum* J. de V. (Coleoptera: Tenebrionidae). J Stored Prod Res **19**:19–24.

Buscarlet LA, Lasceve G, Garcia J (1974) Utilisation de la cinétique de ²²Na pour estimer la ration alimentaire de *Sitophilus granarius* dans différentes conditions de température et d'alimentation. Int J Appl Radiat Isotop **25**:445–453.

Calver MC (1984) A review of ecological applications of immunological techniques for diet analysis. Aust J Ecol **9**:19–25.

Cammen LM (1977) On the use of liquid scintillation counting of ⁵¹Cr and ¹⁴C in the twin tracer method of measuring assimilation efficiency. Oecologia (Berl) **30**:249–251.

Chambers DL (1977) Quality control in mass rearing. Annu Rev Entomol **22**:289–308.

Chambers DL, Calkins CO, Boller EF, Ito Y, Cunningham RF (1983) Measuring, monitoring and improving the quality of mass-reared Mediterranean fruit flies, *Ceratitis capitata* (Wied.). 2. Field tests for confirming and extending laboratory results. Z Ang Entomol **95**:285–303.

Chou YM, Rock GC, Hodgson E (1973) Consumption and utilization of chemically defined diets by *Argyrotaenia velutinana* and *Heliothis virescens*. Ann Entomol Soc Am **66**:627–632.

Cohen AC, Patana R (1984) Efficiency of food utilization by *Heliothis zea* (Lepidoptera: Noctuidae) fed artificial diets or green beans. Can Entomol **116**:139–146.

Dadd RH (1977) Qualitative requirements and utilization of nutrients: Insects. In: CRC Handbook Series in Nutrition and Food. Section D: Nutritional Requirements, Vol. 1. Rechcigl Jr M (ed), CRC Press, Cleveland, pp 305–346.

Daum RJ, Mckibben GH, Davich TB, McLaughlin R (1969) Development of the bait principle for boll weevil control: calco oil red N-1700 dye for measuring ingestion. J Econ Entomol **62**:370–375.

Duke KM, Crossley Jr DA (1975) Population energetics and ecology of the rock grasshopper, *Trimeroptropis saxatilis*. Ecology **56**:1106–1117.

Fraenkel G (1953) The nutritional value of green plants for insects. Trans IX Int Congr Entomol (Amsterdam) **2**:90–100.

Gist CS, Crossley Jr DA (1975) Feeding rates of some cryptozoa as determined by isotopic half-life studies. Environ Entomol **4**:625–631.

Gordon HT (1968) Quantitative aspects of insect nutrition. Zool **8**:131–138.

Gordon HT (1972) Interpretations of insect quantitative nutrition. In: Insect and Mite Nutrition. Rodriguez JG (ed), North-Holland Publishing, Amsterdam, pp 73–105.

Grabstein EM, Scriber JM (1982) Host-plant utilization by *Hyalophora cecropia* as affected by prior feeding experience. Entomol Exp Appl **32**:262–268.

Hamamura Y, Hayashiya K, Naito K, Matsuura K, Nishida J (1962) Food selection by silkworm larvae. Nature **194**:754–755.

Hendricks DF, Graham HM (1970) Oil soluble dye in larval diet for tagging moths, eggs, and spermatophores of tobacco budworms. J Econ Entomol **63**:1019–1020.

Henneberry TJ, Kishaba AN (1966) Cabbage loopers. In: Insect Colonization and Mass Production. Smith CN (ed), Academic Press, New York, pp 461–478.

Holter P (1974) Food utilization of dung-eating *Aphodius* larvae (Scarabaeidae). Oikos **25**:71–79.

Hori K, Endo M (1977) Metabolism of ingested auxins in the bug *Lygus disponsi:* conversion of indole-3-acetic acid and gibberillin. J Insect Physiol **23**:1075–1080.

Jones RL, Perkins WD, Sparks AN (1975) Heliothis zea: effects of population density and a marker dye in the laboratory. J Econ Entomol **68**:349–350.

Kasting R, McGinnis AJ (1965) Measuring consumption of food by an insect with carbon-14 labelled compounds. J Insect Physiol **11**:1253–1260.

Kasting R, McGinnis AJ (1966) Radioisotopes and the determination of nutrient requirement. Ann NY Acad Sci **139**:98–107.

Kester KM, Smith CM (1984) Effect of diet on growth, fecundity and duration of tethered flight of *Nezara viridula*. Entomol Exp Appl **35**:75–81.

Klein I, Kogan M (1974) Analysis of food intake, utilization, and growth in phytophagous insects—a computer program. Ann Entomol Soc Am **67**:295–297.

Kogan M (1972) Intake and utilization of natural diets by the Mexican bean beetle, *Epilachna varivestis*—a multivariate analysis. In: Insect and Mite Nutrition. Rodriguez JG (ed), North Holland Publishing, Amsterdam, pp 107–126.

Kogan M, Cope D (1974) Feeding and nutrition of insects associated with soybeans. 3. Food intake, utilization, and growth in the soybean looper, *Pseudoplusia includens*. Ann Entomol Soc Am **67**:66–72.

Kogan M, Parra JRP (1981) Techniques and application of measurements of consumption and utilization of food by phytophagous insects. In: Current Topics in Insect Endocrinology and Nutrition. Bhaskaran G, Friedman S, Rodriguez JG (eds), Plenum, New York, pp 337–352.

Krebs CJ (1972) Ecology: The Experimental Analysis of Distribution and Abundance. Harper & Row, New York, 694 pp.

Kuramochi K, Nishijima Y (1980) Measurement of the meal size of the horn fly, *Haematobia irritans* (L.) (Diptera: Muscidae) by the use of amaranth. Appl Entomol Zool **15**:262–269.

Leppla NC, Spangler HG (1971) A flight-cage actograph for recording circadian periodicity of pink bollworm moths. Ann Entomol Soc Am **64**:1431–1434.

Liddle L, Seegmiller JE, Laster L (1959) The enzymatic spectrophotometric method for determination of uric acid. J Lab Clin Med **54**:903–913.

Lund RD, Turpin FT (1977) Serological investigation of black cutworm larval consumption by ground beetles. Ann Entomol Soc Am **70**:322–324.

McGinnis AJ, Kasting R (1964a) Colorimetric analysis of chromic oxide used to study food utilization by phytophagous insects. J Agric Food Chem **12**:259–262.

McGinnis AJ, Kasting R (1964b) Comparison of gravimetric and chromic oxide methods for measuring percentage utilization and consumption of food by phytophagous insects. J Insect Physiol **10**:989–995.

McGinnis AJ, Kasting R (1969) Digestibility studies with cellulose-U-C[14] on the pale western cutworm, *Agrotis orthogonia*. J Insect Physiol **15**:5–10.

Michel R, Colin Y, Rodriguez M, Richard JP (1977) Automatic measurement and recording of insect flight activity. Entomol Exp Appl **21**:199–206.

Miller TA (1979) Insect Neurophysiological Techniques. Springer-Verlag, New York, 308 pp.

Moss JI, van Steenwyk RA (1982) Marking pink bollworm (Lepidoptera: Gelechiidae) with cesium. Environ Entomol **11**:1264–1268.

Nabholz JV, Crossley Jr DA (1978) Ingestion and elimination of cesium-134 by the spider, *Pardosa lapidicina*. Ann Entomol Soc Am **71**:325–328.

Parra JRP, Kogan M (1981) Comparative analysis of methods for measurements of food intake and utilization using the soybean looper, *Pseudoplusia includens* and artificial media. Entomol Exp Appl **30**:45–57.

Peters TM, Barbosa P (1977) Influence of population density on size, fecundity, and developmental rate of insects in culture. Annu Rev Entomol **22**:431–450.

Petrusewicz K (ed) (1967) Secondary Productivity of Terrestrial Ecosystems; Principles and Methods. Panstowowe Wydawnictwo Naurowe, Warsaw, Poland, 2 vol, 879 pp.

Petrusewicz K, Macfadyen A (1970) Productivity of Terrestrial Animals—Principles and Methods. Int Biol Proj Handbook 13, Blackwell, Oxford, U.K., 190 pp.

Porres MA, McMurtry JA, March RB (1975) Investigation of leaf sap feeding by three species of phytoseiid mites by labelling with radioactive phosphoric acid ($H_3{}^{32}PO_4$). Ann Entomol Soc Am 68:871–872.

Radford PJ (1967) Growth analysis formulae—their use and abuse. Crop Sci 7:171–175.

Reese JC (1978) Chronic effects of plant allelochemics on insect nutritional physiology. Entomol Exp Appl 24:625–631.

Reese JC (1979) Interactions of allelochemics and nutrients in herbivore food. In: Herbivores, Their Interactions with Secondary Plant Metabolites. Rosenthal GA, Janzen DH (eds), Academic Press, New York, pp 309–330.

Reese JC, Beck SD (1978) Interrelationships of nutritional indices and dietary moisture in the black cutworm (*Agrotis ipsilon*) digestive efficiency. J Insect Physiol 24:473–479.

Rosenthal GA, Dahlman DL (1975) Non-protein amino acid-insect interactions: II. Effects of canaline-urea cycle amino acids on growth and development of the tobacco hornworm, *Manduca sexta* L. (Sphingidae). Comp Biochem Physiol 52A:105–108.

Schmidt DJ, Reese JC (1986) Sources of error in nutritional index studies of insects on artificial diet. J Insect Physiol 32:193-198.

Scriber JM (1977) Limiting effects of low leaf water content on the nitrogen utilization, energy budget, and larval growth of *Hyalophora cecropia* (Lepidoptera: Saturniidae). Oecologia (Berl) 28:269–287.

Scriber JM (1979) Effects of leaf-water supplementation upon post-ingestive nutritional indices of forb-, scrub-, vine, and tree-feeding Lepidoptera. Entomol Exp Appl 25:240–252.

Scriber JM (1984) Host-plant suitability. In: Chemical Ecology of Insects. Bell WJ, Carde RT (eds), Chapman and Hall, Sunderland, MA, pp 159–202.

Scriber FM, Slansky Jr F (1981) The nutritional ecology of immature insects. Annu Rev Entomol 26:183–211.

Shepard M, Waddill VW (1976) Rubidium as a marker for Mexican bean beetles, *Epilachna varivestis* (Coleoptera: Coccinellidae). Can Entomol 108:337–339.

Slansky Jr F (1980) Food consumption and reproduction as affected by tethered flight in female milkweed bugs (*Oncopeltus fasciatus*). Entomol Exp Appl 28:277–286.

Slansky Jr F (1982) Insect nutrition: an adaptationist's perspective. Fla Entomol 65:45–71.

Slansky Jr F, Scriber JM (1982) Selected bibliography and summary of quantitative food utilization by immature insects. Bull Entomol Soc Am 28:43–55.

Southwood TRE (1978) Ecological Methods. Halsted Press, John Wiley & Sons, New York. 524 pp.

Stadler E, Hanson FE (1978) Food discrimination and induction of preference for artificial diets in the tobacco hornworm, *Manduca sexta*. Physiol Entomol 3:121–133.

Stepien ZA, Rodriguez JG (1972) Food utilization in acarid mites. In: Insect and Mite Nutrition. JG Rodriguez (ed), North Holland Publishing, Amsterdam, pp 127–151.

Stimman MW (1974) Marking insects with rubidium: imported cabbage worm marked in the field. Environ Entomol 3:327–328.

Svoboda JA, Kaplanis JN, Robbins WE, Thompson MJ (1975) Recent development in insect steroid metabolism. Annu Rev Entomol 20:205–220.

van Lenteren JC, van der Linder RW, Gluvers A (1976) A "border-line detector" for recording locomotory activities of animals. Oecologia (Berl) 23:133–137.

van Steenwyk RA, Ballmer GT, Page AL, Reynolds HT (1978) Marking pink bollworm with rubidium. Ann Entomol Soc Am 71:81–84.

Viaud P, Le Cain Y (1975) An apparatus for recording animal motor activity. Behavior 52:312–316.

Waldbauer GP (1964) The consumption, digestion and utilization of solanaceous and non-solanaceous plants by larvae of the tobacco budworm, *Protoparce sexta* (Johan.) (Lepidoptera: Sphingidae). Entomol Exp Appl 7:253–269.

Waldbauer GP (1968) The consumption and utilization of food by insects. Adv Insect Physiol 5:229–288.

Waldbauer GP, Cohen RW, Friedman S (1984) Self-selection of an optimal nutrient mix from defined diets by larvae of the corn earworm, *Heliothis zea* (Boddie). Physiol Zool 57:590–597.

Webb JC, Agee HR, Leppla NC, Calkins CO (1981) Monitoring insect quality. Trans ASAE 24:476–479.

Whitham TG (1983) Host manipulation of parasites: Within-plant variation as a defense against rapidly evolving pests. In: Variable Plants and Herbivores in Natural and Managed Systems. Denno RF, McClure MS (eds), Academic Press, New York, pp 15–41.

Williams GC (1966) Adaptation and Natural Selection. Princeton University Press, Princeton, NJ, 307 pp.

Wright PG, Fisher DB, Mittler TE (1985) Measurement of aphid feeding rates on artificial diets using ^{3}H-inulin. Entomol Exp Appl 37:9–11.

Chapter 7

Nutritional and Allelochemic Insect–Plant Interactions Relating to Digestion and Food Intake: Some Examples

Isaac Ishaaya[1]

I. Introduction

Digestive enzymes in insects are generally adapted to the diet on which the species feed (Wigglesworth, 1965). Blowfly larvae that feed on animal tissues exhibit high protease and lipase activities (Hobson, 1931). Polyphagous insects such as *Spodoptera littoralis* larvae secrete high levels of protease, amylase, and invertase (Ishaaya et al., 1971, 1974), whereas hemipterous insects such as aphids and scales, which usually feed on plant fluid, exhibit a relatively high level of invertase activity (Ishaaya and Swirski, 1970, 1976). Nutritional and environmental factors affect digestive enzymes; the digestive proteolytic and amylolytic activities of *Spodoptera littoralis,* for example, are closely correlated with the protein level in the diet or with the environmental temperature (Ishaaya et al., 1971). In some cases digestive enzymes can be used as parameters for assessing antifeeding activity (Ascher and Ishaaya, 1973; Ishaaya and Casida, 1975; Ishaaya et al., 1974, 1977, 1980, 1982), or phagostimulation (Ishaaya and Meisner, 1973). Despite availability of ample information concerning biochemical properties of digestive enzymes in various insects (House, 1974; Wigglesworth, 1974), relatively little is known about their role in insect feeding and insect–host compatability. Trehalase, another carbohydrase that in insects degrades trehalose to glucose for internal energy supply (Wyatt, 1967), is used as a biochemical parameter for assessing the adaptability of the black scale *Saissetia oleae* to various host plants (Ishaaya

---

[1]Division of Entomology, Agricultural Research Organization, The Volcani Center, Bet Dagan, Israel.

and Swirski, 1976). In *Aphis citricola,* two types of aphid trehalase have been identified: one is water-soluble and the other is a membrane-bound enzyme. An increased level of the soluble trehalase in the alate morphs of *Aphis citricola* indicates the importance of this enzyme system in the energy supply needed for aphid flight (Neubauer et al., 1980).

Secondary compounds present in plants such as saponins (Birk, 1969; Cheeke, 1971; Horber, 1972; Bondi et al., 1973; Applebaum and Birk, 1979), tannins (Haslam, 1966, 1974; Swain, 1979), proteinous inhibitors of protein-digesting enzymes (Vogel et al., 1968; Liener and Kalade, 1969; Laskowski and Sealock, 1971; Fritz et al., 1974; Birk, 1976; Richardson, 1977; Ryan, 1979), and gossypol (Abou-Donia, 1976) are widely reviewed. These compounds deter insects and other herbivores from feeding and exert their effect, in many cases, directly on the predator digestive enzymes (Birk et al., 1962, 1963a,b,c; Applebaum, 1964; Applebaum et al., 1964a; Ishaaya, 1965; Ishaaya and Birk, 1965; Laskowski and Sealock, 1971; Norris et al., 1971; Rozental and Norris, 1973; Ryan, 1973; Singleton and Kratzer, 1973; Rozental et al., 1975; Jones and Mangan, 1977; Meisner et al., 1978; Griffiths and Moseley, 1980).

On the other hand biochemical changes occur in plants as a result of insect infestation (Ishaaya and Sternlicht, 1969, 1971; Hori 1973a,b, 1975; Hori and Atalay, 1980). An increase in phenol level and in phenol oxidase and peroxidase activities is observed in lemon buds infested with the citrus bud mite, *Aceria sheldoni* (Ishaaya and Sternlicht, 1969, 1971), and in sugar beet leaves infested with the bug *Lygus disponsi* (Hori, 1973a; Hori and Atalay, 1980). Phenol extract obtained from lemon buds infested with mites greatly inhibits the development of the pest, and strongly infested buds are not suitable hosts for mites (Ishaaya and Sternlicht, 1971). Enhanced activity of oxidative enzymes (Ishaaya and Sternlicht, 1969) may play a role in the formation of quinones which may act as deterrents to further infestation. These data strongly suggest that biochemical changes occurring in plants after infestation may, in some cases, play a defensive role against insects and other herbivores. The reduced level of nutritional factors such as sugars and proteins observed, in some cases after insect attack (Hori, 1973b; Hori and Atalay, 1980), may also deter insects and other herbivores from feeding.

This chapter deals with insect–plant biochemical interactions relating specifically to nutritional factors and to plant allelochemic compounds that interfere with digestion and food intake.

II. Insect–Plant Interactions Relating to Nutritional Factors

The feeding behavior of phytophagous insects has been discussed in several reviews (Beck, 1965; Dethier, 1966, 1970; Schoonhoven, 1968, 1982; Dadd, 1970; Hsiao, 1972; Chapman, 1974; Barton Brown, 1975; McKey, 1979;

Bernays and Simpson, 1982). Plant chemicals are of major importance in regulating insect feeding and are classified according to their effects as positive or negative stimuli (Beck, 1965). These stimuli result from biochemical and physiological processes occurring in the mouth parts or the digestive tract of the insect. This section deals with nutritional factors relating to proteinous and carbohydrate compounds important in digestion and food intake along with relevant biological and biochemical assays.

2.1. Carbohydrates

The role of some sugars as feeding stimulants for insects is well documented (Dethier, 1953; Thorsteinson, 1960; Lapidus et al., 1963; Davis, 1968). Following feeding tests, various sugars have been classified according to their ability to sustain life or promote feeding. In the Colorado beetle larva the biting and feeding response was induced most efficiently by sucrose (Hsiao and Fraenkel, 1968); the tobacco hornworm larva (Yamamoto and Fraenkel, 1960) responded well to sucrose and glucose; the Mexican bean beetle (Augustine et al., 1964) to sucrose, glucose, and fructose; the silkworm larva (Ito, 1960) to sucrose, fructose, and raffinose; and the grasshopper, *Camnula pellucida* (Thorsteinson, 1960), to sucrose, fructose, glucose, maltose, and raffinose. Generally, sucrose, maltose, glucose, and fructose stimulate feeding and are well utilized by insects, whereas cellobiose, lactose, and most of the pentoses fail to support growth (Dethier, 1953; Pielou and Glaser, 1953; Galun and Fraenkel, 1957; Ito, 1960; Meisner et al., 1972a). Electrophysiological and chemoreception studies (Chapter 10, *this volume*) confirm that various sugars do stimulate certain chemoreceptors in various insect species and this may be one of the reasons for their feeding stimulating activity (Schoonhoven, 1968).

Among the di- and trisaccharides, sucrose, maltose, raffinose, and melezitose induce significantly the larval amylase, invertase, and protease activities of *Spodoptera littoralis* larvae. Melibiose and β-lactose affect the various digestive enzymes to some extent, whereas cellobiose has practically no effect (Ishaaya and Meisner, 1973). Among the monosaccharides, glucose and fructose induce digestive enzyme activity to about the same level, but to a greater extent than galactose (Ishaaya and Meisner, 1973). These data confirm that some carbohydrates do stimulate insect digestive enzymes, which in turn induce feeding.

In general, the effect of various sugars on the digestive enzymes of *Spodoptera littoralis* larvae correlates well with larval growth and food intake (Meisner et al., 1972a,b; Ishaaya and Meisner, 1973). Among the di- and trisaccharides, sucrose, raffinose, and melezitose elicited the strongest feeding response and β-lactose and cellobiose the weakest. The excretion index in larvae fed on sugars stimulating digestive enzymes is considerably lower than that obtained in larvae fed on nonstimulatory sugars (Ishaaya and Meisner, 1973).

The physiological effect of di- and trisaccharides is due to the activity of their monosaccharide components and/or to a specific structure. In comparative assays, sucrose, maltose, and raffinose induce the various digestive enzymes to about the same level as their equivalent monosaccharide components (Ishaaya and Meisner, 1973). On the other hand, β-lactose and cellobiose, which have the same components as melibiose and maltose but differ in their configuration, are barely hydrolyzed by insects (Galun and Fraenkel, 1957). Small differences, as between galactose and glucose, result in great physiological differences; yet larger differences, as in the case of fructose and glucose, yield a similar effect. These results indicate that both the monosaccharide components and the sugar configuration affect insect digestive enzymes. It may be suggested that digestive enzymes can be used as parameters for evaluating phagostimulant activity relating to carbohydrate moieties.

Differences in the behavior of scale insects involving reproductivity and voltinism are induced by various host plants (Flanders, 1970). The black scale, *Saissetia oleae,* in a year produces one generation on olive trees growing under nonirrigated conditions, and two generations on irrigated trees (Peleg, 1965). On citrus trees it usually produces one generation (Peleg, 1965) but in certain groves there are two clearly defined generations (Blumberg et al., 1975). Observations carried out in glasshouses revealed that this insect produces two or three generations a year on oleander and four or five generations on potato sprouts (Blumberg and Swirski, 1977). In addition, rearing and development of endoparasites in scale insects are strongly affected by the host plant on which the scale develops (Flanders, 1970; Blumberg and DeBach, 1979). Differences in scale growth on various host plants could result from availability of some nutritional essentials, subsequently inducing physiological processes that, in turn, affect growth and development.

Trehalose plays a significant role in the supply of energy to an insect (Wyatt, 1967), and the activity of trehalase might serve as an indicator of energy reserve resulting from availability of carbohydrate nutrients. The activity of digestive enzymes depends mostly on the presence of a respective substrate in the food (House, 1965). Invertase activity, and to some extent amylase activity (Ishaaya and Swirski, 1970, 1976), are essential digestive enzymes in plant-sucking insects and can be used as additional parameters for assessing the availability of nutrients and the adaptability of the scale to its host plant. Trehalase and digestive enzymes are strongly affected by various host plants (Ishaaya and Swirski, 1976). Trehalase activity in scales reared on potato sprouts is about 3.5- and 4-fold that obtained in scales reared on oleander and citrus plants, respectively. An increase of about 35–60% in invertase activity was obtained in scales reared on potato sprouts as compared with those reared on oleander or citrus plants (Table 1, after Ishaaya and Swirski, 1976). A good cor-

Table 1. Trehalase and invertase activity in young females of *Saissetia oleae* reared on various host plants

Host plant	Enzyme activity relative to that present in scales reared on citrus plants (%)		Generation length (months)
	Trehalase	Invertase	
Citrus	100	100	6–12
Oleander	118	120	4–5
Potato sprouts	395	162	2–3

After Ishaaya and Swirski 1976.

relation is obtained between enzyme activity—especially of trehalase—and scale development (Table 1). The induced activity of these enzymes may result either from the availability of excess substrates in the preferred host plant, in this case sucrose, or from the presence of activators (Hori, 1969). The nutritional factors present in the preferred hosts of the black scale are still to be clarified. The above results suggest that trehalase and to some extent invertase can be used as parameters to assess the adaptability of the black scale to its host plant.

In contrast to the results obtained with *Saissetia oleae,* host plants such as grapefruit, lemon fruit, squash, and potato tubers show no significant effect on Florida or California red scale invertase activity (Ishaaya and Swirski, 1970). This may indicate the adaptability of these species to various host plants. A decrease of about 30% in the invertase activity is found in starved scales, 1 day after removal of the scales from their host, or when they were kept for 24 hr at temperatures (2–8°C) below the threshold for development (Ishaaya and Swirski, 1970). At the high temperature of 32°C the Florida red scale invertase was found to be physiologically better adapted than that of the California red scale; this may reflect the adaptability of the Florida red scale to high temperatures (Ishaaya and Swirski, 1970).

2.2. Proteins

Proteins are the source of amino acids required for the production of tissues and enzymes in insects. Some amino acids, such as arginine, lysine, leucine, isoleucine, tryptophan, histidine, phenylalanine, methionine, valine and threonine, are essential for normal insect growth. Others are essential for some species, such as glycine for some dipterous insects, alanine for *Blattella,* and proline for *Phormia.* Balance between different amino acids is particularly important (Chapman, 1969).

Protein degradation and absorption take place primarily in the midgut.

The diet plays an important part in egg production (Johannson, 1964). Protein is important for yolk production and is essential for oogenesis in *Musca domestica* and in many biting *Diptera* such as mosquitoes and tabanids (Chapman, 1969). The quantity of food is also important. In *Ephestia* the number of eggs laid is related to the amount of flour ingested by the larva, and in *Cimex* the number of eggs laid increases with the size of the blood meal. In *Calliphora* the intake of protein during the early stages of egg development activates the corpora allata to secrete a factor leading to an increase in carbohydrate intake during the period of yolk deposition. The removal of protein metabolites from the blood at this stage leads to a reduction in corpus allatum activity and to a smaller carbohydrate intake (Strangways-Dixon, 1959).

Protein intake stimulates the activity of digestive enzymes in insects. In *Aedes aegypti,* protease activity is increased by 26-fold after a meal of blood but there is only a 2-fold rise after a meal of syrup (Fisk and Shambaugh, 1952). Similarly, invertase activity increases 4-fold after blood feeding, but shows only a slight change after insects feed on sucrose (Fisk and Shambaugh, 1954). In *Spodoptera littoralis,* both protease and amylase activity are affected by the protein level in the diet and are in close accord with larval growth (Ishaaya et al., 1971). Reducing the protein content in an artificial diet from 7.6 to 3.6% resulted in a decrease of about 75% in the proteolytic and amylolytic activities of the midgut wall, with a similar decrease observed in the larval weight (Table 2, after Ishaaya et al., 1971). These results indicate that protein affects digestive enzymes in general and not only protease, the enzyme specifically acting on protein. On clover—which contains about 2.5% protein—amylase activity in *Spodoptera littoralis* larvae is much higher than expected from a similar level of protein in the artificial diet (Ishaaya et al., 1971). This may be due to nonprotein inducing factors present in clover, such as a high level of carbohydrates

Table 2. Effect of protein level in the diet, on larval weight, and on larval protease and amylase activities of the gut wall of *Spodoptera littoralis* larvae

Protein in artificial diet (%)	Larval weight 11 days after emergence (mg)	Protease activity[a]	Amylase activity[a]
1.6	0.6	—	—
3.6	37.8	110	51
5.6	91.3	397	236
7.6	137.2	531	211

[a]Protease activity expressed as OD units $\times$ 10^3 at an absorbancy of 280 nm; amylase activity expressed as OD units $\times$ 10^3 at an absorbancy of 550 nm.
After Ishaaya *et al., 1971.*

and/or a presence of active compounds that may stimulate amylase activity (Applebaum et al., 1964b; Hori, 1969; Ishaaya and Meisner, 1973). These data concur with the concept that certain protein fractions (Fisk and Shambaugh, 1952, 1954) can stimulate digestive enzymes, probably through a hormonal mechanism (Dadd, 1961; Wigglesworth, 1965). In *Calliphora* the ingestion of protein stimulates the median neurosecretory cells to produce a hormone, which in turn acts on cells in the midgut epithelium and results in the release of protease (Thomsen and Moller, 1963).

The relationship between digestive enzymes and larval growth is further emphasized in studies of the effect of environmental temperature on the digestive protease and amylase activities of *Spodoptera littoralis* larvae. At 10°C, which is below the threshold for larval development (Bishara, 1934), these enzymes lose about 90% of their activity as compared with that at 32°C. A gradual increase in larval amylase and protease activity occurs with the increase in the environmental temperature from 10° to 32°C, in accordance with the increase in larval growth (Ishaaya et al., 1971).

Some proteins may inhibit digestive enzymes. Hemoglobin cannot replace casein as a substrate for proteolytic activity of *Tenebrio molitor* larvae. The addition of hemoglobin to the enzyme reaction inhibits the activity of protease enzymes obtained from the midgut of *Tribolium confusum, T. castaneum,* and *Tenebrio molitor* (Birk et al., 1962). The poor digestion of hemoglobin by phytophagous insects may be of significance as regards the specificity of these enzymes, since hemoglobin is an incomplete protein, being deficient in methionine and isoleucine (Frost, 1959). A similar inhibitory effect is noticed when only the globin component is added to the basic diet of *Tribolium* (Birk et al., 1962). The suppression of growth of *Tribolium* larvae by hemoglobin could not be overcome by the addition of methionine and/or isoleucine, the missing amino acids in hemoglobin (Birk et al., 1962). This indicates that the detrimental effect of hemoglobin on larval growth cannot be attributed solely to its deficiency in these two essential amino acids, but rather to a more specific action on the insect proteolytic activity.

Protein may counteract the inhibitory effect of various toxic compounds on insect digestive enzymes. Preincubation of soybean saponins and triorganotins with various proteins counteracts the inhibitory effect of these compounds on insect proteolytic and amylolytic activity (Ishaaya and Birk, 1965; Ishaaya and Casida, 1975).

2.3. Biological and Enzyme Assays

In this section, our attention is focused on describing biological and biochemical techniques relating to the effects of various nutritional factors on digestion and food intake in insects. Proteases, amylases, and invertases are used for determining phagostimulation and deterrent activities. The trehalase system and digestive enzymes are used as parameters for insect–host compatability in scale insects. In addition, various proteins are found to counteract the inhibitory effects of various compounds on insect digestive enzymes.

A. Phagostimulation and Enzyme Assays

The phagostimulatory effect of various sugars on digestive enzymes and on feeding rate is tested on thin (0.6 mm) rectangular (6 × 3 cm) lamellae of Styropor (foamed polystyrene) (Meisner et al., 1972a; Ishaaya and Meisner, 1973). The lamellae are weighed and then painted with solutions of 50% ethanol containing 0.25 M of a sugar or other phagostimulant compound. The lamellae are left to dry for 48 hr and then reweighed. The weight of the dry sugar deposit on each lamella can thus be calculated. Lamellae of similar weight, but painted with 50% ethanol only, served as controls. Fifth instar larvae of Spodoptera littoralis, weighing between 170 and 190 mg, are starved for 3 hr and introduced singly into petri dishes (15 cm diameter) containing one lamella each. After 48 hr at 27°C the larvae are taken for determination of gut protease, amylase, and invertase activities. The quantity of treated lamella, and consequently the amount of the sugar consumed per larva, can be determined. The excretion index (see also Chapter 6, this volume) is calculated according to Meisner et al. (1972b) and expressed as (weight of fecal pellets/weight of lamellae consumed) × 100.

Gut enzyme solution from Spodoptera littoralis is prepared from midgut walls and midgut content (Ishaaya et al., 1971) or from the whole midgut (Ishaaya and Meisner, 1973; Ascher and Ishaaya, 1973; Ishaaya et al., 1974). The midgut of fifth instar larvae (200–250 mg) is collected by exposing the alimentary canal and sectioning it first slightly posterior to the cardiac sphincter and again slightly anterior to the pyloric sphincter. Slight pressure at one end of the excised midgut results in the extrusion of the midgut contents. Washed midgut walls or contents representing 1 g larval weight is homogenized in 10 ml of distilled water using a chilled glass Teflon tissue grinder. The suspensions are centrifuged in the cold at 10,000 × g for 15 min, and the supernatants are used as enzyme solutions. In some cases the whole midgut (gut walls and contents) is used for enzyme preparation.

Protease activity is determined under optimal experimental conditions (Ishaaya et al., 1971) in a reaction mixture of 0.2 ml 0.2 M glycine-NaOH buffer (pH 11.0), 0.4 ml 1.5% casein solution in 0.01 M phosphate buffer, at pH 8.0 and 0.2 ml enzyme solution. Enzyme activity is terminated after 60 min of incubation at 37°C by addition of 1.2 ml of 5% trichloroacetic acid. The reaction mixture is then filtered through Whatman filter paper

No. 1 and the filtrate is taken for enzyme activity evaluation. The proteolytic activity is expressed in optical density (OD) units $\times$ 10^3 at an absorbance of 280 nm. For standardization, the activity can be expressed in milligrams fresh larval or gut weight or milligrams protein.

Amylase and invertase activities are determined under optimal experimental conditions (Ishaaya et al., 1971, 1974; Ishaaya and Meisner, 1973), using 3,5-dinitrosalicylic acid reagent for determining the free aldehydic groups of glucose formed after starch or sucrose digestion. This reaction is based on the reduction of the dinitrosalicylic acid by the aldehydic groups of glucose units in basic medium. The reduced dinitrosalicylic acid can be measured spectrophotometrically at an absorbance of 550 nm. Amylase activity is determined in a reaction mixture of 0.4 ml 0.05 M glycine-NaOH buffer (pH 9.5), 0.2 ml 1% starch solution, and 0.2 ml enzyme solution. Invertase activity is determined in a reaction mixture of 0.4 ml 0.05 M phosphate buffer (pH 7.0), 0.2 ml 4% sucrose solution, and 0.2 ml enzyme solution. After 30 min incubation at 37°C, enzyme activity is terminated by addition of 1.6 ml of the 3,5-dinitrosalicylic acid reagent. The reaction mixture is heated for 5 min at 100°C, followed by immediate cooling in an ice bath and dilution with 1.6 ml distilled water. Enzyme activity is expressed in OD units $\times$ 10^3 at an absorbance of 550 nm. For standardization, the activity can be expressed in milligrams fresh larval or gut weight or milligrams protein.

The preparation of the dinitrosalicylic acid reagent used for the amylase and invertase reactions is based on that described by Noelting and Bernfeld (1948): 1 g 3,5-dinitrosalicylic acid, 20 ml NaOH 2 N, and 50 ml H_2O are mixed in a glass container, with a magnetic stirrer, until all the dinitrosalicylic acid is dissolved. Potassium sodium tartrate (30 g) is added and the solution is mixed again until it becomes clear. Distilled water is then added to bring the final volume to 100 ml. The reagent is now ready for use. (It should be kept in the dark, and can be used for 3 months.)

According to standard curves obtained from a direct measurement of tyrosine and glucose under experimental conditions similar to those used for protease, amylase, and invertase reactions, protease activity can be determined in mg tyrosine/mg protein/hr, and amylase and invertase activities in mg glucose/mg protein/hr.

B. Insect–Host Compatability Assays in Scale Insects

These assays are carried out with the black scale, *Saissetia oleae* (Ishaaya and Swirski, 1970). Trehalase, amylase, and invertase activities are used for determining the suitability of these scales to feed on various host plants.

The rearing of the black scale is carried out by the modified procedure of Flanders (1942), using sprouts of potato tubers detached from soil before infestation (Ishaaya and Swirski, 1976; Blumberg and Swirski, 1977). The rearing assays are kept in glasshouses at 22°–28°C. The biological and biochemical aspects of second instars and young females are compared with those reared, under similar temperature conditions, on oleander and citrus plants.

The Florida and California red scales are reared on grapefruit, lemon, squash, and potato tubers. Invertase and amylase activities of adult females are assayed to determine the suitability of these scales for various host plants and environmental temperatures (Ishaaya and Swirski, 1970).

(a) Enzyme preparation from the black scale Saissetia oleae and enzyme assays. A total of 500 to 1000 second instar or young black scale females are removed from their host and homogenized in a chilled tissue grinder containing 10–20 ml of distilled water. The homogenate is centrifuged at 20°C for 20 min at 40,000 × g. The supernatant is filtered through glass wool to remove fatty materials and freeze-dried with a Virtis lyophylizer. A light-colored powder of 1–1.5 mg for 1000 second instar scales, and of 10–15 mg for 1000 young females, is obtained and used for enzyme activity determination. The enzyme extract containing 65–70% protein according to the Lowry method (Lowry et al., 1951) is dissolved in H_2O and used as enzyme solution. The enzyme powder can be kept in a freezer ($-20°C$) and used for at least 2 weeks.

Invertase, amylase, and trehalase assays based on the digestion of sucrose, starch, and trehalose are done by a spectrophotometric method using the 3,5-dinitrosalicylic acid reagent to determine the free aldehyde groups of glucose formed after sucrose, starch, or trehalose digestion (Ishaaya and Swirski, 1976). Optimum conditions for enzyme reaction, i.e., pH, initial velocity, enzyme and substrate concentrations are determined in a series of preliminary experiments, in which individual factors are varied and all the others are kept at the optimum.

The invertase reaction under optimal conditions consists of 0.2 ml 4% sucrose, 0.1 ml 0.2 *M* acetate buffer (pH 5.5), and 0.1 ml 0.2% enzyme solution; the amylase reaction consists of 0.1 ml 2% starch, 0.1 ml 0.2 *M* phosphate buffer (pH 6.0), and 0.2 ml 0.2% enzyme solution; the trehalase reaction consists of 0.2 ml 3% trehalose, 0.1 ml 0.2 *M* acetate buffer (pH 5.5), and 0.1 ml 0.2% enzyme extract. After 60 min incubation at 37°C of either invertase or amylase reaction, enzyme activity is terminated by addition of 0.8 ml 3,5-dinitrosalicylic acid reagent (for preparation, see previous section). The reaction mixture is heated for 5 min at 100°C and followed by immediate cooling in an ice bath. The activity is determined in extinction units (E) at an absorbance of 550 nm as described in the previous section. According to the standard curve obtained from a direct reaction of glucose with dinitrosalicylic acid reagent, under conditions similar to those of the enzyme reaction, 1 E unit = 0.4 ml glucose. For standardization, the activity can be expressed in mg glucose/mg protein/ hr (Ishaaya and Swirski, 1976).

(b) Enzyme preparation from armored scales and enzyme assays. One hundred adults of the Florida red scale (*Chrysomphalus aonidum*) and the California red scale (*Aonidiella aurantii*) are removed from their host and homogenized in a chilled tissue grinder containing 1 ml distilled water or buffer. The homogenate is centrifuged for 15 min at 2500 × *g* and the supernatant is used as the enzyme solution. Because of the nature of the protective shields, the Florida red scales are removed from their shields and taken for homogenization, whereas the California red scales, which are strongly bound to their shields, are taken whole for homogenization (Ishaaya and Swirski, 1970).

Invertase and amylase activities are determined by a procedure similar to that described in the previous section. Under optimal experimental conditions, the reaction mixture for amylase consists of 100 μl 1% starch, 50 μl 0.05 *M* phosphate buffer (pH 6.0), and 50 μl enzyme solution; the reaction mixture of invertase is similar to that of amylase, using 1% sucrose instead of 1% starch. After 10 min incubation at 37°C for the amylase assay and 60 min for the invertase assay, the enzyme activity is terminated by addition of 0.8 ml of the 3,5-dinitrosalicylic acid reagent. The reaction mixtures are heated for 5 min at 100°C and then diluted by addition of 2.4 ml H_2O. Amylase and invertase activities are determined as described in the previous section.

C. Protein–Inhibitor Interactions

Proteins may counteract the enzyme inhibitory effect of various secondary compounds such as saponins (Ishaaya and Birk, 1965) and triorganotins (Ishaaya and Casida, 1975). Determination of the counteractivity of protein on various digestive enzyme inhibitors is done in vitro by incubating the inhibitor with the protein and/or with the enzyme for 5–30 min prior to the start of reaction. In case of counteraction, incubation of protein with the inhibitor prior to the start of reaction decreases the inhibitory effect of the test compound, whereas incubation of the enzyme with the inhibitor accentuates the activity of the compound on digestive enzymes, as in the case of saponins and triorganotins (Ishaaya and Birk, 1965; Ishaaya and Casida, 1975). Protein inhibitor interaction will be discussed further in the next sections.

III. Insect–Plant Interactions Relating to Allelochemic Compounds

Plants produce a diverse set of chemicals that are toxic, in various degrees, to plant pathogens, insects, and mammals, and are therefore of potential selective advantage in deterring their enemies. Toxic chemicals can be of

high molecular weights such as phytohemagglutinins (lectins), polysaccharides, and proteinous substances. Some of these compounds inhibit
growth and development of insect species (Applebaum et al., 1970; Janzen
et al., 1976) whereas others inhibit proteolytic and amylolytic activities
(Birk et al., 1962; Shainkin and Birk, 1970; Strumayer, 1972; Applebaum
and Birk, 1972); and serve as antifeedants for insects and mammals. A
diverse range of allelochemic compounds of low molecular weights is
present in plants and plays important defensive roles against insects and
other herbivores; included among these compounds are saponins (Applebaum and Birk, 1979), tannins and lignins (Swain, 1979), gossypol and
other terpenoids (Abou-Donia, 1976; Mabry and Gill, 1979), alkaloids (Bell,
1978), nonprotein amino acids (Rosenthal and Bell, 1979), and cyanogenic
compounds (Conn, 1973, 1979). These compounds may serve as powerful
toxicants to deter insects and other herbivores from feeding and in some
cases they serve as a starting point for developing novel insecticides, as
in the case of the recent synthetic pyrethroids (Elliott et al., 1978). This
section deals with plant allelochemic compounds affecting specifically
digestion and food intake in insects.

3.1. Gossypol

Gossypol [1,1',6,6',7,7'-hexahydroxy-3,3'-dimethyl-5,5'-diisopropyl (2,2'-
binaphthalene)-8,8'-dicarboxaldehyde] (see Fig. 1), a substance occurring
naturally in the pigment glands of cotton, affects the feeding and growth
of a number of phytophagous insects (Bottger et al., 1964; Lukefahr et
al., 1968; Shaver and Parrott, 1970; Meisner et al., 1972, 1973, 1977a,b).
The influence of gossypol on insect development is assayed either by its
incorporation in a basic diet (Lukefahr and Martin, 1966; Maxwell et al.,
1967; Shaver et al., 1970) or by offering strains of cotton with various
gossypol contents to insects (Wene and Sheets, 1966; Brader, 1967; Singh
and Weaver, 1972; Meisner et al., 1977b). Third instars of *Spodoptera
littoralis* larvae (40–50 mg) fed on cotton leaves of low gossypol content
for 3 days gain about three times as much weight as those fed on leaves
of high gossypol content. The detrimental effect of gossypol increases

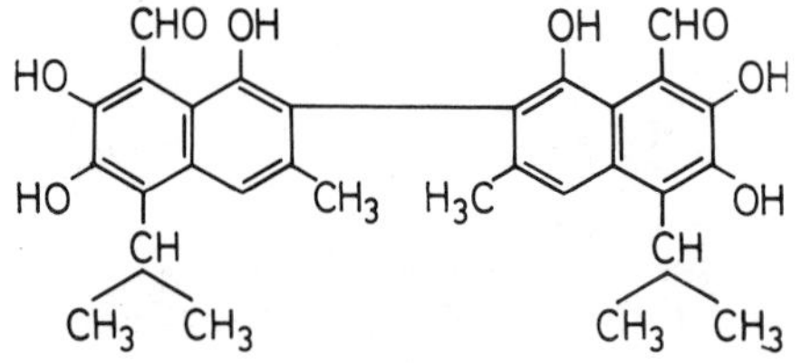

Figure 1. Gossypol structure.

Table 3. Protease activity in 90–110-mg-weight *Spodoptera littoralis* larvae fed for 2 days on a semisynthetic diet treated with various concentrations of gossypol acetate

Gossypol acetate added to the semisynthetic diet (%)	Average weight per larva after 2 days (mg)	Protease activity relative to control (%)
0 (control)	545.6 ± 23.6	100
0.25	491.0 ± 23.4	89.1
0.50	391.6 ± 24.4	54.9

After Meisner *et al.*, 1978.

Neonate and young larvae are affected much more than are older larvae (Meisner et al., 1977a). A decrease of about 40% in larval amylase and protease activities occurs 1 day after feeding on leaves with a high gossypol content. A gradual decrease in larval weight gain and protease activity is observed in larvae fed an artificial diet containing increasing levels of gossypol (Table 3, after Meisner et al., 1978). The affinity of gossypol to protein is well established and seems to resemble that of tannins (Feeny, 1970). When gossypol acetate is incubated with the enzyme or its casein substrate, prior to enzyme reaction, there is a marked inhibition of the proteolytic activity (Table 4, after Meisner et al., 1978). The interaction of gossypol is rapid, so the inhibitory effect obtained after 5 min preincubation is similar to that after 30 min. These results indicate that gossypol may affect larval proteolytic activity by reacting either with the proteinous substrate or with the enzyme itself.

The possible interaction of gossypol with cottonseed protein has been demonstrated by Lyman et al. (1959) and Damaty and Hudson (1975).

Table 4. Effect of preincubation of gossypol with the enzyme protein or the substrate casein on the protease activity

Gossypol in preincubation mixture (mg)	Preincubation		Enzyme activity relative to control (%)
	Mixture	Time (min)	
0.25	Enzyme + gossypol	5	69.8
	Enzyme + gossypol	30	73.9
	Casein + gossypol	5	70.7
	Casein + gossypol	30	74.1
1.5	Enzyme + gossypol	5	45.6
	Enzyme + gossypol	30	44.9
	Casein + gossypol	5	48.9
	Casein + gossypol	30	56.7

After Meisner *et al.*, 1978.

Baliga and Lyman (1957) showed that with the formation of the gossypol–
protein complex, lysine availability decreases by about one-half. Incubation of gossypol with pepsinogen at a 2:1 or 3:1 ratio resulted in a complete inhibition of the autocatalytic conversion of pepsinogen to pepsin
(Tanksley et al., 1970). The protein level in the diet affects both protease
and amylase activities of *Spodoptera littoralis* larvae (Ishaaya et al., 1971),
so that reduced availability of protein as a result of interaction with gossypol in situ may affect the activities of these enzymes.

3.2. Saponins

Saponins are glycosides containing a polycyclic aglycone moiety of either
triterpenoid or steroidal structure attached to a mono- or oligosaccharide
chain (Fig. 2). The aglycone moieties are collectively termed sapogenins.
Triterpenoid sapogenins contain 30 carbon atoms, most of which are pen-
tacyclic compounds derived from oleanan (Steiner and Holtzem, 1955),
whereas steroidal sapogenins contain usually 27 carbon atoms and exhibit
a steroidal structure. The main constituents of carbohydrates identified
in triterpenoid saponins are glucose, galactose, arabinose, xylose, rhamnose, quinovose, and glucuronic and galactoronic acids; and in steroidal
saponins the main constituents are glucose, galactose, rhamnose, xylose,
and arabinose (Shoppee, 1964). Further information concerning saponin
structure and composition can be obtained from excellent surveys and
reviews published during the last two decades (Boiteau et al., 1964; Basu
and Rastogi, 1967; Birk, 1969; Cheeke, 1971; Horber, 1972; Bondi et al.,
1973; Agarwal and Rastogi, 1974).

Saponins are widely distributed throughout the plant kingdom and have
been identified in at least 400 species belonging to more than 80 different
families (Birk, 1969). Plants that contain saponins include spinach, beet
root, sugar beet, beech, asparagus, crocus, horse chestnut, alfalfa, and
soybeans (George, 1965). They also occur in clover (Walter, 1957, 1960)
and other pasture plants (Walter, 1961). A variety of saponins that have
been identified in soybean and alfalfa plants differ in their aglycone and

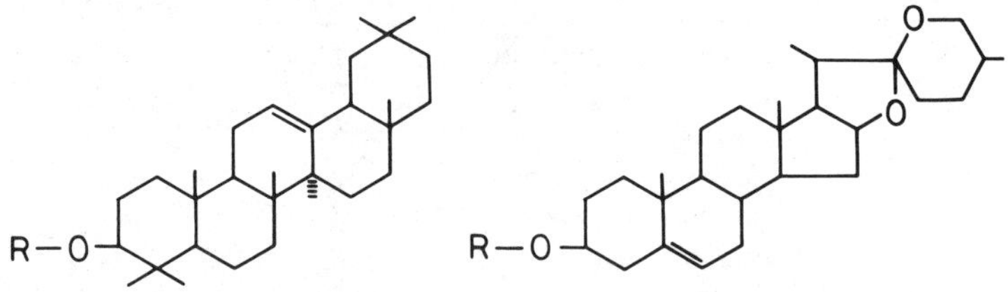

Figure 2. Basic structure formulae of triterpenoid saponins (oleanan type) and of
steroid saponins. R represents sugar moiety.

carbohydrate moieties (Birk, 1969; Applebaum and Birk, 1979). Medicagenic acid is a major sapogenin present in alfalfa but not in soybean plants (Gestetner et al., 1970).

Soybean saponins inhibit various mammalian proteases such as trypsin, chymotrypsin, and papain as well as digestive protease from the midgut of *Tribolium castaneum, Tenebrio molitor,* and *Dermestes maculatus* (Ishaaya and Birk, 1965; Ishaaya, 1965). The inhibitory effect of soybean saponins on *T. castaneum* proteolytic activity is more accentuated after incubation with the enzyme prior to the start of reaction and fully abolished after incubation with the proteinous substrate such as casein, hemoglobin, and soybean protein (Table 5, after Ishaaya and Birk, 1965). These results indicate that the inhibitory effect of soybean saponins is nonspecific and results from a protein (i.e., enzyme)–saponin interaction. Alfalfa saponins inhibit α-chymotrypsin and digestive proteases of *T. castaneum* to the same extent as do soybean saponins (Ishaaya, 1965; Birk, 1969). The nonspecific type of inhibition, which has been proved for soybean saponins, is also valid for the inhibition of these enzymes by alfalfa saponins, since it could be overcome, fully or partly, by preincubation of alfalfa saponins with casein. Cholesterol, which is a useful complexing agent with alfalfa saponins but not with soybean saponins, could not abolish the inhibitory effect of alfalfa saponins on enzymes when preincubated with the saponin

Table 5. Effect of preincubation of soybeans saponins with *Tribolium castaneum* proteolytic enzymes or with their proteinous substrates on enzyme activity

Substrate	Preincubation[a] of	time (min)	Enzyme activity relative to control (%)
Casein	—		100
	Enzyme + saponins	5	10
		30	0
	Substrate + saponins	5	89
		30	89
Hemoglobin	—		100
	Enzyme + saponins	5	78
		30	24
	Substrate + saponins	5	104
		30	100
Soybean protein	Enzyme + saponins	5	45
		30	42
	Substrate + saponins	5	94
		30	93

[a]The enzyme or the substrate protein is incubated for 5 or 30 min prior to the start of reaction. Enzyme activity is expressed as percent of that of the untreated reaction. After Ishaaya and Birk, 1965.

prior to the start of the reaction (Ishaaya, 1965). Alfalfa saponins seem to have two sites of interaction, one with protein and the other with cholesterol, whereas soybean saponins interact only with protein.

Alfalfa saponins inhibit larval growth and development of *Tribolium castaneum* more strongly than do soybean saponins (Table 6, after Ishaaya et al., 1969). Prolongation of the larval stage is proportional to larval weight inhibition and both effects are reversed by the addition of cholesterol to the diet containing alfalfa saponins (Table 6). The fact that the growth impairment caused by alfalfa saponin is counteracted by cholesterol is probably due to the complex formed between the two. It also indicates that the inhibitory effect of alfalfa saponin may arise from complexes formed with larval dietary sterols. This is also supported by the relatively lower inhibitory effect of soybean saponins—which do not combine with cholesterol—on *T. castaneum* (Table 6). The detrimental effect of alfalfa saponins is apparently due to the presence of medicagenic acid (Gestetner et al., 1970). Free medicagenic acid is highly toxic towards *T. castaneum*, whereas the soya sapogenins are harmless. The toxicity of medicagenic acid can be reversed by cholesterol, stigmasterol, β-sitosterol, or campesterol (Shany et al., 1970).

Water resorption in the hindgut of *Locusta migratoria* is impaired and

Table 6. Effect of soybean and alfalfa saponins and saponin–cholesterol interaction on larval weight of *Tribolium castaneum*

Dietary saponins (%)			Larval weight relative to control[a]
Soybean	Alfalfa	Cholesterol	
0.5			88
1.0			81
5.0			54
10.0			40
	0.25		72
	0.5		56
	1.0		44
	0.5	0.5	74
	0.5	1.0	80
	0.5	2.0	89
	1.0	0.5	66
		0.5	99
		1.0	95

[a]Larval weight is determined 18 days after the start of the reaction and expressed as percent of untreated control. Average weight of the untreated larvae = 2.06 mg. After Ishaaya et al., 1969.

hemolymph volume is decreased when this insect is fed on a diet containing alfalfa saponins (Applebaum and Birk, 1979).

Insect growth impairment caused by saponins may result from a non-specific interaction with protein, thereby inhibiting digestive enzymes and food intake, and/or interaction with sterols present in the diet and insect tissues, and thus affecting hormone production. The ability of saponins to interfere with water resorption in the hindgut of insects may lead to loss of liquid in insect tissues and to physiological disturbances. The wide distribution of saponins in a variety of plants led to the suggestion that saponins, along with other secondary substances that have been formed in plants during their parallel evolution with insects (Fraenkel, 1959), impart resistance against some species of insects (Applebaum, and Birk, 1972, 1979).

3.3. Tannins

Tannins are phenolic compounds widely distributed among varieties of plants, often in high concentrations. Phenolic compounds of low molecular weight that occur in plants may undergo polymerization by either autoxidation or enzyme-catalyzed oxidation to yield tannin-like compounds (Swain, 1979). Tannins can be divided into two main groups according to their molecular weight and biological activity: condensed tannins and hydrolyzable tannins (Figs. 3 and 4). Condensed tannins are phenolic polymers with a molecular weight of 1000–2500 or more, act through their ability to combine with protein (and other hydrogen-bonding polymers), and thus inhibit enzymes and reduce the availability of protein and other nutrients in foods (Swain, 1978). Various plants produce hydrolyzable tannins that in some cases are more effective than the condensed tannins as antifungal agents or feeding deterrents for insects and other herbivores. Furthermore, the hydrolyzable tannins are biodegradable and thus can be recycled by the plant if necessary (Swain, 1979).

The wide range and content of tannins in various plants (Singleton and Kratzer, 1969) accentuate their role in plant defensive systems against

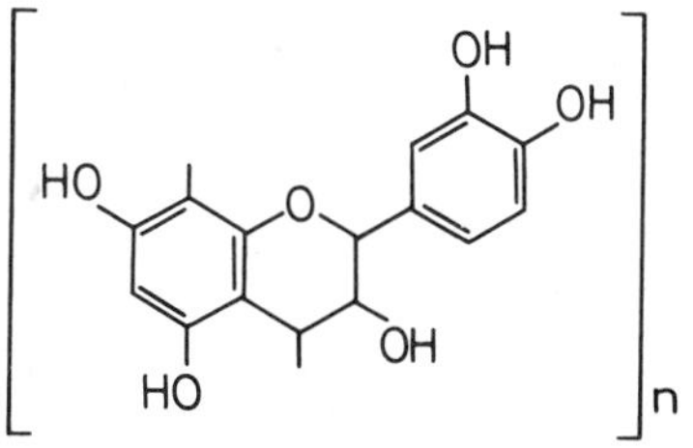

Figure 3. Procyanidin (condensed tannins), n = 2–6.

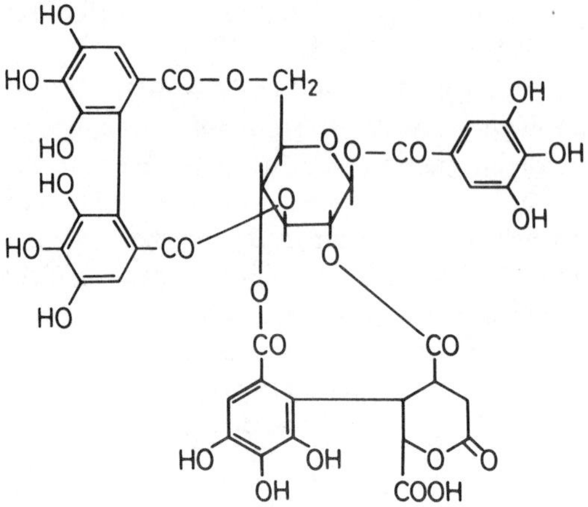

Figure 4. Chebulagic acid (hydrolysable tannin).

pathogens (Levin, 1971, 1976; Harborne, 1977; Swain, 1977, 1978) and insects (Feeny, 1968, 1970, 1976; Chan et al., 1978; Klocke and Chan, 1982) whose diets include condensed tannins. In addition, a relatively high level of condensed tannins in some cotton strains is found to be associated with the plant's resistance to spider mites (Lane and Schuster, 1981).

It was originally proposed that the biochemical mode of action of tannins in insects involves a reduced availability of dietary protein (Feeny, 1970, 1976) or a decreased activity of digestive enzymes (Goldstein and Swain, 1965; Klocke and Chan, 1982); but, the reader should refer to Chapter 5 (*this volume*) for reinterpretations of the original data. Tannins may also interact with dietary amino acids and decrease their availability to the insect (Delort-Laval and Viroben, 1969). In some cases they may exhibit a direct toxic action (Feeny, 1970) or a repellent effect reducing palatability (Lipke and Fraenkel, 1956). Condensed tannins incorporated into an artificial diet inhibit growth of *Heliothis zea* larvae in accordance with the concentration of dietary tannin (Klocke and Chan, 1982). Larvae feeding on a tannin-treated diet exhibit a decreased activity of the digestive protease and amylase enzymes along with a reduced level of protein and sugars in larval hemolymph (Klocke and Chan, 1982). These results concur with those of earlier studies (Goldstein and Swain, 1965; Singleton and Kratzer, 1973) in which tannins were shown to inhibit digestive enzymes by a direct interaction with the enzyme protein. It is not clear whether reduced digestive enzyme activity is the main reason for the larval growth impairment, since the addition of glucose and casein hydrolysate does not reverse the larval growth retardation occurring when a tannin-supplemented diet is consumed (Klocke and Chan, 1982). Tannins may interact with the gut wall protein and affect permeability and assimilation of nutrients. In "choice experiments," condensed tannins are found to deter insects from feeding (Bernays and Chapman, 1978; Jones and Firn, 1979); possibly due to their effect on the insect's peripheral chemoreceptors (Bernays, 1981).

Insect growth inhibition by tannins represents probably a complex interaction that includes a lowered food intake, decreased activity of midgut digestive enzymes, and a reduced level of protein and sugars in the hemolymph, all of which may result from tannin–protein interactions occurring in the insect gut lumen, the peripheral chemoreceptor sites, or the gut wall proteins.

3.4. Proteinous Inhibitors and Nonprotein Amino Acids

Early studies indicated the ability of soybean soluble fractions to inhibit trypsin activity (Bowman, 1944; Ham et al., 1945). Shortly thereafter, a proteinous inhibitor of this enzyme was isolated from raw soybean in crystalline form by Kunitz (1946). Then, the effect of soybean inhibitors on growth of rats, chicks and insects was reported (Liener, 1953; Almquist and Merritt, 1953; Lipke and Fraenkel, 1954). Extensive studies followed, resulting in the isolation and characterization of soybean trypsin inhibitor, SBTI (Kunitz, 1947), a trypsin and chymotrypsin inhibitor, AA (Birk, 1961; Birk and Gertler, 1962; Birk et al., 1963b); and a *Tribolium* larval protease inhibitor (Birk et al., 1962, 1963a). In studies using purified soybean protease inhibitors, *Tribolium* proteases seem to differ in their inhibition specificities from the mammalian trypsin and chymotrypsin enzymes (Applebaum and Birk, 1972). It is suggested that this property may serve for selecting legume seed varieties containing higher concentrations of the specific *Tribolium* protease inhibitor for deterring storage pests (Applebaum and Birk, 1972). The possible involvement of protease inhibitors in plant protection received considerable support from the finding that Colorado potato beetle infestation induces a rapid accumulation of protease inhibitors in the leaves of potato and tomato plants (Green and Ryan, 1972). Accumulation of protease inhibitors is observed also in tomato plants after wounding, accounting over 0.02% of the leaf tissue homogenate (Melville and Ryan, 1972; Green and Ryan, 1973). A wide spectrum of protease inhibitors of the proteinous type is found in a variety of plants (Bryant et al., 1976; Walker-Simmons and Ryan, 1977; Ryan, 1979), strengthening the concept that these proteins, along with other secondary compounds, play an important defensive role against insects and other herbivores.

The nonprotein amino acids are widely distributed among plants and most of them are extremely important precursors of plant metabolites (Rosenthal and Bell, 1979). The majority of these compounds are α-amino acids of the L-configuration, and occur usually in a free state as α-glutamyl derivatives (Robinson, 1976). Several nonprotein amino acids present in legume seeds are lathyrogenic or neurotoxic to various vertebrates (Ressler, 1964; Murti et al., 1964), and others, such L-canavanine and L-canaline, exhibit a relatively high insecticidal activity (Rosenthal and Bell, 1979). Canavanine is the guanidinooxy structural analog of arginine, characterized by the replacement of the terminal methylene group with oxygen, and canaline is the hydrolytic product of canavanine (Rosenthal and Bell, 1979).

$$H_2N—C(NH)—NH—O—CH_2—CH_2—CH(NH_2)—COOH$$

Canavanine

$$H_2N—O—CH_2—CH(NH_2)—COOH$$

Canaline

A pronounced insecticidal property of canavanine has been demonstrated in the fruit fly (Harrison and Holiday, 1967) and in the boll weevil (Vanderzant and Chremos, 1971). A dietary concentration of 3–45 mM L-canavanine causes significant developmental aberrations and decreased survival of the tobacco hornworm, *Manduca sexta,* and of the southern army worm, *Prodenia eridania* (Rehr et al., 1973; Dahlman and Rosenthal, 1975). The antimetabolic properties of canavanine seem to result from the ability of canavanine to replace arginine in the nascent polypeptide chain; this substitution may cause structural and protein alterations (Attias et al., 1969; Prouty et al., 1975). The possible substitution of canavanine for arginine in histones alters the histone property and ultimately affects genome expression. In this regard, canavanine has been shown to disrupt histone synthesis in HeLa cells (Ackermann et al., 1965) and in cell cultures of hamsters and mice (Hare, 1969). According to this concept, canavanine may affect protein production and enzyme properties in insects and thus, in turn, affect biochemical processes in digestion and metamorphosis. Further studies are needed to evaluate the importance of canavanine in biochemical functions and the properties of various enzyme systems. Other uncommon amino acids, such as β-cyanoalanine and 2,4-diaminobutyric acid, are found at high levels in some legume seeds and exert a detrimental effect on *Tribolium* and *Callosobruchus* (Applebaum and Birk, 1979). Most of the nonprotein amino acids are toxic to humans and to farm animals (Rosenthal, 1977; Rosenthal and Bell, 1979); thus, further studies are needed to make possible the practical use of these compounds as a factor of resistance in plants against insects, without side effects on the environment.

3.5. Biological and Biochemical Assays

In this section our attention is focused on determining biological and biochemical assays relating to the effects of various secondary compounds on digestion and food intake. Assays with gossypol, saponins, tannins, and proteinous inhibitors on insect growth and larval digestive enzymes are described.

A. Assays with Gossypol

Feeding assays are carried out with cotyledons of high and low gossypol content or with artificial diets containing various amounts of gossypol. Assays with cotyledons are carried out with fourth instar *Spodoptera littoralis* larvae. After 3 hr of starvation, each larva is introduced into a petri dish (7 cm diameter) containing a layer of sawdust. Freshly collected cotyledons are provided daily, two per larva. The amount of food intake, larval weight, and digestive enzyme activities are determined after different intervals (Meisner et al., 1978). A semisynthetic diet is prepared according to the method of Shorey and Hale (1965) as modified by Kehat and Gordon (1975). Acetone solutions with various concentrations of gossypol acetate are added to alfalfa meal and mixed thoroughly. The mixture is placed in a fume cupboard for 1–2 days to evaporate the acetone and then stirred in an electric mixer with the other ingredients of the diet. The alfalfa meal in the control diet is mixed with acetone alone. The diets are distributed in small plastic cloth-covered containers. For each treatment, 30 larvae weighing 90–110 mg each are used, one larva per container. After 48 hr of feeding, larval weight gain and protease and amylase activities are determined. Enzyme preparation from larval gut assays is as described in the previous section.

Protease inhibition in vitro by gossypol is carried out according to Meisner et al. (1978), using gossypol solution in ethanol (12.5 μg/ml). Levels of 0.02 and 0.12 ml gossypol containing 0.25 and 1.5 mg are added to the 1-ml substrate solution (casein) or to the 1-ml enzyme-buffer solution and incubated for 5 and 30 min at 37°C prior to the initiation of the reaction. The enzyme activity is carried out as described in the previous section and expressed as a percentage of the appropriate controls containing the ethanol used to prepare the gossypol solution. The ethanol concentrations used in these assays have no effect on the enzyme activity.

B. Assays with Saponins

These assays are carried out with *Tribolium castaneum* larvae (Ishaaya and Birk, 1965; Ishaaya et al., 1969). The larvae are reared on wheat flour containing 5% dried yeast as a basic diet. For the bioassay, 10 g of diet is mixed with soybean or alfalfa saponins in 12 ml acetone. Following evaporation of the solvent and thorough mixing, the diet is distributed in portions of 2 g in test vials 2.5 cm in diameter. Ten white larvae, 0–3 hr after hatching, are placed in each test vial and kept at 28°C to determine larval growth, pupation, and emergence. The addition of cholesterol to the diet reverses the detrimental effect of alfalfa saponins (Ishaaya et al., 1969) and addition of protein to enzyme reaction reverses the inhibitory effect of soybean saponins on digestive enzymes (Ishaaya and Birk, 1965). Each in vivo treatment is carried out with five to ten replicates of ten larvae each. Larval weight, pupation, and emergence relative to untreated control are then determined.

(a) Larval enzyme preparation and enzyme assays. The gut of fourth instar (1.2 ± 0.1 mg) or last instar (2.0 ± 0.1 mg) larva is dissected free from other tissues by cutting off the larval head with a razor blade and removing the alimentary canal on a wet filter paper, using a fine forceps assisted by slight pressure at the posterior end of the body. The enzyme solutions are prepared at 0–3°C by homogenizing 20–50 guts in distilled water (50-fold relative to larval weight for the protease assay, and 200-fold for the invertase and amylase assays).

Protease activity is assayed colorimetrically (280 nm) on the basis of liberated tyrosine (1 E unit = 0.35 mg of tyrosine) under optimal conditions (Birk et al., 1962; Ishaaya et al., 1977, 1980) involving 0.8 ml of 1.5% wt/vol casein solution in 0.2 M phosphate buffer (pH 6.6), 0.4 ml of H_2O, and 0.4 ml enzyme solution, incubated for 60 min at 37°C. Addition of 0.4 ml of 15% (wt/vol) trichloroacetic acid solution terminated the protease activity, after which the mixture was centrifuged at 10,000 × g for 10 min and filtered through Whatman No. 1 filter paper. The enzyme activity is expressed in mg tyrosine/mg larval weight/hr, or in mg tyrosine/mg protein/hr.

Invertase and amylase activities are assayed colorimetrically, under optimal conditions, on the basis of liberated glucose (Ishaaya and Casida 1975; Ishaaya et al., 1977). The reaction mixtures consist of 0.2 ml of 4% (wt/vol) sucrose (for invertase) or 0.2 ml 4% (wt/vol) starch (for amylase), 0.1 of 0.2 M phosphate buffer (pH 5.5), and 0.1 ml of enzyme solution. After 60 min incubation at 37°C, 0.8 ml of 3,5-dinitrosalicylic acid reagent (for preparation, see p. 299) is added, the mixture is heated for 5 min at 100°C, and then immediately cooled in an ice bath. The absorbance at 550 nm is determined in extinction (E) units. Under conditions similar to those for the enzyme assay, direct reaction of glucose with the dinitrosalicylic acid reagent gives 1 E unit = 0.4 mg of glucose. Enzyme activity is expressed in mg glucose/mg larval weight/hr or in mg glucose/mg protein/hr.

(b) Enzyme inhibition by saponins. The effect of saponins on larval digestive enzymes can be assayed in vivo by determining enzyme activity after feeding on a saponin-supplemented diet or in vitro by determining the direct effect of saponins on the enzyme activity. For determining the counteractivity of protein on the inhibitory effect of saponins on digestive enzymes, saponin is incubated with the enzyme or with the proteinous substrate (in the case of protease) for 5 and 30 min. The reaction is then

initiated by adding the substrate or the enzyme solution (Ishaaya and Birk, 1965). In the case of invertase and amylase, in which the substrates are carbohydrates, saponins may be incubated with proteins such as casein, soybean protein, and hemoglobin prior to the addition of enzyme solution (Ishaaya and Birk, 1965). The enzyme activity is then determined as described above.

C. Assays with Tannins

The effect of tannins on the larval digestive protease and invertase activities is studied with *Heliothis zea* as described by Klocke and Chan (1982). Larval enzyme solution is prepared by a slight modification of the method described by Ishaaya et al. (1971). Larvae are chilled on ice for 15 min followed by rapid dissection of the midguts. Guts are cleaned of their attachments and cut between the pyloric and the cardiac valves. After removal of the peritrophic membrane and its content, midgut walls are homogenized in ice-cold distilled water in a chilled glass Teflon tissue grinder. Following centrifugation in the cold at 20,000 $\times$ *g* for 20 min, the supernatant is used as the enzyme solution.

Protease activity is based on the method described for *Spodoptera littoralis* (Ishaaya et al., 1971) and used by Klocke and Chan (1982). The reaction mixture consists of 0.5 ml 0.075 *M* glycine-NaOH buffer at pH 11.0, 1.0 ml 1.5% casein solution, and 0.5 ml enzyme solution incubated at 37°C for 60 min.

Invertase activity is based on the method described by Ishaaya and Swirski (1970) and used by Klocke and Chan (1982). The reaction mixture consists of 100 μl phosphate buffer at pH 7.0, 200 μl 0.0075 *M* sucrose, and 25–50 μl enzyme solution, brought to a final volume of 800 μl with distilled water. The reaction mixture is incubated for 30 min at 37°C and terminated by the addition of 1.6 ml 3,5-dinitrosalicylic acid reagent (for preparation, see p. 299). The mixture is heated for 5 min at 100°C and immediately cooled in an ice bath. Enzyme activity is determined as described in the previous section.

For studying the in vitro inhibition of protease and invertase activities (Klocke and Chan, 1982), tannins are preincubated for 30 min with the enzyme extract, centrifuged at 1500 $\times$ *g* for 5 min, followed by determination of the residual enzyme activity in the supernatant.

D. Assays with Proteinous Inhibitors

The effects of proteinous inhibitors on larval digestive enzymes can be tested in vivo and in vitro as described in the previous sections for other secondary compounds. For in vivo assays, dietary proteinous inhibitors are given to the larvae for determinations of protease, amylase, or invertase activity at various intervals. For the in vitro assays, the inhibitors are incubated with the enzyme for 5 and/or 30 min prior to the enzyme reaction. Enzyme assays can be adapted from those described above for *Spodoptera littoralis, Tribolium castaneum,* and *Heliothis zea* larvae.

IV. Conclusions

The biochemical aspects of plant nutritional and allelochemic factors affecting insect digestion and food intake are reviewed. The activities of digestive enzymes such as protease, amylase, and invertase are useful parameters for evaluating detrimental effects occurring as a result of these factors.

Nutritional factors such as proteins and sugars may stimulate or inhibit digestive enzymes in insects. Allelochemicals such as gossypol, soybean saponins, and tannins interact with nutrients (mainly proteins) or with larval digestive enzymes, resulting in deleterious effects. In many cases digestive enzymes are primarily affected and can be used as parameters for antifeeding activity. The trehalase-trehalose system, which is important for internal energy supply in insect (Wyatt, 1967), can be used, in some cases, as an additional parameter for assessing the adaptability of insects to various host plants (Ishaaya and Swirski, 1976).

Assays for determining digestive enzymes and trehalase activities in various insects along with in vivo and in vitro effects of various nutritional and secondary compounds are discussed.

Acknowledgments. I wish to thank Dr. A. Navon and Dr. M. Weissenberg for stimulating discussions and critical reading of the manuscript, and Mrs. Sara Yablonski for expert technical assistance. I also acknowledge with thanks the permission granted for reprinting the material presented in the Tables by Pergamon Press (Tables 1 and 2), Entomological Society of America (Tables 3 and 4), The Institute of Food Technologist (Table 5) and the Society of Chemical Industry (Table 6). This review constitutes a contribution No. 673-E, 1983 series, from the Agricultural Research Organization, The Volcani Center, Bet Dagan, Israel.

References

Abou-Donia MB (1976) Physiological effects and metabolism of gossypol. Residue Rev **61**:125–166.
Ackermann WW, Cox DC, Dinks S (1965) Control of histone and DNA synthesis with canavanine, puromycin, and poliovirus. Biochem Biophys Res Commun **19**:745–750.
Agarwal SK, Rastogi RP (1974) Triterpenoid saponins and their genins. Phytochemistry **13**:2623–2645.

Almquist HJ, Merritt JB (1953) Effect of crystalline trypsin on the raw soybean growth inhibitor. Proc Soc Exp Biol Med **83**:269.

Applebaum SW (1964) Physiological aspects of host specificity in the Bruchidae — I. General considerations of developmental compatibility. J Insect Physiol **10**:783–788.

Applebaum SW, Birk Y (1972) Natural mechanisms of resistance to insects in legume seeds. In: Insect and Mite Nutrition. Rodriguez JG (ed), North-Holland Publishing, Amsterdam pp 629–636.

Applebaum SW, Birk Y (1979) Saponins. In: Herbivores, Their Interaction with Secondary Plant Metabolites. Rosenthal GA, Janzen DH (eds), Academic Press, New York, pp 539–566.

Applebaum SW, Birk Y, Harpaz I, Bondi A (1964a) Comparative studies on proteolytic enzymes of *Tenebrio molitor* L. Comp Biochem Physiol **11**:85–103.

Applebaum SW, Harpaz I, Bondi A (1964b) Amylase secretion in the larvae of *Prodenia litura* F. (Insecta). Comp Biochem Physiol **13**:107–111.

Applebaum SW, Tadmor U, Podoler H (1970) The effect of starch and of a heteropolysaccharide fraction from *Phaseolus vulgaris* on development and fecundity of *Collosobruchus chinensis* (Coleoptera: Bruchidae). Entomol Exp Appl **13**:61–70.

Ascher KRS, Ishaaya I (1973) Antifeeding and protease and amylase inhibiting activity of fentin acetate in *Spodoptera littoralis*. Pestic Biochem Physiol **3**:326–336.

Attias J, Schlesinger MJ, Schlesinger S (1969) The effect of amino acid analogues on alkaline phosphatase formation in *Escherichia coli* K-12—IV. Substitution of canavanine for arginine. J Biol Chem **244**:3810–3817.

Augustine MG, Fisk FW, Davidson RH, Lapidus JB, Cleary RW (1964) Host-plant selection by the Mexican bean beetle, *Epilachna varivestis*. Ann Entomol Soc Am **57**:127–134.

Baliga BP, Lyman CM (1957) Preliminary report on the nutritional significance of bound gossypol in cottonseed meal. J Am Oil Chem Soc **34**:21–24.

Barton Brown L (1975) Regulatory mechanism in insect feeding. In: Treherne JG, Berridge MJ, Wigglesworth VB (eds), Adv. Ins. Physiol Academic Press, London, Vol II, pp 1–116.

Basu N, Rastogi RP (1967) Triterpenoid saponins and sapogenins. Phytochemistry **6**:1249–1270.

Beck SD (1965) Resistance of plant to insects. Annu Rev Entomol **10**:207–232.

Bell EA (1978) Toxin in seeds. In: Biochemical Aspects of Plant and Animal Coevolution. Harborne JB (ed), Academic Press, London, pp 143–161.

Bernays EA (1981) Plant tannins and insect herbivores: an appraisal. Ecol Entomol **6**:353–360.

Bernays EA, Chapman RF (1978) Plant chemistry and acridoid feeding behavior. In: Biochemical Aspects of Plant and Animal Coevolution. Harbone JB (ed), Academic Press, New York, pp 99–141.

Bernays EA, Simpson SJ (1982) Control of food intake. Adv Insect Physiol **16**:59–118.

Birk Y (1961) Purification and some properties of a highly active inhibitor of trypsin and α-chymotrypsin from soybeans. Biochim Biophys Acta **54**:378–381.

Birk Y (1969) Saponins. In: Toxic Constituents of Plant Foodstuffs, 1st edit. Liener IE (ed), Academic Press, New York, pp 169–210.

Birk Y (1976) Proteinase inhibitors from plant sources. In: Methods in Enzymology Vol. 45, Part B, Lorand L (ed), Academic Press, New York, pp 695–739.

Birk Y, Gertler A (1962) Effect of mild chemical and enzymatic treatment of soybean meal and soybean trypsin inhibitors on their nutritive and biochemical properties. J Nutr 75:379–387.

Birk Y, Gertler A, Khalef S (1963a) Separation of a *Tribolium* protease inhibitor from soybean on a calcium phosphate column. Biochim Biophys Acta 67:326–328.

Birk Y, Gertler A, Khalef S (1963b) A pure trypsin inhibitor from soybeans. Biochem J 87:281–284.

Birk Y, Harpaz I, Ishaaya I, Bondi A (1962) Studies on the proteolytic activity of the beetles *Tenebrio* and *Tribolium*. J Insect Physiol 8:417–429.

Bishara I (1934) The cotton worm, *Prodenia litura* F., in Egypt. Bull Soc R Entomol Egypt 18:288–420.

Blumberg D, DeBach P (1979) Development of *Habrolepis rouxi* Compere (Hymenoptera: Encyrtidae) in two armoured scale hosts (Homoptera: Diaspididae) and parasite egg encapsulation by California red scale. Ecol Entomol 4:299–306.

Blumberg D, Swirski E (1977) Mass breeding of two species of *Saissetia* (Hom.: coccidae) for propagation of their parasitoids. Entomophaga 22:147–150.

Blumberg D, Swirski E, Greenberg S (1975) Evidence for bivoltine populations of the Mediterranean black scale *Saissetia oleae* (Olivier) on citrus in Israel. Israel J Entomol 10:19–24.

Boiteau P, Pasich B, Rakoto Ralsimamangu A (1964) Les triterpenoides en physiologie végétale et animale. C.N.R.S., Gauthier-Villars, Paris, 1370 pp.

Bondi A, Birk Y, Gestetner B (1973) Forage saponins. In: Chemistry and Biochemistry of Herbage, Vol 1, Butler GW, Bailey RW (eds), Academic Press, New York, pp 511–528.

Bottger GT, Sheehan ET, Lukefahr MJ (1964) Relation of gossypol content of cotton plants to insect resistance. J Econ Entomol 57:283–285.

Bowman DE (1944) Fractions derived from soybeans and navy beans which retard tryptic digestion of casein. Proc Soc Exp Biol Med 57:139–140.

Brader L (1967) La faune des côtoniers sans glandes dans la partie meridional du Tchad. 1. Les altises. Coton Fibres Trop 22:171–181.

Bryant J, Green TR, Gurusaddaiah T, Ryan CA (1976) Proteinase inhibitor II from potatoes: isolation and characterization of its protomer components. Biochemistry 15:3418–3424.

Chan BG, Waiss AC, Jr, Binder RG, Elliger CA (1978) Inhibition of lepidopterous larval growth by cotton constituents. Entomol Exp Appl 24:94–100.

Chapman RF (1969) The Insects: Structure and Function. The English Universities Press, London, p 72.

Chapman RF (1974) The chemical inhibition of feeding by phytophagous insects: a review. Bull Entomol Res 64:339–363.

Cheeke PR (1971) Nutritional and physiological implications of saponins: a review. Can J Anim Sci 51:621–632.

Conn EE (1973) Cyanogenetic glycosides. In: Toxicants Occurring Naturally in Foods. Strong FM (ed), National Academy of Science, Washington, DC, pp 299–308.

Conn EE (1979) Cyanide and cyanogenic glycosides. In: Herbivores: Their Interaction with Secondary Plant Metabolites. Rosenthal GA, Jenzen DH (eds), Academic Press, New York, pp 387–412.

Dadd RH (1961) Evidence for humoral regulation of digestive secretion in the beetle, *Tenebrio molitor*. J Exp Biol **38**:259–266.

Dadd RH (1970) Arthropod nutrition. In: Chemical Zoology: Arthropoda Vol. 5, Florkin M, Scheer BT (eds), Academic Press, New York, pp 39–95.

Dahlman DL, Rosenthal GA (1975) Non-protein amino acid-insect interactions — I. Growth effects and symptomology of L-canavanine consumption of tobacco hornworm, *Manduca sexta* (L.). Comp Biochem Physiol **51A**:33–36.

Damaty S, Hudson BJF (1975) Interaction of gossypol with cottonseed protein: potentiometric studies. J Sci Food Agric **26**:1667–1672.

Davis GRF (1968) Phagostimulation and consideration of its role in artificial diets. Bull Entomol Soc Am **14**:27–30.

Delort-Laval J, Viroben G (1969) Nutrition. C R Acad Sci Paris **269**:1558–1561.

Dethier VG (1953) Chemoreception. In: Insect Physiology. Roeder KD (ed), John Wiley, New York, pp 544–576.

Dethier VG (1966) Feeding behavior. In: Insect Behavior. Haskell PT (ed), Symp R Entomol Soc Lond **3**:46–58.

Dethier VG (1970) Chemical interactions between plants and insects. In: Chemical Ecology. Sondheimer E, Simeone JB (eds), Academic Press, New York, pp 83–102.

Elliott M, Janes NF, Potter C (1978) The future of pyrethroids in insect control. Annu Rev Entomol **23**:443–469.

Feeny P (1968) Effect of oak leaf tannins on larval growth of the winter moth, *Operophtera brumata*. J Insect Physiol **14**:805–817.

Feeny P (1970) Seasonal changes in oak leaf tannins and nutrients as a cause of spring feeding by winter moth caterpillars. Ecology **51**:565–581.

Feeny P (1976) Plant apparency and chemical defense. Recent Adv Phytochem **10**:1–40.

Fisk FW, Shambaugh GF (1952) Protease activity in adult *Aedes aegypti* mosquitoes as related to feeding. Ohio J Sci **52**:80–88.

Fisk FW, Shambaugh GF (1954) Invertase activity in adult *Aedes aegypti* mosquitoes. Ohio J Sci **54**:237–239.

Flanders SE (1942) Propagation of black scale on potato sprouts. J Econ Entomol **35**:687–689.

Flanders SE (1970) Observations on host plant induced behavior of scale insects and their endoparasites. Can Entomol **102**:913–926.

Fraenkel GS (1959) The raison d'être of secondary plant substances. Science **129**:1466–1470.

Fritz H, Tschesche H, Green L, Truscheit E (eds) (1974) Proteinase Inhibitors. 5th Bayer Symposium, Springer-Verlag, Berlin.

Frost DV (1959) Methods of measuring nutritive values of proteins, protein hydrolysates and amino acid mixtures. The repletion method. In: Protein and Amino Acid Nutrition. Albanese AA (ed), Academic Press, New York, pp 225–279.

Galun R, Fraenkel G (1957) Physiological effects of carbohydrates in the nutrition of a mosquito, *Aedes aegypti* and two flies, *Sarcophaga bullata* and *Musca domestica*. J Cell Comp Physiol **50**:1–23.

George AJ (1965) Legal status and toxicity of saponins. Food Cosmet Toxicol 3:85–91.

Gestetner B, Shany S, Tencer Y, Birk Y, Bondi A (1970) Lucerne saponins. II. Purification of saponins from lucerne tops and roots and characterisation of the isolated fractions. J Sci Food Agric 21:502–507.

Goldstein JL, Swain T (1965) The inhibition of enzymes by tannins. Phytochemistry 4:185–192.

Green TR, Ryan CA (1972) Wounds-induced proteinase inhibitor in plant leaves: A possible defense mechanism against insects. Science 175:776–777.

Green TR, Ryan CA (1973) Wounds-induced proteinase inhibitor in tomato leaves — Some effects of light and temperature on the wound response. Plant Physiol 51:19–21.

Griffiths DW, Moseley G (1980) The effect of diet containing field beans of high or low polyphenolic content on the activity of digestive enzymes in the intestines of rats. J Sci Food Agric 31:255–259.

Ham WE, Sandstedt RM, Mussehl FE (1945) The proteolytic inhibiting substance in the extract from unheated soy bean meal and its effect upon growth in chicks. J Biol Chem 161:635–642.

Harborne JB (1977) Introduction to Ecological Biochemistry. Academic Press, New York, NY.

Hare JD (1969) Reversible inhibition of DNA synthesis by the argenine analogue canavanine in hamster and mouse cells in vitro. Exp Cell Res 58:170–174.

Harrison BJ, Holliday R (1967) Senescence and the fidelity of protein synthesis in Drosophila. Nature (Lond) 213:990–992.

Haslam E (1966) The Chemistry of Vegetable Tannins. Academic Press, New York.

Haslam E (1974) The Shikimic Acid Pathway. Halsted Press, New York.

Hobson RP (1931) Studies on the nutrition of blowfly larvae. 1. Structure and function of the alimentary tract. J Exp Biol 8:109–123.

Horber E (1972) Alfalfa saponins significant in resistance to insects. In: Insect and Mite Nutrition. Rodriguez JG (ed), North-Holland Publishing, Amsterdam, pp 611–627.

Hori K (1969) Effect of various activators on the salivary amylase of the bug Lygus disponsi. J Insect Physiol 15:2305–2317.

Hori K (1973a) Studies on the feeding habits of Lygus disponsi Linnavuori (Hemiptera: Miridae) and the injury to its host plant. III. Phenolic compounds, acid phosphatase and oxidative enzymes in the injured tissue of sugar beet leaf. Appl Entomol Zool 8:103–113.

Hori K (1973b) Studies on the feeding habits of Lygus disponsi Linnavuori (Hemiptera: Miridae) and the injury to its host plant. IV. Amino acids and sugars in the injured tissue of sugar beet leaf. Appl Entomol Zool 8:138–142.

Hori K (1975) Plant growth regulating factor, substances reacting with Salkovski reagent and phenoloxidase activities in vein tissue injured by Lygus disponsi Linnavuori (Hemiptera: Miridae) and surrounding mesophyll tissues of sugar beet leaf. Appl Entomol Zool 10:130–135.

Hori K, Atalay R (1980) Biochemical changes in the tissue of Chinese cabbage injured by the bug Lygus disponsi. Appl Entomol Zool 15:234–241.

House HL (1965) Digestion. In: The Physiology of Insecta, Vol 3, Rockstein M (ed), Academic Press, New York, pp 815–858.

House HL (1974) Nutrition. In: The Physiology of Insecta, 2nd edit, Vol 5, Rockstein M (ed), Academic Press, New York, pp 63–117.

Hsiao TH (1972) Chemical feeding requirements of oligophagous insects. In: Insects and Mite Nutrition. Rodriguez JG (ed), North-Holland Publishing, Amsterdam, pp 225–240.

Hsiao TH, Fraenkel G (1968) The influence of nutrient chemicals on the feeding behavior of the Colorado beetle, *Leptinotarsa decemlineata* (Coleoptera: Chrysomelidae). Ann Entomol Soc Am **61**:44–54.

Ishaaya I (1965) Significance of Soybean Saponins in Animal Nutrition and Their Effect on Digestive Enzymes. Ph.D thesis, The Hebrew University of Jerusalem, Israel, 97 pp.

Ishaaya I, Ascher KRS, Shuval G (1974) Inhibitory effect of the antifeeding compound AC-24,055 4'(-3,3-dimethyl-1-triazeno) acetanilide on digestive enzymes of *Spodoptera littoralis* larvae. Pestic Biochem Physiol **4**:19–23.

Ishaaya I, Ascher KRS, Yablonski S (1982) Differential effect of triphenyltin chloride and its cyclohexyl derivatives on growth and development of *Tribolium confusum*. Phytoparasitica **10**:205–208.

Ishaaya I, Birk Y (1965) Soybean saponins IV. The effect of proteins on the inhibitory activity of soybean saponins on certain enzymes. J. Food Sci **30**:118–120.

Ishaaya I, Birk Y, Bondi A, Tencer Y (1969) Soybean saponins IX. Studies of their effect on birds, mammals and cold blooded organisms. J Sci Food Agric **20**:433–436.

Ishaaya I, Casida JE (1975) Phenyltin compounds inhibit digestive enzymes of *Tribolium confusum* larvae. Pestic Biochem Physiol **5**:350–358.

Ishaaya I, Holmstead RL, Casida JE (1977) Triphenyl derivatives of group IV elements as inhibitors of growth and digestive enzymes of *Tribolium castaneum* larvae. Pestic Biochem Physiol **7**:573–577.

Ishaaya I, Meisner J (1973) Physiological effect of sugars on various digestive enzymes of *Spodoptera littoralis* larvae. J Comp Physiol **86**:117–124.

Ishaaya I, Moore I, Joseph D (1971) Protease and amylase activity in larvae of the Egyptian cotton worm, *Spodoptera littoralis*. J Insect Physiol **17**:945–953.

Ishaaya I, Sternlicht M (1969) Growth accelerators and inhibitors in lemon buds infested by *Aceria sheldoni* (Ewing) (Acarina: Eriophyidae). J Exp Bot **20**:796–804.

Ishaaya I, Sternlicht M (1971) Oxidative enzymes, ribonuclease, and amylase in lemon buds infested with *Aceria sheldoni* (Ewing) (Acarina: Eriophyidae). J Exp Bot **22**:146–152.

Ishaaya I, Swirski E (1970) Invertase and amylase activity in the armored scales *Chrysomphalus aonidum* and Aonidiella aurantii. J Insect Physiol **16**:1599–1606.

Ishaaya I, Swirski E (1976) Trehalase, invertase and amylase activities in the black scale, *Saissetia oleae,* and their relation to host adaptability. J Insect Physiol **22**:1025–1029.

Ishaaya I, Yablonski S, Ascher KRS, Casida JE (1980) Triphenyl and tetraphenyl derivatives of group V elements as inhibitors of growth and digestive enzymes of *Tribolium confusum* and *Tribolium castaneum* larvae. Pestic Biochem Physiol **13**:164–168.

Ito T (1960) Effect of sugars on feeding of larvae of the silkworm, *Bombyx mori*. J Insect Physiol 5:95–107.

Janzen DH, Juster HB, Liener IE (1976) Insecticidal action of the phytohemag-glutinin in black beans on a bruchid beetle. Science 192:795–796.

Johannson AS (1964) Feeding and nutrition in reproductive processes in insects. Symp R Entomol Soc Lond 2:43–55.

Jones CG, Firn RO (1979) Some allelochemicals of *Pteridium aquilinum* and their involvement in resistance to *Pieris brassicae*. Biochem System Ecol 7:187–192.

Jones WT, Mangan JL (1977) Complexes of the condensed tannins of sainfoin (*Onobrychis viccifolia* Scop) with fraction 1 leaf protein and with submaxillary mucoprotein, and their reversal by polyethylene glycol and pH. J Sci Food Agric 28:126–136.

Kehat M, Gordon D (1975) Mating, longevity, fertility and fecundity of the cotton leafworm, *Spodoptera littoralis* Boisd. (Lepidoptera: Noctuidae). Phytoparasitica 3:87–102.

Klocke JA, Chan BG (1982) Effects of cotton condensed tannin on feeding and digestion in the cotton pest, *Heliothis zea*. J Insect Physiol 28:911–915.

Kunitz M (1946) Crystalline soybean trypsin inhibitor. J Gen Physiol 29:149–154.

Kunitz M (1947) Crystalline soybean trypsin inhibitor—II. General properties. J Gen Physiol 30:291–310.

Lane HC, Schuster MF (1981) Condensed tannins of cotton leaves. Phytochemistry 20:425–427.

Lapidus JB, Cleary RW, Davidson RH, Fisk FW, Augustine MG (1963) Chemical factors influencing host selection by the Mexican bean beetle *Epilachna varivestis* Muls. J Agric Food Chem 11:462–463.

Laskowski M, Jr, Sealock RW (1971) Protein proteinase inhibitors - Molecular aspects. In: The Enzymes, 3rd edit, Vol 3, Boyer PD (ed), Academic Press, New York, pp 375–473.

Levin D (1971) Plant phenolics: an ecological perspective. Am Nat 105:157–181.

Levin D (1976) The chemical defenses of plants to pathogens and herbivores. Annu Rev Ecol Syst 7:121–159.

Liener IE (1953) Soyin, a toxic protein from the soybean. I. Inhibition of rat growth. J Nutr 49:527–539.

Liener IE, Kalade ML (1969) Protease inhibitors. In: Toxic Constituents in Plant Foodstuffs. Liener IE (ed), Academic Press, New York, pp 8–68.

Lipke H, Fraenkel GS (1954) Growth inhibitors—effect of soybean inhibitors on growth of *Tribolium confusum*. J Agric Food Chem 2:410–414.

Lipke H, Fraenkel GS (1956) Insect nutrition. Annu Rev Entomol 1:17–44.

Lowry OH, Rosebrough NJ, Farr AL, Randall RJ (1951) Protein measurement with the Folin phenol reagent. J Biol Chem 193:267–275.

Lukefahr MJ, Cowan CB Jr, Bariola LA, Houghtaling JE (1968) Cotton strains resistant to the cotton fleahopper. J Econ Entomol 61:661–664.

Lukefahr MJ, Martin DF (1966) Cotton plant pigments as a source of resistance to the bollworm and tobacco budworm. J Econ Entomol 59:176–179.

Lyman CM, Baliga BP, Slay MW (1959) Reactions of proteins with gossypol. Arch Biochem Biophys 84:486–497.

Mabry TJ, Gill JE (1979) Sesquiterpene lactones and other terpenoids. In: Herbivores: Their Interaction with Secondary Plant Metabolites. Rosenthal GA, Janzen DH (eds), Academic Press, New York, pp 501–537.

Maxwell FG, Jenkins JN, Parrott WL (1967) Influence of constituents of the cotton plant on feeding, oviposition and development of the bollweevil. J Econ Entomol **60**:1294–1297.

McKey D (1979) The distribution of secondary compounds within plants. In: Herbivores: Their Interaction with Secondary Plant Metabolites. Rosenthal GA, Janzen DH (eds), Academic Press, New York, pp 56–134.

Meisner J, Ascher KRS, Flowers HM (1972a) The feeding response of the larva of the Egyptian cotton leafworm, *Spodoptera littoralis* Boisd., to sugars and related compounds. I. Phagostimulatory and deterrent effects. Comp Biochem Physiol **42A**:899–914.

Meisner J, Ascher KRS, Zur M (1977) Phagodeterrency induced by pure gossypol and leaf extracts of a cotton strain with high gossypol content in the larva of *Spodoptera littoralis*. J Econ Entomol **70**:149–150.

Meisner, J, Flowers HM, Ascher KRS, Ishaaya I (1973) The feeding response of the larva of the Egyptian cotton leafworm, *Spodoptera littoralis* Boisd., to sugars and related compounds. III. Biochemical and enzymological aspects of sucrose consumption. Comp Biochem Physiol **44A**:793–806.

Meisner J, Ishaaya I, Ascher KRS, Zur M (1978) Gossypol inhibits protease and amylase activity of *Spodoptera littoralis* Boisduval larvae. Ann Entomol Soc Am **71**:5–8.

Meisner J, Zur M, Kabonci E, Ascher KRS (1977b) Influence of gossypol content of leaves of different cotton strains on the development of *Spodoptera littoralis* larvae. J Econ Entomol **70**:714–716.

Melville JC, Ryan CA (1972) Chemotrypsin inhibitor I from potatoes: large scale preparation and characterization of its subunit components. J Biol Chem **247**:3445–3453.

Murti VVS, Seshadri TR, Venkitasubramanian TA (1964) Neurotoxic compounds of the seeds of *Lathyrus sativus*. Phytochemistry **3**:73–78.

Neubauer I, Ishaaya I, Aharonson N, Raccah B (1980) Activity of soluble and membrane-bound trehalase in apterous and alate morphs of *Aphis citricola*. Comp Biochem Physiol **66B**:505–510.

Noelting G, Bernfeld P (1948) Sur les enzymes amylolytiques III. La β-amylase: dosage d'activité et contrôle de l'absence d'α-amylase. Helv Chim Acta **31**:286–290.

Norris DM, Ferkovich SM, Baker JE, Rozental JM, Borg TK (1971) Energy transduction in quinone inhibition of insect feeding. J Insect Physiol **17**:85–97.

Peleg BA (1965) Observation on the life cycle of the black scale, *Sassetia oleae* Bern., on citrus and olive trees in Israel. Isr J Agric Res **15**:21–26.

Pielou DP, Glaser RF (1953) Survival of *Macrocentrus ancylivorus* Roh., a parasite of the oriental fruit moth, on different concentrations of various sugar solutions. Can J Zool **31**:121–124.

Prouty WF, Karnovsky MJ, Goldberg AL (1975) Degradation of abnormal proteins in *Escherichia coli*. J Biol Chem **250**:1112–1122.

Rehr SS, Bell EA, Janzen DH, Feeny PP (1973) Insecticidal amino acids in legume seeds. Biochem Syst Ecol **1**:63–67.

Ressler C (1964) Neurotoxic amino acids of certain species of *Lathyrus* and vetch. Fed Proc 23:1350–1353.

Richardson M (1977) The proteinase inhibitors of plants and micro-organisms. Phytochemistry 16:159–169.

Robinson T (1976) D-amino acids in higher plants. Life Sci 19:1097–1102.

Rosenthal GA (1977) The biological effects and mode of action of L-canavanine, a structural analogue of L-arginine. Q Rev Biol 52:155–178.

Rosenthal GA, Bell EA (1979) Naturally occurring toxic nonprotein amino acids. In: Herbivores: Their Interaction with Secondary Plant Metabolites. Rosenthal GA, Janzen DH (eds), Academic Press, New York, pp 353–385.

Rozental JM, Norris DM (1973) Chemosensory mechanism in American cockroach olfaction and gustation. Nature (Lond) 244:370–371.

Rozental JM, Singer G, Norris DM (1975) Affinities of certain quinone repellents for detergent-solubilized proteins from *Periplaneta americana* antennae. Biochem Biophys Res Commun 65:1040–1046.

Ryan CA (1973) Proteolytic enzymes and their inhibitors in plants. Annu Rev Plant Physiol 24:173–196.

Ryan CA (1979) Proteinase inhibitors. In: Herbivores: Their Interaction with Secondary Plant Metabolites. Rosenthal GA, Janzen DH (eds), Academic Press, New York, pp 599–618.

Schoonhoven LM (1968) Chemosensory bases of host plant selection. Annu Rev Entomol 13:115-136.

Schoonhoven LM (1982) Biological aspects of antifeedants. Entomol Exp Appl 31:57-69.

Shainkin R, Birk Y (1970) α-Amylase inhibitors from wheat: isolation and characterization. Biochem Biophys Acta 221:502–513.

Shany S, Gestetner B, Birk Y, Bondi A (1970) Lucerne saponins III. Effect of lucerne saponins on larval growth and their detoxification by various sterols. J Sci Food Agric 21:508–510.

Shaver TN, Lukefahr MJ, Garcia JA (1970) Food utilization, ingestion and growth of larvae of the bollworm and tobacco budworm on diets containing gossypol. J Econ Entomol 63:1544–1546.

Shaver TN, Parrott WL (1970) Relationship of larval age to toxicity of gossypol to bollworms, tobacco bollworms, and pink bollworms. J Econ Entomol 63:1802–1804.

Shoppee CW (1964) Chemistry of the Steroids, 2nd edit. Butterworths, London.

Shorey HH, Hale LL (1965) Mass rearing of nine noctuid species on a simple artificial medium. J Econ Entomol 58:522–524.

Singh ID, Weaver JB (1972) Growth and infestation of bollweevils on normal-glanded, glandless, and high gossypol strains of cotton. J Econ Entomol 65:821–824.

Singleton VL, Kratzer FH (1969) Toxicity and related activity of phenolic substances of plant origin. J Agric Food Chem 17:497–512.

Singleton VL, Kratzer FH (1973) Plant phenolics. In: Toxicants Occurring Naturally in Foods, 2nd edit. Committee on Food Protection (eds), National Academy of Sciences, Washington, DC, pp 309–345.

Steiner M, Holtzem H (1955) In: Moderne Methoden der Pflanzenanalyse. Vol 3. Paech K, Tracey MV (eds), Springer-Verlag, Berlin, p 58.

Strangways-Dixon J (1959) Hormonal control of selective feeding in female *Calliphora erythrocephala* Meig. Nature (Lond) **184**:2040–2041.

Strumayer DH (1972) Protein amylase inhibitors in the gliadin fraction of wheat and rye flour: possible factors in celiac disease. Nutr Rep Int **5**:45–52.

Swain T (1977) Secondary compounds as protective agents. Annu Rev Plant Physiol **28**:479–501.

Swain T (1978) Plant-animal coevolution; a synoptic view of the paleozoic and mezozoic. In: Biochemical Aspects of Plant and Animal Coevolution. Harborne JB (ed), Academic Press, London, pp 3–19.

Swain T (1979) Tannins and lignins. In: Herbivores: Their Interaction with Secondary Plant Metabolites. Rosenthal GA, Janzen DH (eds), Academic Press, New York, pp 657–682.

Tanksley TD, Neumann H, Lyman CM, Pace CN, Prescott JM (1970) Inhibition of pepsinogen activation by gossypol. J Biol Chem **245**:6456–6461.

Thomsen E, Moller I (1963) Influence of neurosecretory cells and of corpus allatum on intestinal protease activity in the adult *Calliphora erythrocephala* Meig. J Exp Biol **40**:301–321.

Thorsteinson AJ (1960) Host selection in phytophagous insects. Annu Rev Entomol **5**:193–218.

Vanderzant ES, Chremos JH (1971) Dietary requirement of the boll weevil for arginine and the effect of arginine analogues on growth and on the composition of the body amino acids. Ann Entomol Soc Am **64**:480–485.

Vogel R, Trautschold I, Werle E (1968) Natural Proteinase Inhibitors. Academic Press, New York.

Walker-Simmons M, Ryan CA (1977) Wound-induced accumulation of trypsin inhibitor activities in plant leaves. Plant Physiol **59**:437–439.

Walter ED (1957) Isolation of a saponin, hederin and its sapogenin, hederagenin, from bur clover (*Medicago hispida*). J Am Pharm Assoc **46**:466–467.

Walter ED (1960) Note on saponins and their sapogenins from strawberry clover. J Am Pharm Assoc **49**:735–736.

Walter ED (1961) Isolation of oleanolic acid and saponin from trefoil (*Lotus corniculatus*, var. Viking) J Pharmacol Sci **50**:173.

Wene GP, Sheets LW (1966) Comparative susceptibility of long- and short-staple cotton varieties to bollworm injury in Arizona. J Econ Entomol **59**:1538–1539.

Wigglesworth VB (1965) The Principles of Insect Physiology, 6th edit. Methuen, London, p 438.

Wigglesworth VB (1974) The Principles of Insect Physiology. Chapman and Hall London.

Wyatt GR (1967) The biochemistry of sugars and polysaccharides in insects. Adv Insect Physiol **4**:287–360.

Yamamoto RT, Fraenkel GS (1960) Assay of the principal gustatory stimulant for the tobacco hornworm, *Protoparce sexta*, from solanaceous plants. Ann Entomol Soc Am **53**:499–503.

Chapter 8

Chemical Methods for Isolating and Identifying Phytochemicals Biologically Active in Insects

Isao Kubo[1] and Frederick J. Hanke[1]

I. Introduction

In contrast with the other contributions to this volume, we will be concerned primarily with the new chemical methods for isolating and identifying phytochemicals biologically active in insects.

Host-plant resistance to insect attack is due largely to chemical factors. In fact, it has long been speculated (Brues, 1946; Dethier, 1954) that phytophagous insects attack any available plant not containing repellent or toxic factors. The elucidation of these factors is important not only for understanding evolutionary and ecological aspects of plant–insect relationships, but also in practical pest control. For example, plant breeding programs may genetically enhance the chemical defenses of some plant species, and it is possible that synthesized defense compounds discovered from certain plants may be applied artificially to other plant species.

Although the study of natural plant products reveals an immense array of structural types, all such chemicals cannot be treated in this limited space; instead the phytoecdysteroids will be emphasized as a good example. Today it is known that insect hormones are widely distributed in the plant kingdom, and the structures of over 40 phytoecdysteroids have been elucidated. The phytoecdysteroids are polyhydroxysteroids containing a 5β-H-7-ene-6-one system commonly hydroxylated at sites 2β,

[1]Division of Entomology and Parasitology, College of Natural Resources, University of California, Berkeley, California 94720, U.S.A.

3β, 14α, and usually *20R, 22R* (Fig. 1). Other functional groups include extra hydroxyls, ethers, and lactones (Hikino, 1981).

II. Early Isolation of Phytoecdysteroids

The first isolation of ecdysteroids from plants was a rather remarkable coincidence. While investigating the chemical constituents of the leaves of *Podocarpus nakaii,* a traditional Chinese medicinal plant, Nakanishi et al. isolated three polyhydroxy steroids (Fig. 1), ponasterone A, B, and C (Nakanishi et al., 1966). The similarity of these structures to ecdysone was duly noted, and a subsequent bioassay showed strong molting hormone activity. At the same time Takemoto and co-workers were investigating the bioactive principles from the roots of *Achyranthes fauriei* and they isolated 20-hydroxyecdysone and inokosterone (Takemoto et al., 1967). Thus, the compounds shown in Figure 1 were the first ecdysteroids to be found in plants, with their discovery coming only 1 year after the

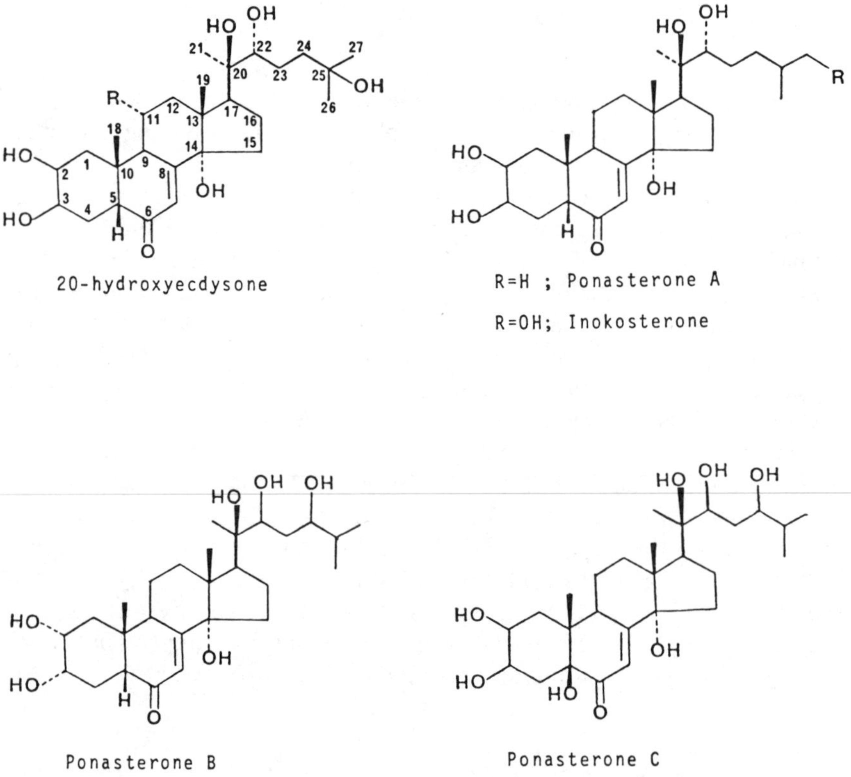

20-hydroxyecdysone

R=H ; Ponasterone A

R=OH; Inokosterone

Ponasterone B

Ponasterone C

Figure 1. Structures of the earliest isolated phytoecdysteroids.

structural elucidation of ecdysone from the silkworm *Bombyx mori* (Huber and Hoppe, 1965).

III. Bioassay

Natural product chemists have devoted many years to isolation and structure determination of plant constituents. However, the most scientific approach is to make selections and isolations based on bioassay information. Only in this way can quantitatively minor, but often very active, compounds be located.

To detect phytoecdysteroids the plant extracts are frequently incorporated into an artificial diet optimized for several economically important pest insects (Chan et al., 1978). For example, a methanolic extract from the leaves of *Ajuga remota* (Labiatae) was dissolved in solvent and added to a non-nutritive filler (α-cellulose) and evaporated to dryness. This was

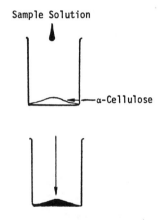

1. Solid Nutrients (Casein, Sugar, Salt, Wheat germ)
2. Buffered Vitamin Solution (B-vitamin, Vitamin C)
3. 4% Agar

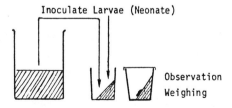

Figure 2. Artificial diet feeding bioassay for lepidopterous larvae. Samples are placed on α-cellulose which is then incorporated into an agar-based meridic artificial diet.

added to the components of a meridic artificial diet that included solid nutrients (casein, sucrose, wheat germ, Wesson salts), vitamins (C and B-complex), and 4% agar. Newly hatched larvae of the pink bollworm, *Pectinophora gossypiella,* and of the fall armyworm, *Spodoptera frugiperda,* were placed singly on portions of the diet in plastic vials (Kubo and Klocke, 1983). Additional bioassays were conducted with the silkworm, *B. mori,* by incorporating dissolved *A. remota* extracts directly into dried mulberry powder (Nihon Nohsan), evaporating the solvent to dryness, and adding a 2% agar solution (Kubo et al., 1983a) (Fig. 2).

Physiological analysis of test insects fed *A. remota* extracts revealed a developmental disruption in which the insects died in the pharate condition following initiation of molting (apolysis), but before completion of molting (ecdysis) (Figs. 3–5). The insect molting cycle is visibly initiated when the cuticular epithelium separates from the overlying cuticle in the process of apolysis. This step is followed by digestion and resorption of the old endocuticle, emergence from the old exocuticle (ecdysis), and then expansion and sclerotization of the new cuticle. The *A. remota* extract apparently inhibits ecdysis. Figure 3 exemplifies this as the newly molted *B. mori* larva died while encased by the old cuticular skin and head capsule (pharate condition). The effects of this pharate condition are to prevent feeding due to the masking of the mouth parts by the head capsule and also to prevent locomotory and excretory functions because the whole

Figure 3. A molting cycle failure of the silkworm *Bombyx mori* caused by ingestion of the crude methanol extract of *Ajuga remota* root. The insect underwent normal apolysis, but failed to complete ecdysis. Thus, it could not remove its head capsule or its trunk exuviae. ×11.

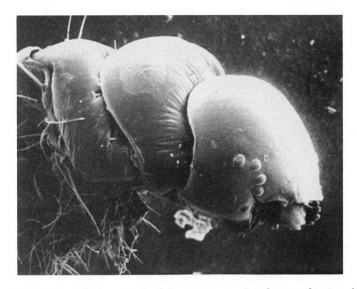

Figure 4. Electron micrograph of a fall armyworm, *Spodoptera frugiperda,* after ingestion of the crude methanol extract of *Ajuga remota* roots. This insect has three head capsules that mask its functional mouth parts. The insect eventually starved to death. ×28.5.

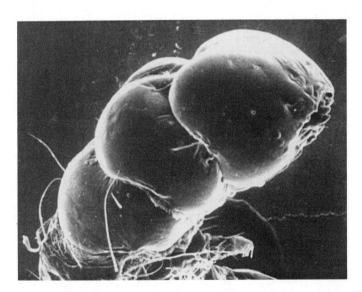

Figure 5. Electron micrograph of a pink bollworm, *Pectinophora gossypiella,* after ingestion of the crude methanol extract of *Ajuga remota* roots. This insect has three head capsules that mask its functional mouth parts. The insect eventually starved to death. ×84.75.

body is trapped by the retention of the entire cuticular skin (Kubo et al., 1981).

Figures 4 and 5 are electron micrographs of *S. frugiperda* and *P. gossypiella,* respectively. Both insects have three head capsules because they underwent two failed molting cycles before death. That is, even though feeding became impossible after the first inhibited ecdysis because the adhering second head capsule covered the mouth parts, these larvae could continue to synthesize a third head capsule.

We have utilized the molting-cycle-failure bioassay for the initial detection of phytoecdysteroid in several other plant species including *A. reptans, A. chameacistus* (Kubo et al., 1983a), *Podocarpus gracilior* (Kubo et al., 1984a), and many others. These detections were substantiated by subsequent isolation and structure elucidation of phytoecdysteroids.

Several other bioassays have been developed for the detection of these compounds (Karlson and Shaaya, 1964; Kaplanis et al., 1966; Imai et al., 1969; and others) but since these bioassays entail unnatural application of the phytoecdysteroids (i.e., injection, immersion, ligation, etc.), they can be used only for detection and not for understanding the ecological or physiological phenomena involved in plant–insect interactions. Since our molting-cycle-failure bioassay is based on ingestion of dietary phytoecdysones, we believe the effects observed in this bioassay to be comparable to those of ingestion by insects of phytoecdysteroids in the host plant (see Berenbaum, *this volume,* for discussion of the ecological relevance of bioassays with plant compounds).

IV. Isolation

Separation of phytoecdysteroids has been accomplished by a number of methods.

4.1 High-Performance Liquid Chromatography (HPLC)

HPLC fractionation followed by molting-cycle-failure bioassay resulted in the isolation of 20-hydroxyecdysone, cyasterone, and ajugasterone C as the active principles from the *A. remota* extract (Fig. 6) (Kubo et al., 1983a).

HPLC is an excellent method for the analysis of phytoecdysteroids. It is also very useful for their isolation. As it is nondestructive the natural products themselves can be isolated. However, a disadvantage of this method is that the hundreds of milligrams of phytoecdysteroids needed for bioassays are not readily isolated because of the inherently low sample capacity of laboratory size prep-scale columns (10 mm × 300 mm).

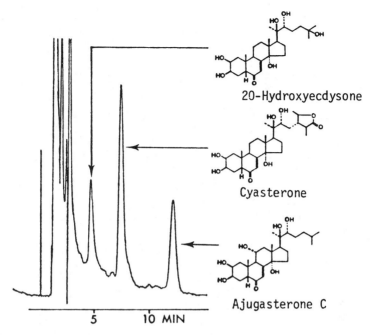

20-Hydroxyecdysone

Cyasterone

Ajugasterone C

5   10 MIN

Figure 6. C_{18} reversed-phase HPLC of *Ajuga remota* extract. Conditions: H_2O/ CH_3CN/CH_3OH (82:18:1.8 by vol); 1.5 ml/min; 254 nm.

4.2 Droplet Countercurrent Chromatography (DCCC)

DCCC, on the other hand, is an especially efficient method for the preparative separation of phytoecdysteroids with its high sample capacity of several grams, and it requires only 1–2 liters of solvent. In this way HPLC and DCCC tend to complement each other.

DCCC was developed in 1971 by Tanimura, Ito, and co-workers, and is based on the Craig countercurrent distribution (CCC) principle (Tanimura et al., 1970). Unlike CCC, DCCC has inherently higher resolution and the apparatus is far less cumbersome.

Separation in DCCC is accomplished by pumping droplets of a mobile phase through columns of a stationary liquid phase. Solutes will move through the system at various rates in accordance with their partition coefficients between the two immiscible liquid phases. Theoretically each droplet remains discrete in its travel through the system.

The apparatus consists of about 300 vertical glass columns connected in series "head to tail" by capillary Teflon tubing. Depending on the separation needs, the mobile phase can be heavier or lighter than the stationary phase. When lighter, the mobile phase is pumped in at the bottom of the first column (ascending mode droplets) and when heavier, it is delivered at the top (descending mode droplets) (Fig. 7).

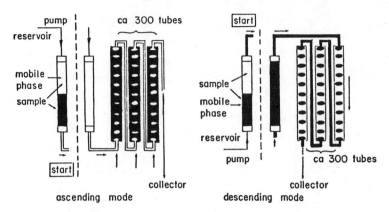

Figure 7. Diagram of ascending and descending modes of droplet countercurrent chromatography (DCCC).

A quick empirical method for selecting a solvent system and its pumping mode (ascending or descending) is to observe the thin-layer chromatography (TLC) behavior of the solute with the water-saturated organic layer. If the R_f values of the desired compounds are >0.5, the less polar (organic phase layer) is used as the mobile phase; when the R_f values are <0.5, the more polar (aqueous layer) is used as the mobile phase. However, since DCCC depends only on partition coefficients rather than adsorption, separation is not always predicted by TLC behavior.

For example, with DCCC, the bioactive *Ajuga* phytoecdysteroids were isolated rapidly and nondestructively (Fig. 8). Roughly 500 mg of the crude *A. remota* extract was separated by DCCC using a solvent system of chloroform/methanol/water (13:7:4 vol/vol) in the ascending mode with the eluent collected in 200 4-ml fractions. The fraction tubes were evaporated in vacuo and the weight of the residue was plotted versus fraction number to give the chromatogram shown. From the 500 mg of crude extract, the following pure bioactive compounds were obtained directly by DCCC: 20-hydroxyecdysone, 31 mg; ajugasterone C, 36 mg; and cyasterone, 82 mg (Kubo et al., 1983b). DCCC has now made it possible to obtain abundant quantities of phytoecdysteroids.

Traditionally these compounds have been difficult to isolate especially on a preparative scale. As shown in Figure 8 there is a baseline resolution of 20-hydroxyecdysone from ajugasterone C although both of these phytoecdysteroids have five hydroxyl groups and they give overlapping spots on TLC. The structures of these previously known compounds were determined mainly by spectroscopic data.

When the isolated phytoecdysteroids were fed to the pink bollworm *P. gossypiella* in artificial diets, the same type of ecdysis inhibition caused

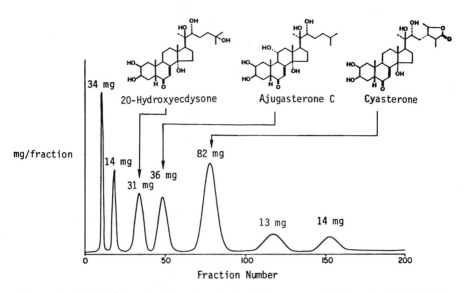

Figure 8. DCCC of the ethyl acetate extract of *Ajuga remota* (500 mg) with CHCl₃/
CH₃OH/H₂O (13:7:4, by vol) by the ascending method.

by the crude extract was observed. Thus, these two phytoecdysteroids
are responsible for the observed bioactivity.

An example of an even larger scale separation is shown in Figure 9.
The methanol extract of rootbark from *Vitex madiensis* was partitioned
with ethyl acetate to give 2.65 g of crude oil. This was in turn applied in
a single injection onto a DCCC utilizing the solvent system and mode
mentioned above for *A. remota*. This single 2.65 g injection gave 1.04 g
pure 20-hydroxyecdysone and 0.38 g pure ajugasterone C (Kubo et al.,
1984b). It should be pointed out that DCCC can also be used very effectively for nonpreparative work. The high yields of ecdysteroids from the
rootbark of *V. madiensis* prompted us to investigate the leaves and fruits
of the related *V. thyrisflora*. Preliminary TLC of the crude methanol extract
showed no ecdysteroids in detectable amounts from either the fruits or
the leaves. HPLC techniques could be used very effectively to detect
their presence; however, laborious preconcentrating steps, such as solvent–solvent partitioning followed by open column chromatography, are
usually necessary to make sure all components in an injected mixture are
compatible with the particular solvent system and column and also to
increase the percentage of ecdysteroids. Because of the inherently low
sample capacity of HPLC a low percentage of ecdysteroids in a small
injected sample could be below the detectable amount. In a high sample
capacity system as DCCC, however, a large enough sample can be applied

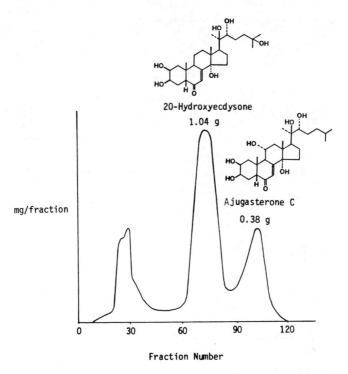

Figure 9. DCCC of the ethyl acetate extract of *Vitex madiensis* (2.65 g) with CHC1₃/CH₃OH/H₂O (13:7:4, by vol) by the ascending method; 1.5 ml/fraction.

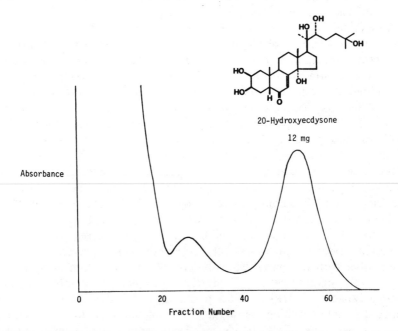

Figure 10. DCCC of the methanol extract of the fruit of *Vitex thyrsiflora* (2.05 g) with CHC1₃/CH₃OH/H₂O (13:7:4, by vol) by the ascending method. UV detection at 254 nm.

so that even a low percentage of ecdysteroids are present in a large amount and, due to the low solvent use, also in concentrations detectable by a UV monitor. A 2.05-g portion of the crude methanol extract of the fruits of *V. thyrsiflora* was applied in a single injection onto a DCCC utilizing the solvent system and mode mentioned above for *A. remota*. Fractions were monitored at 254 nm to give the chromatogram seen in Figure 10. In this way 12 mg (0.58%) of 20-hydroxyecdysone was isolated from 2.05 g of the crude methanol extract without any preliminary purification (Kubo et al., 1984c). An even more useful example is seen in Figure 11 where we applied the analytical use of DCCC to an isolation of ecdysone and 20-hydroxyecdysone from the pupae of the silkworm *B. mori*. It is typically much more difficult to work with animal extracts than plant extracts. The results of injecting the ethyl acetate soluble portion (0.65 g) of a methanol extract of 405 (654 g) male pupae is shown in Figure 11 as monitored at 254 nm. 20-Hydroxyecdysone and ecdysone were found at only 0.12% (0.8 mg) and 0.22% (1.4 mg), respectively (Kubo et al., 1984d). The efficiency of this separation is most clearly seen when compared to the original isolation of 20-hydroxyecdysone and ecdysone (0.33 mg and 25 mg, respectively) from an extract of 500,000 g of pupae (Butenandt and Karlson, 1954; Karlson, 1956).

V. Identification

The procedures for the identification of compounds fall into two types.

5.1 Comparison to Authentic Compounds

One method is that of direct comparison to an authentic sample of compound. This could utilize chemical, physical, or spectral techniques but would most often consist of a comparison of chromatographic properties.

A. Gas Chromatography (GC)

GC is an excellent technique for identification of stable and volatile compounds and has been used extensively. Recent advances in this technique are in improved capillary column design and electronics to make instruments more durable, easier to operate, and give better resolution and sensitivity. However, GC is unsuitable for large, nonvolatile, and heat-sensitive compounds such as the phytoecdysteroids.

B. HPLC

Of far more importance is HPLC. Improvements to HPLC are similar to GC improvements. New electronics make operations easier, especially

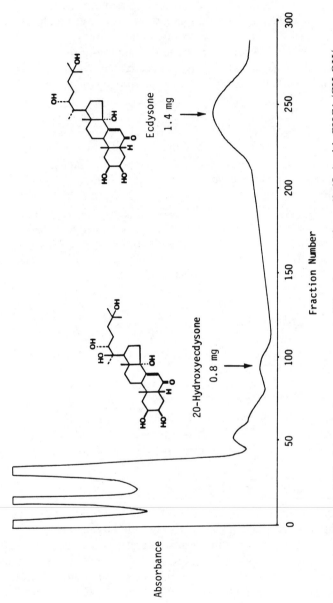

Figure 11. DCCC of the ethyl acetate extract of *Bombyx mori* larva (0.65 g) with $CHC1_3/CH_3OH/$ H_2O (13:7:4, v/v) by the ascending method. 1.0 ml/fraction. UV detection at 254 nm.

for gradient elution. Micro columns are available with 3 μm packing and use less solvent with faster separations. Of primary importance to compound identification is detection. New precolumn derivitization with 1-anthroyl nitrile has enabled us recently to lower the detection limit in a mixture of phytoecdysteroids to below 50 pg per ecdysteroid (Kubo and Komatsu, 1986). Considering the silkworm *B. mori* can contain over 600 ng per whole animal, this technique can be used to investigate ecdysteroid concentrations on a single insect's individual organs and body parts. The ability to monitor differences in concentrations within an individual will greatly simplify studies as levels of ecdysteroids often will vary among individual specimens in a population depending on an individual's particular stage of development.

5.2 Spectral Characterization

The other method of compound identification is structural characterization. Spectral techniques have been used as a primary source of information for 15–20 years. As the instruments involve complex electronics it is not surprising that large advances have been made within the last few years. While some improvements have been made to ease operation and increase reliability, others have spawned a new generation of spectroscopy.

A. Nuclear Magnetic Resonance Spectroscopy

The most noticeable advances in spectroscopy have occurred in Fourier Transform Nuclear Magnetic Resonance (FT NMR) spectroscopy. Instruments have become increasingly easier to operate because of availability of new software and increased stability of fields. There is also a trend of decreasing cost as a result of lower power and helium consumption and initial purchase price. No doubt these effects will expand availability to novel uses such as *in vivo* studies. Increased computer capabilities have increased research into the new areas of Two-Dimensional NMR (2D-NMR) and alternative one-dimensional pulse sequences.

Two-dimensional FT NMR was first proposed by Jeener in 1971 but it was not until 1975 that it was applied (Freeman and Morris, 1979). Until very recently two-dimensional techniques remained experimental and available to only a few select groups. Today many new spectrophotometers are commercially available with enough computer capability and software to perform most of the hetero- and homonuclear two-dimensional experiments described in the literature (Benn and Günther, 1983). These programs have simplified operations so most two-dimensional work can be considered routine on these instruments. Reading how these programs work can be confusing even to people with a reasonable background in NMR although there is at present an outstanding book on the subject (Bax, 1982).

To interpret spectra is more straightforward and we shall concentrate on the latter. Two-dimensional experiments can complement one-dimensional experiments such as heteronuclear decoupling experiments but also can create new spectra not normally possible such as a homonuclear decoupled ^1H spectrum. It should be pointed out that several one-dimensional NMR experiments can be displayed as two-dimensional as in a sequence of spin relaxation experiments. However to be a true two-dimensional experiment there must be a second frequency dimension. Two time variables must be monitored. These variables form a two-dimensional matrix of signals. Just as a Fourier transform turns a single time domain signal into a one-dimensional spectrum with one frequency axis, the Fourier transform on the matrix forms two frequency axes in addition to the normal intensity axis. What time domain information is collected depends on the type of experiment to be run. Many are possible.

A general type is J-resolved spectroscopy. These can be hetero- or homonuclear. In J-resolved 2D-NMR one axis is a decoupled spectrum containing only chemical shift data. The other contains only magnitudes of couplings without any chemical shift information. Thus, the wings of the coupled peaks, rather than overlapping to the right and left, go in back and in front of the chemical shift and in this way are more clearly visible. In heteronuclear J-resolved 2D-NMR, one axis might give a decoupled carbon spectrum and the other give proton–carbon couplings. In homonuclear J-resolved two-dimensional spectra, one axis might be a decoupled proton spectrum (a very nice trick!) and the other give proton–proton couplings. Spectra are often shown as viewed from above and show peaks as circles. Concentric circles give information of peak height similar to what one would find in a topographical map of a mountain. The J-resolved spectrum gives couplings and chemical shifts. Another type of two-dimensional spectrum called homonuclear shift-correlated (or COSY) spectrum gives chemical shifts, couplings, and also correlates the origins of the couplings. A coupled spectrum is located on each axis and a coupled spectrum runs down a diagonal with couplings shown to each side. A line perpendicular to an axis (45° angle to diagonal) is drawn from a particular shift. Off-diagonal multiplets along this line show couplings, and, when a new line is drawn perpendicular to the first and through the intersected off diagonal multiplets, it will intersect a second chemical shift on the other axis which is coupled to the original chemical shift. A peak with multiple couplings will have several off-diagonal multiplets along the first line drawn and this procedure can be repeated to correlate all couplings.

Below we give a few examples of the application of these NMR techniques to phytoecdysteroid research. An example of J-resolved 2D-NMR is seen in Figure 12 (Kubo et al., 1984c). Here we see the ^1H homonuclear COSY spectrum of 20-hydroxyecdysone at 300 MHz in methanol-d_4. This one spectrum contains the same information as do many one-dimensional selective spin-decoupled spectra and more. Lines have been drawn to

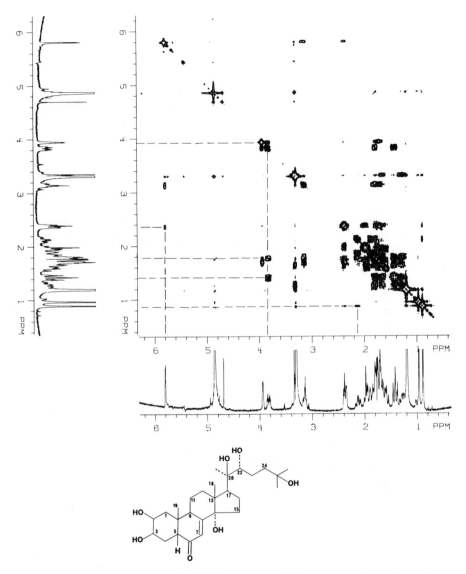

Figure 12. The ^{1}H homonuclear COSY spectrum of 20-hydroxyecdysone in methanol-d$_4$ at 300 MHz.

show the couplings of H-2 (3.83 δ) to H-3 (3.94 δ), H-1 ax (1.42 δ), and H-1 eq even though H-1 eq is hidden in a complex multiplet at 1.75 δ. It is not uncommon to observe long-range couplings not generally seen in one-dimensional spectra. An example is seen here in Figure 13 where a very small and unexpected coupling through the carbonyl carbon can clearly be seen for the vinyl H-7 proton (5.80 δ) and H-5 (2.37 δ). This

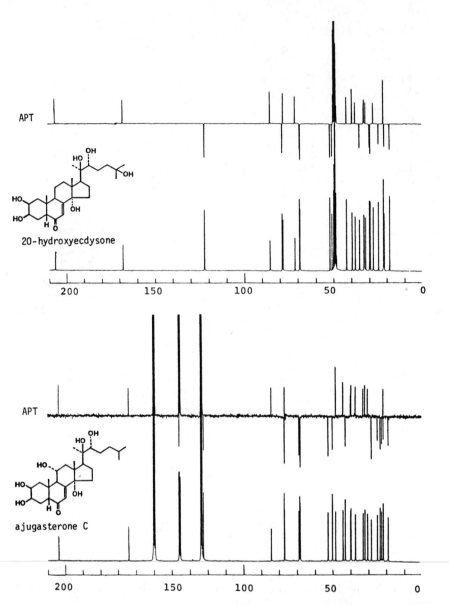

Figure 13. The Attached Proton Test (APT) and completely decoupled ^{13}C NMR spectrum of 20-hydroxyecdysone and ajugasterone C.

coupling is not obvious in one-dimensional spectra at 300 MHz without careful and time-consuming decoupling experiments (Kubo et al., 1977). Another example of long-range couplings can be seen by the coupling between the methyl at 0.88 δ and H-12 ax (2.13 δ). This coupling identifies the 0.88 δ methyl as C-18 and the singlet at 0.96 δ as C-19, though once again no coupling is seen in the one-dimensional spectra at 300 MHz. There has been some very recent work on better utilization of such long-range couplings through modifications of existing two-dimensional pulse sequences. Applications have been made to long-range ^{1}H–^{1}H couplings (Batta and Liptak, 1984) and even long-range ^{13}C–^{1}H couplings (Seto et al., 1984). Also, more information would be clear at the higher available fields.

It is also possible to perform heteronuclear correlated two-dimensional spectroscopy. This is most commonly done with ^{13}C–^{1}H but has also been done with ^{15}N–^{1}H. In ^{13}C–^{1}H spectra one axis represents a decoupled ^{13}C spectrum while the other a coupled ^{1}H spectra. With the ^{13}C as the horizontal axis, ^{1}H multiplets line up in columns above the carbon with which they correlate. A line perpendicular to the carbon axis and through a ^{13}C chemical shift will intersect a line drawn perpendicular to the proton axis and through a ^{1}H chemical shift at these multiplets and thereby correlate carbons and hydrogens. When performed in conjunction with two-dimensional homonuclear COSY spectra, information on the connectivity of hydrogen can be extrapolated to give carbon connectivity data.

Carbon connectivity data can also be acquired from a single spectrum in what is called the INADEQUATE experiment. Technically a two-dimensional display of a one-dimensional pulse sequence, it fits very nicely with our previous discussion. The net result is a single spectrum with the horizontal axis being a decoupled ^{13}C spectrum and the vertical axis representing the single quantum residual. While this has no simple meaning, the spectra evaluation is straightforward. A spectrum consists of a series of rows of doublet pairs representing ^{13}C–^{13}C satellites. Lines perpendicular to the decoupled axis and through a chemical shift will intersect a doublet in a row. The other doublet in that row is used to correlate two carbons next to one another in the compound. If a carbon has 2 neighbors, the perpendicular line will intersect two doublets in two separate rows.

2D-NMR will not replace one-dimensional NMR for a number of reasons. Not all instruments can perform two-dimensional spectra, as it takes a large computer to handle data accumulation and reduction. It takes much longer to perform the various individual two-dimensional experiments than single one-dimensional experiments and instrument time is not always available. Also, in many cases information is very clear from one-dimensional spectra and it would be a waste of time and resources to perform two-dimensional experiments in all instances. In addition, several one-dimensional experiments have been developed that can give needed information much faster and with less computer capacity than is necessary with two-dimensional experiments. Sometimes these can be used to enhance the sensitivity of insensitive nuclei but for the most part are used to distinguish the multiplicities of carbon since peak overlap often complicates information from normal off-resonance ^{13}C spectra. The following experiments get around this quite cleverly.

There are three predominantly popular one-dimensional multipulse sequences of the many reported variations in recent literature. They go by the acronyms of INEPT (Insensitive Nuclei Enhanced by Polarization Transfer), DEPT (Distortionless Enhancement by Polarization Transfer), and APT (Attached Proton Test). All are of the spin-echo type of pulse sequence with the first two utilizing special pulse sequences for the ^1H nuclei to cause a transfer of magnetization from the much more abundant ^1H nuclei to the less abundant ^{13}C nuclei and thus enhancing the normally weak ^{13}C signal. A more exact discussion of the pulse sequences will be omitted in favor of a discussion of the interpretation of spectra.

There are two main types of INEPT spectra (Doddrell and Pegg, 1980). By making a small change in delay times, one type gives a decoupled ^{13}C spectrum where only CH-type carbons are enhanced and others are nulled. In the other type, CH and CH_3 carbons are enhanced, CH_2 carbons are inverted and enhanced, and C-type quaternary carbons are nulled. By comparing these two INEPT spectra to a decoupled spectrum all types of carbon can be assigned multiplicities. Even peaks that overlap completely can sometimes be simplified by having both a peak up and down.

DEPT spectra are similar to INEPT spectra. However there is a minor change in the pulse sequences (Pegg et al., 1982). Because of this change it is claimed this sequence is relatively distortionless in comparison to INEPT spectra and, as such, spectra can be edited successfully. Initially three subspectra are taken and these are subtracted to plot three new spectra, one with only CH-type carbon peaks, one with only CH_2-type carbon peaks, and one with only CH_3-type carbon peaks. All are enhanced by polarization transfer and appear in a normal phase. Comparison to a normal decoupled spectra will identify C-type quaternary carbons.

An APT spectrum is a single spectrum consisting of normal phase CH_3-type and CH-type carbons and inverted C-type and CH_2-type carbons (Patt and Schoolery, 1982). Slow relaxing singlets show up in normal intensity, and although there is no enhancement by polarization transfer, it is claimed this enhancement is only moderately greater than the NOE enhancement retained in this experiment. This program might be used in conjunction with a rapid pulsing of decoupled spectra to null the slow relaxing C-type quaternary carbons and distinguish C-type from CH_2-type carbons.

An example of APT type spectra can be seen in Figure 13 (Kubo et al., 1984b) which includes APT and decoupled ^{13}C spectra for the two similar ecdysteroids, 20-hydroxyecdysone and ajugasterone-C. (It should be pointed out those APT spectra have been arbitrarily phased so that multiplet identification is the direct opposite of the norm previously mentioned. C- and CH_2-type carbons are shown in normal phase while CH- and CH_3-type are inverted.) The only difference between these two structures is a transposition of a single hydroxy group between C-25 and C-11. These compounds can best be distinguished by examining of the hydroxy-containing carbons between 60 and 90 ppm. The spectra of 20-hydroxyecdysone has three normal phase and three inverted peaks corresponding to its three tertiary and secondary hydroxy groups, respectively. Ajugasterone-C has two normal phase and four inverted peaks corresponding to its two tertiary and four secondary hydroxy groups, respectively. In this example two APT spectra could distinguish these two compounds and could be acquired much faster than conventional one-dimensional spectra or two-dimensional spectra.

There are potential problems with all three of these one-dimensional spin-echo techniques. When dealing with compounds containing groups with abnormal coupling constants ambiguous results can occur although it has been shown DEPT spectra have a lower dependence on coupling constants than INEPT, as does APT. It is difficult to choose the best of the above J-modulated spin-echo experiments. Some are faster, others have more information, whereas others may be run only after instrumentation modification. Also the choices of experiments change rapidly in this young field of research as is evident by the recent introduction by Wesener et al. (1984) of the TANDEM-SEFT sequence. In this report it is claimed this approach provides the more general approach with less sophisticated instrumentation needed than with INEPT or DEPT and allowing for the selective detection of C-type quarternary carbons. With so little information now available on this new technique we decline to comment.

B. Mass Spectroscopy (MS)

MS has been a very useful tool in structure analysis of natural products for decades. Its great sensitivity and its coupling to GC have been of particular use to the analysis of complex mixtures commonly found in phytochemical studies. The development of Tandem MS (Maquestiau and Flammang, 1983), the coupling of two mass spectrometers, has allowed a better understanding of mass spectral fragments. Analysis is now possible on positive and negative ions. However, the recent advances of most interest to studies on biologically active chemicals are most likely the soft ionization techniques used for nonvolatile or unstable compounds.

Often the very chemical groups of a molecule that give it is biological activity, such as hydroxy or aldehyde groups, also make it thermally unstable due to the inherent reactivity of these groups. For example, in ec-

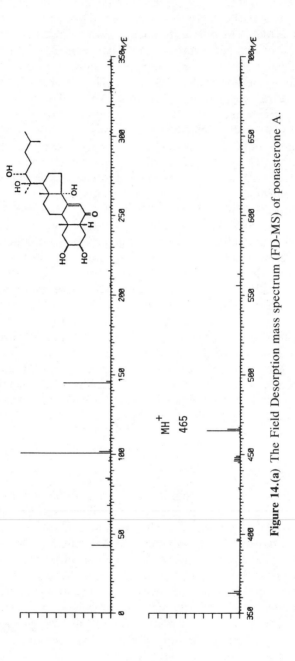

Figure 14.(a) The Field Desorption mass spectrum (FD-MS) of ponasterone A.

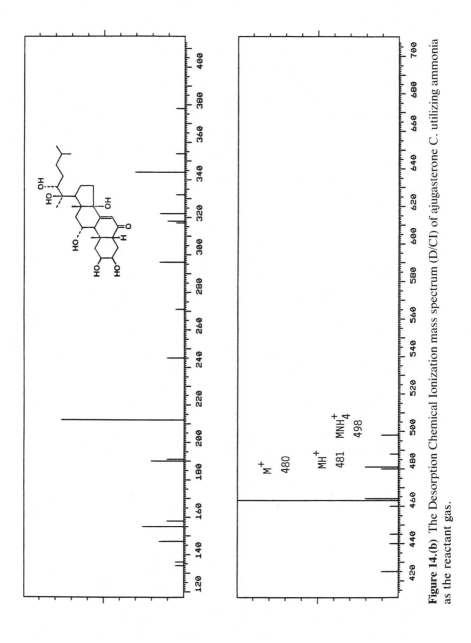

Figure 14.(b) The Desorption Chemical Ionization mass spectrum (D/CI) of ajugasterone C. utilizing ammonia as the reactant gas.

dysteroids conventional electron impact (EI) MS commonly provides useful fragment information but no parent ion (M^+) due to the ease of dehydration of alcohol groups. Though fragment information is very useful, often the M^+ ion is of greatest interest, as it can be used to solve for the molecular formula. Several techniques have been used to ionize compounds "softly" and without transferring extra energy to cause complete fragmentation. These have been very successful. A complete discussion here is not necessary as there are several reviews in the literature (Daves, 1979). Varied techniques are now in common use although availability of any one technique would depend on the individual facility. Some examples of soft ionization mass spectra of ecdysteroids are shown in Figures 14a and b.

The first example shows a Field Desorption mass spectrum (FD-MS) of the phytoecdysteroid ponasterone A. In this technique the sample is placed onto a surface and subjected to an increasing electrostatic field. Immediately upon ionization the ion will leave this area and proceed to the detection area of the instrument. Since only enough energy can be transferred to cause ionization, little energy is available for fragmentation. The lack of fragmentation can clearly be seen. As with most soft ionization techniques, the parent ion at m/z 464 is identified from the protonated species (MH^+) at m/z 465.

A more advanced technique was used for the spectrum of the phytoecdysteroids ajugasterone-C called Desorption Chemical Ionization (D/CI) (Arpino and Devant, 1979). Little fragmentation is present and the parent ion of m/z 480 is identified from the MH^+ at m/z 481 and the ammonium complex MNH_4^+ at m/z 498 characteristic when ammonia is used as the reactant gas. Though the information is similar in both these examples, D/CI-MS is generally capable of more structural information than FD-MS. For example, it has been used effectively to sequence the glycosides of saponins (Hostettmann et al., 1981; Kubo and Hanke, 1986).

VI. Conclusions

The preceding text was designed to introduce new chemical methods for isolating and identifying phytochemicals biologically active in insects. Although a degree in chemistry is not a prerequisite for utilizing these techniques a more complete understanding of the techniques than is presented here is probably necessary.

A large amount of work with biologically active phytochemicals can be done by either a chemist or a biologist. The beginning steps for the isolation of compounds, such as solvent partitioning, TLC, and open or flash column chromatography, are best learned by experience. The classically trained chemist will not have a large advantage over a person with a biological background who understands the basics of chromatography. HPLC systems closely follow the principles of open-column systems. Problems with HPLC systems tend to be mechanical in nature, such as leaks, or are involved with columns damaged because the compatibility of the column's support and either the sample or the solvent was not checked properly.

The identification of compounds by spectroscopic means tends to be more technical than their isolation. The acquisition and interpretation of IR and UV data is rather straightforward. However, the operation of NMR instrumentation is much more complex and should be left to only those well trained. The interpretation of poorly run NMR data can be difficult or impossible. Although the interpretation of NMR spectra can usually be understood much more easily than how it is acquired there are many subtleties, such as deceptively simple ^{1}H–^{1}H NMR coupling patterns or transient peaks appearing in a ^{1}H–^{13}C COSY 2D-NMR experiment, that a novice might not be aware of. Mass spectral instrumentation is typically very sensitive and it is rather simple to cause several thousand dollars in damages. It is best to have as few operators as possible. The interpretation of mass spectra, such as fragmentation mechanisms, is easily understood. Problems with MS interpretation are often due to an overinterpretation. It should be considered a rule that a great deal of caution be exhibited when interpreting any type of spectra. Spectroscopy is an area of chemistry particularly susceptible to the overextending of skills.

References

Arpino PJ, Devant G (1979) Analyse par spectrométrie de masse en désorption/ionisation chimique (D/IC) de substances organiques peu volatiles. Analysis 7:348–354.

Batta G, Liptak A (1984) Long-range ^{1}H–^{1}H spin-spin couplings through the interglycosidic oxygen and the primary structure of oligosaccharides as studied by 2D-NMR. J Am Chem Soc **106**:248–250.

Bax A (1982) Two-Dimensional NMR in Liquids. Reidel Publishing, Amsterdam.

Benn R, Günther H (1983) Modern pulse methods in high-resolution NMR spectroscopy. Angew Chem Int (Engl) **22**:350–380.

Brues CT (1946) Insect Dietary. Harvard University Press, Cambridge.

Butenandt A, Karlson P (1954) Uber die Isolierang eines Metamorphosehormons der Insekten in kristallisierter Form. Z Naturforsch B **96**:389–391.

Chan BG, Waiss AC Jr, Stanley WL, Goodban AE (1978) A rapid diet preparation method for antibiotic phytochemical bioassay. J Econ Entomol **71**:366–368.

Daves GD Jr (1979) Mass spectrometry of involatile and thermally unstable molecules. Acc Chem Res **12**:359–365.

Dethier VG (1954) Evolution of feeding preferences in phytophagous insects. Evolution **8**:33–54.

Doddrell DM, Pegg DT (1980) Assignment of proton-decoupled carbon-13 spectra of complex molecules by using polarization transfer spectroscopy. A superior method to off-resonance decoupling. J Am Chem Soc **102**:6388–6390.

Freeman R, Morris GA (1979) Two-dimensional Fourier transformation in NMR. Bull Magnet Reson 1:5–26.

Hikino H (1981) Isolation and bioassay of phytoecdysteroids. In: Advances in Natural Products Chemistry. Natori S, Ikekawa N, Suzuki M (eds), Kodansha, Tokyo, pp 458–471.

Hostettmann K, Doumas J, Hardy M (1981) Desorption/chemical ionization mass spectrometry of naturally occurring glycosides. Helv Chim Acta 64:297–303.

Huber R, Hoppe W (1965) Zur chemie des Ecdysones VII. Die Kristall und Molekulstrukturanalyse des Insectenverpuppungs-hormons Ecdyson mit der Automatisierten Faltmolekulmethode. Chem Ber 98:2403–2424.

Imai S, Toyosato T, Sakai M, Sato Y, Fujioka S, Murata E, Goto M (1969) Screening results of plants for phytoecdysones. Chem Pharm Bull (Tokyo) 17:335–339.

Kaplanis JN, Tobor CA, Thompson MJ, Robbins WE, Shortino TJ (1966) Assay for ecdysone (molting hormone) activity using the housefly, *Musca domestica*. Steroids 8:625–631.

Karlson P (1956) Biochemical studies on insect hormones. Vitam Hormones 14:227–266.

Karlson P, Shaaya E (1964) Ecdysone titer during insect development. A method for determining ecdysone content. J Insect Physiol 10:797–804.

Kubo I, Hanke FJ (1986) Submitted.

Kubo I, Klocke JA (1983) Isolation of phytoecdysones as insect ecdysis inhibitors and feeding deterrents. In: ACS Symposium Series 208. Plant Resistance to Insects. Hedin P. (ed), American Chemical Society, Washington, DC, pp 208–229.

Kubo I, Klocke JA, Asano S (1981) Insect ecdysis inhibitors from the East African medicinal plant *Ajuga remota* (Labiatae). Agric Biol Chem 45:1925–1927.

Kubo I, Klocke JA, Asano S (1983a) Effects of ingested phytoecdysteroids on the growth and development of two lepidopterous larvae. J Insect Physiol 29:307–316.

Kubo I, Klocke JA, Ganjian I, Ichikawa N, Matsumoto T (1983b) Efficient isolation of phytoecdysones from *Ajuga* plants by high-performance liquid chromatography and droplet counter-current chromatography. J Chromatogr 257:157–161.

Kubo I, Klocke JA, Matsumoto T (1984a) Multichemical resistance of the conifer *Podocarpus gracilior* (Podocarpaceae) to insect attack. J Chem Ecol 10:547–559.

Kubo I, Komatsu S (1986) Micro analysis of prostaglandins and ecdysteroids in insects by high-performance liquid chromatography and fluorescence labeling. J. Chromatogr, in press.

Kubo I, Matsumoto A, Ayafor JF (1984b) Efficient isolation of a large amount of 20-hydroxyecdysone from *Vitex madiensis* (Verbenaceae) by droplet counter-current chromatography. Agric Biol Chem 48:1683–1684.

Kubo I, Matsumoto A, Hanke FJ (1984c) The ^{1}H-NMR assignment of 20-hydroxyecdysone. Agric Biol Chem 49:243–244.

Kubo I, Matsumoto A, Asano S (1984d) Efficient isolation of ecdysteroids from the silkworm, *Bombyx mori* by droplet counter-current chromatography (DCCC). Insect Biochem 15:45–47.

Kubo I, Nakanishi K, Kamikawa T, Isobe T (1977) The structure of inflexin. Chem Lett 99–102.

Maquestiau A, Flammang R (1983) Tandem mass spectroscopy of natural products. In: Tandem Mass Spectrometry. McLafferty FW (ed), John Wiley & Sons, New York.

Nakanishi K, Koreeda M, Sasaki S, Chang ML, Hsu HY (1966) Insect hormones. The structure of ponasterone A, an insect moulting hormone, from the leaves of *Podocarpus nakaii* Hay. J. Chem. Soc., Chem Commun 915–917.

Patt SL, Schoolery JN (1982) Attached proton test for carbon-13 NMR. J Magnet Reson 46:535–539.

Pegg DT, Doddrell DM, Bendall MR (1982) Proton-polarization transfer enhancement of a heteronuclear spin multiplet with preservation of phase coherency and relative component intensities. J Chem Phys 77:2745–2752.

Seto H, Furihata K, Otake N, Itoh Y, Takahashi S, Haneishi T, Ochuchi M (1984) Application of long range J C-H resolved 2D spectroscopy (LRJR) in structural elucidation of natural products. The structure of oxiraperityn. Tetrahed Lett 337–340.

Takemoto T, Ogawa S, Nishimoto Y, Hirayama H, Taniguchi T (1967) Isolation of the insect-molting hormones from mulberry leaves. Yakugaku Zasshi 87:748.

Tanimura T, Pisano JJ, Ito Y, Bowman RL (1970) Droplet Countercurrent Chromatography. Science 169:54.

Wesener JR, Schmitt P, Günther H (1984) Spin-echo ^{13}C NMR spectroscopy for the analysis of deuterated carbon compounds. J Am Chem Soc 106:10–13.

Chapter 9

Techniques for Evaluating Plant Resistance to Insects

Ward M. Tingey[1]

I. Introduction

The development and deployment of crop varieties defended or tolerant against insect attack is a major tactic in pest management. In agricultural terms, plant resistance to insects is a property that enables a plant to avoid, tolerate, or recover from the injurious effects of insect feeding and oviposition. Plants vary considerably in their mechanisms for defense against insects. Defensive strategies range from disruption of insect behavior and development to repair or replacement of organs and tissues damaged through insect attack. The purpose of this chapter is to present research methods and considerations appropriate for: (1) evaluation and development of insect-resistant cultivars, and (2) analysis of plant traits conferring resistance. Additional information of this type is provided by Chesnokov (1962), Dahms (1972), Smith (1978), Tingey and Pillemer (1977), and Maxwell and Jennings (1980).

II. Sources of Germplasm

Over the past several decades, concern about erosion of genetic resources has prompted increased efforts in exploration, collection, preservation, and distribution of wild and semicultivated crop germplasm. As a result, a number of excellent and expanding germplasm collections have been

[1]Department of Entomology, Cornell University, Ithaca, New York 14853, U.S.A.

established in the United States and abroad (Table 1). These collections are valuable sources of diverse plant material for those interested in the study and development of plant resistance to insects. In addition to the germplasm resources listed in Table 1, many collections are maintained by research scientists at public institutions and by the private sector of the plant breeding, horticulture, and seed industries (see Creech and Reitz, 1971; Harlan and Starks, 1980).

III. Sources of Insects

The principal sources of insects for studies of plant resistance are field populations and laboratory (or glasshouse) colonies. Each source has its own particular advantages and disadvantages for consideration. In practice, insects from either source can be used effectively if their inherent or potential limitations are recognized.

3.1. Field Populations

The principal advantages for collection of test insects from field populations are convenience and low cost. Large populations can be gathered in a relatively short period of time using infested plant samples or simple tools such as sweep nets or motorized suction devices ranging from the portable D-vac® sampler (Dietrick, 1961) to tractor-powered suction equipment (Stern, 1969). Many sedentary Homoptera such as aphids and scale insects, leaf-mining and plant-tunnelling species of various taxa (Diptera, Lepidoptera, Coleoptera), and spidermites can be collected in large numbers using naturally infested plant samples. Using motorized vacuum equipment, Stern et al. (1965) collected millions of plant bugs (*Lygus* spp.) from alfalfa for use in dispersal studies. A frequently overlooked advantage of field collections is that they represent a gene pool characteristic of the pest in its agricultural environment.

The disadvantages associated with use of field-collected populations are varied and their particular importance to the success of a research program is best determined by each research worker on a case-by-case basis. A major limitation is the seasonality and unpredictability of natural field populations. Mass field collection may be inappropriate or impractical if the target pest is not available in sufficient quantities at the proper time. Other problems largely outside the control of the researcher include unwanted infection of the target species by pathogenic microorganisms, plant pathogens (in the case of vector species), and parasitoids. Finally, mass field collection is usually burdened by accompanying nontarget arthropods, plant debris, and sometimes soil. The sorting and separation of the desired species from extraneous material may be laborious and introduce significant mortality.

Table 1. Major U.S. and foreign collections of plant germplasm

Location	Principal holdings
A. National Collections (United States, USDA)	
1. National Seed Storage Laboratory, Ft, Collins, CO 80523	Major reserve holdings of wheat, oats, barley, sorghum, soybeans, flax, tobacco, cotton, safflower, and sesame. Reserve collections of many other crop species.
2. National Small Grains Collection, Beltsville, MD 20705	Wheat, barley, oats, rye
3. Regional Plant Introduction Stations	
Northeastern Region Geneva, NY 14456	*Vegetables:* broccoli, cauliflower, celery, lettuce, pea (green), squash, tomato *Field and forage:* birdsfoot trefoil, clovers (perennial), tall oatgrass, timothy
Southern Region Experiment, GA 30212	*Vegetables:* bean (castor, mung, urd, velvet), collards, eggplant, gourd, kale, melon (musk, water), okra, pea (chick, cow, mung, velvet), peanut, pepper (sweet, tabasco), shallot, turnip *Field, forage, oilseed:* alfalfa (annual), clover (annual), grasses (bermuda, grama, bluestem, Indian, rhodes, dallis, carpet, pampa, sudan), pearl millet, sesame, sorghum, teosinte, vetch)
North Central Region Ames, Iowa 50010	*Vegetables:* beets (garden, sugar), carrot, cucumber, dill, rhubarb, parsley, pumpkin, spinach, tomato *Field, forage, grain, oilseed:* alfalfa (perennial), clover (sweet), grass (brome, canary), lespedeza, maize, millet (foxtail, proso), mustard (white, black, yellow), rape, redtop, sunflower
Western Region Pullman, WA 99163	*Vegetables:* bean (broad, common, lima, scarlet, runner), cabbage (common, Chinese), garlic, lentil, lettuce *Field, forage, oilseed:* Kentucky bluegrass, orchardgrass, reed canarygrass, ryegrass, safflower, wheatgrass (crested, western)
4. Potato Introduction Station, Sturgeon Bay, WI 54235	Extensive working collection of wild and cultivated potato germplasm.

Table 1. (*Continued*)

Location	Principal holdings
5. Regional Soybean Laboratory, University of Illinois, Urbana, IL 61801	Extensive working collection of soybean germplasm
B. Foreign National Collections	
1. Australia	
Australian Wheat Collection, Tamworth N.S.W., Australia	Wheat
Waite Agricultural Research Institute Adelaide, Australia	Barley
2. Colombia	
Instituto Colombiana Agropecurio Medellin, Colombia	Maize, beans
3. Czechoslovakia	
Research Institute of Cereals Havlivkovo, Czechoslovakia	Barley, oats, rye
4. England	
Plant Breeding Institute Cambridge, England	Wheat and small grains
5. Guyana	
Central Agricultural Station Non Repos, Guyana	Rice
6. Italy	
Laboratorio del Germoplasmo Bari, Italy	Wheat and small grains
7. Japan	
National Institute of Agricultural Sciences Hiratsuka, Japan	Reserve and working collections of many crop species
Ohara Institute for Agricultural Biology Okayama University Kurashiki, Japan	Barley

8.	Mexico	
	Instituto Nacional de Investigaciones Agricolas (INIA), Chapingo, Mexico	Maize
9.	Sweden	
	Swedish Seed Association Svalov, Sweden	Barley, oats, rye
10.	Turkey	
	Crop Research and Introduction Center Izmir, Turkey	Wheat and small grains
11.	USSR	
	Vavilov All-Union Institute of Plant Industry, Leningrad, USSR	Reserve and working collections of many crop species
C.	International Agricultural Research Centers	
1.	International Center for Tropical Agriculture (CIAT) Cali, Colombia	Beans, cassava
2.	International Crops Research Institute for the Semi-Arid Tropics (ICRISAT) Hyderabad, India	Sorghum
3.	International Institute of Tropical Agriculture (IITA) Ibadan, Nigeria	Beans, cassava, cowpea
4.	International Maize and Wheat Improvement Center (CIMMYT) Mexico, D.F., Mexico	Maize, wheat
5.	International Potato Center Lima, Peru	White potato, sweet potato
6.	International Rice Research Institute (IRRI) Manila, Philippines	Rice
7.	East African Agricultural Research Organization Muguga, Kenya	Sorghum

3.2. Laboratory Colonies

Mass rearing via plant material or artificial diet ensures continuous availability of test insects. Numerous artificial diets have been developed for
insect rearing (Singh, 1977). Many of these are available commercially
(BioServe Inc., Frenchtown, NJ, U.S.A.) and can be useful in rearing
large quantities of test insects. Laboratory or glasshouse insect culture
provides an opportunity to eliminate or at least manage pathogenic microorganisms, parasitoids, and other biotic mortality factors. In some
cases, laboratory rearing of vector species, e.g., aphids, may be necessary
to ensure freedom from plant pathogens.

The major disadvantage of laboratory culture is its cost. Laboratory
culture of insects can be time- and labor-intensive; artificial diets in the
large quantities needed for mass rearing are costly. Furthermore, special
facilities are usually required for control of temperature, humidity, and
illumination, for isolation, and for air handling. Finally, laboratory-reared
insects may differ so radically from natural populations in genetic, behavioral, and physiological characteristics as to limit their usefulness in
screening germplasm and assessment of plant resistance (Schoonhoven,
1967; Guthrie et al., 1974).

3.3. Special Considerations

Several specific considerations in selection of insects for use in bioassays
merit attention. First, the behavioral and developmental biology of test
insects may vary depending on their age (Saxena, 1967), sex (Kinzer et
al., 1972; Hammond et al., 1979), and biotype (Cartier et al., 1965). For
example, young adults of the alfalfa weevil, *Hypera postica* (Gyllenhal)
consume far more foliage than do older individuals (Koehler and Pimentel,
1973), while the reverse is true for the Mexican bean beetle, *Epilachna
varivestis* (Mulsant) (Smith, 1978). Smith et al. (1976) reported that extracts
of white clover attracted more female cloverhead weevils, *Hypera meles*
(F.), than males of this species; feeding rates on tomato foliage by female
Colorado potato beetles, *Leptinotarsa decemlineata* (Say), were greater
than those of males (Schalk and Stoner, 1976).

The use of nontarget host plants for rearing or pretest access to other
hosts can influence the results of behavioral studies (Jermy et al., 1968;
Smith, 1978) as can inattention to circadian and diel rhythms. Obviously,
short-term bioassays focused on evaluation of insect behavior should be
timed to coincide with peak activity periods (Leppla, 1976). Variation
arising from these sources can be minimized by consistency in insect rearing, selection of test insects, pretest conditioning, and timing of bioassays.

Finally, parasitism and infection by pathogens can significantly alter
insect behavior and development. As noted previously, laboratory rearing
may be necessary for management of parasitism and disease, although

judicious use of selective insecticides can be helpful in controlling parasites (Peterson, 1963). Antibiotic chemicals and fungicides originally developed for control of plant pathogens can sometimes aid in management of disease in insect field populations and laboratory colonies (Nanne and Radcliffe, 1971; Horton et al., 1980). Chlorothalonil and maneb, for example, are two widely available and broad-spectrum agricultural fungicides useful in suppression of entomopathogenic fungi, particularly species of *Beauveria* and *Entomophthora* (Soper et al., 1974; Clark et al., 1982). A number of agricultural pesticides and brief descriptions of their useful selective properties for management of arthropod populations are provided in Table 2.

IV. Manipulating and Handling Insect Populations

4.1. Unconfined Field Infestations

Screening and assessment of resistance are frequently accomplished through use of field trials. However, experimental reliance on unmanaged populations of the target pest introduces an element of uncertainty. Infestations may fail to develop at the desired time or magnitude because of disruption in pest migration patterns or because of unpredictable biological and climatic mortality factors. These problems can often be minimized through use of special techniques to enhance or conserve the target pest population. The techniques described below are advantageous in regulating density, timing, and duration of insect infestations. It should be noted, however, that density-enhanced populations may risk collapse because of density-dependent mortality factors, particularly those associated with parasitism and disease. For example, Tingey and van de Klashorst (1976) utilized preinfested transplants of Chinese cabbage to encourage populations of the green peach aphid, *Myzus persicae* (Sulzer), in potato research plots. Although this technique dramatically boosted early to midseason aphid densities on potatoes, the aphid population later collapsed because of an uncontrollable epizootic of the entomopathogenic fungus, *Entomophthora aphidius* Hoffman.

Mass collection from laboratory colonies or from field populations followed by release into field plots is one of the simplest techniques for regulating timing, density, or duration of a pest population. As noted previously, preinfested hosts can be transplanted into field plots and provide some assurance of adequate infestation levels. (Tingey and van de Klashorst, 1976). These techniques can be effective in providing an earlier peak in pest infestation than would normally occur.

Trap hosts can be interplanted within field plots for the purpose of attracting, concentrating, and magnifying populations of the target pest (Stern, 1969; Stride, 1969; Tingey et al., 1982). Later timely removal of

Table 2. Pesticides with properties useful in experimental regulation and management of arthropod populations

Common name	Trade name®	Manufacturer	Useful properties and comments
Insecticides			
Bacillus thuringiensis	Dipel Bactospeine Thuricide	Abbott Laboratories Biochem Products Sandoz Inc.	Specific control of many Lepidoptera. Slow-acting and may require repeat applications.
carbaryl	Sevin	Union Carbide Corp.	Controls many arthropods. Multiple applications are often helpful in producing outbreaks of aphids and spider mites.
diflubenzuron	Dimilin	Uniroyal Chemical Co.	Controls many defoliating species of Coleoptera and Lepidoptera. Acts as larvicide, ovicide, and growth regulator. Generally inactive against aphids and other sucking species. Slow-acting and multiple applications often required.
methiocarb	Mesurol	Mobay Chemical Co.	Controls slugs, snails, and suppresses some non-aphid species. Useful bird repellent.
methyl parathion	Methyl parathion Penncap-M	Monsanto Corp. Pennwalt Corp.	Controls many predacious and parasitic arthropod species. Multiple applications may be useful in producing outbreaks of aphids, spider mites, and some lepidopterous species.
tetraethyl pyrophosphate (TEPP)	Kilmite	Miller Chemical Co.	Broadly toxic to many arthropod species. Extremely short residual activity (<48 hours). Useful in removal of unwanted arthropods from plants prior to bioassay or caging.

Acaricides			
cyhexatin	Plictran	Dow Chemical Co.	Specific control of phytophagous mites; long residual activity. May be phytotoxic at highest rates.
dienochlor	Pentac	Zoecon Corp.	Specific control of spider mites. Long residual activity; seldom phytotoxic even at highest rates.
Fungicides			
chlorothalonil	Bravo	Diamond Shamrock Co.	Aids in suppression of entomopathogenic fungi, particularly species of *Beauveria* and *Entomophthora.* Widely used for control of fungal diseases in vegetable and field crops.
mancozeb	Dithane M-45	Rohm and Haas Co.	
	Manzate 200	E.I. duPont de Nemours	
dinocap	Karathane	Rohm and Haas Co.	Aids in control of spider mites. Excellent for control of powdery mildew on plants in glasshouse culture.
thiram	Thylate	E.I. duPont de Nemours	Foliar applications useful in controlling plant damage by deer and rabbits. General agricultural fungicide.
triphenyltin hydroxide	Du-Ter	Griffin Corporation	Suppresses feeding in some defoliating species of Coleoptera and Lepidoptera. Generally inactive against aphids. Labelled for control of fungal diseases in vegetable and field crops. May produce phytotoxicity at rates required for antifeedant activity.
	Super-Tin		
Herbicides			
dinoseb	Premerge	Dow Chemical Co.	Toxic on contact with many arthropod species. Widely used as preemergent herbicides and as preharvest desiccants.
	Dow General	Dow Chemical Co.	
	Sinox General	FMC Corp.	

the trap host by mechanical means, pest attack, or senescence can force the target pest onto segregating germplasm for subsequent evaluation of resistance (Laster and Meredith, 1974).

Insecticides with selective properties can be powerful tools in conserving and enhancing pest populations. The effectiveness of this technique depends on the availability of an insecticide (or use of a specific dosage) relatively inactive against the target pest, but toxic to nontarget species including competing pests and natural enemies (Eveleens et al., 1973; Shepard et al., 1977). For example, the insecticides carbaryl and methyl parathion are toxic to many species of predators and parasitoids and may encourage populations of aphids and spider mites (Peterson, 1963). A number of additional insecticides and other pesticides with useful properties for selectively conserving and managing insect populations are described in Table 2.

Other considerations for attraction, conservation, and enhancement of target pests include use of semiochemicals, baits, and electromagnetic radiation. In conjunction with appropriate trapping devices, many of these behavior-altering techniques can be utilized in reducing populations of competing pests, predators, and parasitoids (see Mitchell, 1981; Shorey and McKelvey, 1977).

Finally, cultural practices can be manipulated to encourage target pests or to stress nontarget species. Tillage and irrigation procedures can sometimes be modified to encourage pest populations (Hollingsworth and Berry, 1982). The phenological synchrony of the host plant and natural populations of the target pest can often be improved by adjustment of field planting dates (Shands et al., 1972b).

4.2. Confinement Techniques

Spatial confinement procedures are widely used in the study of plant resistance to insects and are particularly appropriate for field, laboratory, and glasshouse trials involving highly mobile species. The benefits of caged confinement are threefold: (1) pest infestation density is easily regulated, (2) immigration and emigration of the pest are eliminated or minimized, and (3) non target arthropods are excluded from the test environment.

Caging devices are as varied as the plant species and target pests for which they are used. Although caging materials and designs are too numerous and varied for extended discussion here, a few common examples are discussed.

Saran mesh (Lumite®; Chicopee Manufacturing Co., Cornelia, GA, U.S.A.) is relatively resistant to deterioration by ultraviolet radiation, photochemical smog, moisture, and temperature extremes. Consequently, this material has been widely used in the construction of screen cages for field use. Cages can be of various sizes, depending on the number and size of the plants to be caged. Large rectangular designs require support; electrical conduit and plastic pipe (PVC, CPVC) are useful alternatives to wood for framing. Cylindrical cages of saran mesh or conventional zinc-coated ferrous screening are popular, relatively inexpensive, and self-supporting if the diameter and height do not exceed 1 m.

Polyester organdy has also been widely used in construction of insect confinement cages. Polyester fabrics must be fastened to a supporting frame in most cases. The ends of polyester organdy sleeve cages can be fitted with Velcro® closures to provide convenience and ease of use (Tingey et al., 1973).

Cellulose sheet or tubing is very useful for construction of small cages (Searls and Harris, 1936). This material is available commercially for use in dialysis procedures and as a casing for packaged meat products. Dialyzing tubing (Fig. 1) is an excellent caging material because of its light weight, transparent properties, and acceptable permeability to water vapor (Kring, 1970; Tingey et al., 1975).

Numerous types of clip-on leaf cages have been developed using rigid and semirigid thermoplastic tubing and sheet materials (Kaloostian 1955; Noble 1958; Hughes et al. 1966). An example of a commonly used clip-on cage is provided in Figure 2. The reader is referred to Adams and van Emden (1972) for other examples of specific designs. Many other tubular materials ranging from drinking straws to gelatin capsules can be adapted for use in insect confinement and their ultimate versatility for this purpose is limited only by the imagination of the researcher.

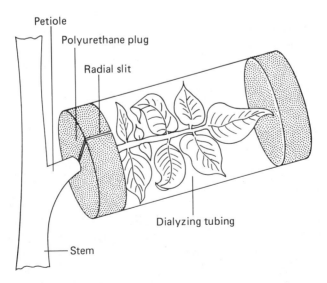

Figure 1. A dialyzing tubing leaf cage. (Courtesy of M.B. Dimock.)

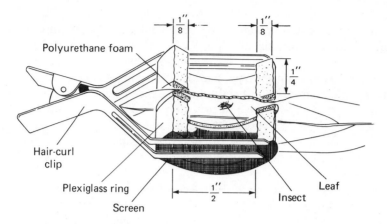

Figure 2. A clip-on leaf cage. From Adams and van Emden (1972).

Although useful in many situations, cages have several limitations. They may be costly to construct and maintain; placement and removal can be time- and labor-intensive. Certain types of materials used in cage construction such as polystyrene, adhesives prone to slow release of organic solvents, and other media, e.g., foam rubber, may be toxic to some insects and plants (Chada, 1962; Hutt and White, 1972). Cages also subject both insect and plant to an unnatural environment. Cage-induced changes in light intensity, temperature, relative humidity, and air movement can directly alter insect behavior and development or exert an indirect influence on insect biology through modification of plant growth processes (Fye et al., 1969; van Steenwyk and Stern, 1975). Finally, cages may confine unwanted arthropods on test plants. Sheltered from their predators and parasitoids and spatially confined, extraneous species can seriously jeopardize the outcome of caged experiments. Short-lived or selective insecticides are useful in eliminating unwanted species from caged plants and should generally be applied following cage placement and prior to introduction of the target pest (see Table 2).

4.3. Handling Considerations

In addition to the generalized handling procedures discussed previously, several specific techniques deserve mention. Brushes and aspirators of various sizes and designs are useful for handling and transfer of individuals or small groups of sedentary insects such as aphids (see van Emden, 1972). For quantification and transfer of large numbers of sedentary species, volumetric and gravimetric procedures may be appropriate (Pielou, 1961). Mechanical foliage brushing devices (Leedom Engineering, Twain Harte, CA, U.S.A.) are also useful for this purpose. Eggs and egg masses can be removed from the surface on which they are deposited (McMillian and Wiseman, 1972), the substrate can be detached for handling and transfer, or eggs can be applied by aerosol and fluid techniques (Thewka and Puttler, 1970; Widstrom and Burton, 1970; Shands et al., 1972a; Palmer et al., 1977; Fery et al., 1979). Larval stages of some lepidopterous species can be dispensed onto plants volumetrically if cushioned and suspended in a dry, flowable medium. A portable dispenser developed at the International Maize and Wheat Improvement Center (CIMMYT) and calibrated to deliver a uniform quantity of a larvae/corncob grits mixture greatly streamlines the artificial infestation of maize with fall armyworm, *Spodoptera frugiperda* (J. E. Smith); sugarcane borer, *Diatraea saccharalis* (F.); southwestern cornborer, *D. grandiosella* Dyar and *D. lineolata* (Walker) (Mihm et al., 1978; Davis and Williams, 1980; Wiseman and Widstrom, 1980).

Easily disturbed and highly mobile species pose a challenge to successful handling and transfer. For some species, the use of an air suction platform may be useful in restrainment. This device can be constructed using an inexpensive prefabricated thermoplastic or polystyrene filter unit (Falcon, Oxnard, CA, U.S.A.; Nalge Company, Rochester, NY, U.S.A.); small-bodied insects are effectively restrained on the upper surface of the filter membrane by a vacuum applied to the lower compartment of the filter assembly. Exposure to low temperature and use of anesthesia are also highly effective in insect restrainment. Carbon dioxide is the most widely used anesthetic gas (Edwards and Patton, 1965); humidification by passage of the gas stream through water can sometimes reduce the hazards of CO_2 anesthesia.

V. Handling and Manipulating Plants

5.1. Horticultural Considerations

Researchers studying plant resistance to insects should strive to achieve plant growth and development representative of the normal cropping system. This objective can be achieved relatively easily in field research where agronomic operations such as fertilization, irrigation, tillage, disease and weed management, and use of cover crops follow commercial practice. Production of field-representative plants through use of glasshouse or laboratory culture is much more difficult. Although it is seldom possible to duplicate field environments in the glasshouse or laboratory, precautions can be taken to minimize differences in horticultural conditions and resulting plant growth.

Plant growth and development are highly subject to variation in light quality and quantity. High-intensity discharge (HID) lamps (sodium, metal halide) provide outstanding supplemental illumination because of their excellent spectral quality and high intensity (Duke et al. 1975). These lamps generally produce plant growth superior to that achieved by the traditional combination of incandescent and fluorescent illumination. Obviously, supplemental lighting should be cycled to approximate the photoperiod of the normal cropping environment.

Temperature is also a major determinant of plant growth and development and efforts should be made to obtain representative cycles in glasshouse and laboratory environments. Forced air exchange, evaporative cooling, or refrigeration may be necessary to produce temperature cycles characteristic of field conditions.

In some cases, use of specialized materials and procedures such as tempered irrigation water, controlled release fertilizers, and artificial soil media may be helpful in optimizing plant growth. A detailed discussion of these and other horticultural techniques is beyond the scope of this chapter and the reader is referred to the technical manual of Lorenz and Maynard (1980) and the textbook by Hartman and Kester (1975).

5.2. Plant Age and Growth Stage

Studies of plant resistance to insects are ideally carried out using plants or plant organs of the same age and growth stage as subject to insect attack in the agricultural environment. If insect attack normally occurs late in crop development or near maturation, research on identification and nature of resistance can be lengthy, costly, and space-intensive. To speed this process, some researchers utilize immature or seedling stage plants, especially during early mass screening cycles. This can be a highly efficient approach, economical in time, numbers of test insects, space, and labor and has been very successfully employed in screening and development of insect-resistant cultivars of several forage and grain crops including alfalfa, rice, and sorghum (Sorensen et al., 1972; Starks and Burton, 1977; Pathak and Saxena, 1980). However, prospective users should be aware of several potential limitations of the seedling screening method for crops normally subject to insect attack at late stages of development. First, this approach may be ineffective if resistance is not manifest during the seedling growth stage (Abernathy and Thurston, 1969). Second, and perhaps more important, the expression of resistance may diminish with increase in plant age and growth stage (Klun and Robinson, 1969; Kindler and Staples, 1969).

5.3. Excised Plant Organs

The successful screening and identification of insect-resistant germplasm is an exercise in probability, dependent on efficient evaluation of large populations of segregating genotypes. In an effort to streamline this often laborious process, excised plants or plant organs have been successfully utilized in several cases (Sams et al., 1975). However, researchers should be aware that excision profoundly alters many fundamental plant physiological processes and that expression and magnitude of resistance can diminish following organ detachment (Mackinnon, 1961; Thomas et al., 1966). In other instances, excision has been associated with an apparent (induced) resistance, not representative of the intact organ (van Emden and Bashford, 1976; see also Chapters 4 and 5, *this volume*).

5.4. Altered Plant Tissue

In the search for and identification of chemical and physical plant factors conferring defense against insects, bioassay techniques employing plant extracts, tissue fractions, or otherwise altered plant organs must be developed. These techniques can range in sophistication from simple observational approaches to complex bioassays of behavioral and metabolic physiology. Similarly, fractions of suspected biological activity can be presented to the target pest in a variety of ways, depending on the insect response under study. At preliminary stages of investigation, three general methods of presenting plant fractions or putative allelochemicals to the target insect may prove useful. These include surface coating or infiltration of host or nonhost tissue (Sinden et al., 1978), application to allelochemically inert substrates such as filter paper (Kennedy et al., 1981), and incorporation in artifical diet (Raman et al., 1979) or surrogate ovipositional substrates (Sparks, 1970; Hower and Ferrer, 1970; Villacorta et al., 1971).

With regard to evaluation of morphological resistance traits, epidermal factors such as trichomes can sometimes be removed mechanically (Gibson, 1976) or by use of selective solvents (Kennedy and Yamamoto, 1979). Trichomes can also be occluded by proper selection of tightly woven fabrics or screens (Wire Cloth Enterprises, Inc., Pittsburgh, PA, U.S.A.), which when placed over the epidermis restrict intimate insect contact but allow diffusion of plant volatiles and access for feeding by insects with piercing/sucking mouthparts (Pillemer and Tingey, 1978). Detailed information regarding preparation and bioassay of plant allelochemicals is provided in Chapters 4–7, *this volume*.

5.5. Special Considerations

Researchers should strive to duplicate horticultural and agronomic regimens representative of the normal cropping environment because environmental temperature (Sosa and Foster, 1976), light intensity (Roberts and Tyrrell, 1961), soil fertility (Kindler and Staples, 1970), and agricultural chemicals (van Emden, 1969) can alter significantly the expression and magnitude of resistance. Numerous specific examples and practical suggestions for dealing with these phenomena are discussed by Tingey and Singh (1980).

VI. Measurement of Resistance

6.1 Insect Assessment

A. *Populations*

The analysis of insect population levels, age structure, and phenology is a fundamental requirement in most studies of plant resistance to insects. Measurement of insect infestation levels is particularly useful in screening large quantities of germplasm. Coupled with data on age structure and phenology, this information can be valuable in defining the cumulative suppressive impact of resistance on dynamics of the pest population. Numerous types of population sampling techniques have been developed, some highly specific to individual species or taxons (Heathcote 1972; Southwood 1978). In general, however, population sampling methods can be assigned to one of three categories: (1) direct observation, (2) sweepnet and vacuum sampling, and (3) trapping.

Direct observation methods are adaptable to many insect–plant associations and employ direct counting of insects, usually in situ, although plants and plant organs can be excised and stored for later examination. Cooling (1–5°C) usually improves the storage life and condition of excised plant samples and aids in immobilizing the pest. Direct observation methods are best suited to relatively sedentary species or life stages because of the disturbance introduced by handling of plants. A useful type of direct observation involves shaking of the plant and collection of dislodged insects on a ground cloth or container (Shepard et al., 1974). Insect counts can be expressed in terms of number per plant (Leigh et al., 1970), plant organ, area of plant tissue, area of plant stand, or units of observation time (Radcliffe and Lauer, 1966). In some cases, it may be advisable to express counts relative to the vertical axis of the plant using node position or arbitrarily defined zones (Mack and Smilowitz, 1981). Selection of specific examination sites should be based on consideration of the pests' preferred feeding and resting sites (Irwin et al., 1979). Typical considerations of this type include insect preference for foliage, fruiting organs, or underground parts; young versus mature or senescing tissue; and abaxial versus adaxial surfaces. Obviously, sampling should coincide with insect activity periods and with the normal period of phenological interaction between the insect and plant.

Sweepnet and vacuum methods are particularly useful for the sampling of easily disturbed and highly mobile species. These techniques are best suited for collection of species and life stages occupying the upper plant canopy. Sweepnet methods are sometimes limited by plant rigidity, height, and growth habit. Sweepnet data may be difficult to interpret because of individual variation in technique; moreover, this method can be injurious to plants if used carelessly. Sweepnet data are usually expressed in terms of numbers of insects per 180° sweep, per plant, or per unit area. Vacuum sampling methods (D-vac®) (Dietrick, 1961) are generally less damaging to plants and introduce less operator variability than the use of sweepnets. Vacuum methods are best suited for collection of light-bodied life stages and species. Vacuum-collected samples can be quantified in terms of numbers of insects per plant, or per unit area of plant canopy. Detailed information on efficiency of sweepnet and vacuum methods relative to direct observation for specific insect–plant associations is provided by Pedigo et al. (1972), Shepard et al. (1974), Marston et al. (1976), and Mayse et al. (1978).

Disturbance of highly mobile species just prior to sampling can be a serious problem in use of sweepnet and vacuum methods. The sampling site should be approached in such a way as to avoid casting a shadow before collection. In the case of vacuum equipment, careless positioning of the downstream air blast can be extremely disturbing to insect populations. Finally, sweepnet and vacuum methods are relatively nonselective with regard to collection of nontarget anthropods, plant debris, and soil. These problems can be corrected to some extent by careful monitoring of operator technique. Unwanted anthropods can sometimes be minimized in vacuum samples by use of screens or sieves in the collection head. Post-collection separation and sorting may require use of Berlese, flotation, or related techniques (Marston and Hennessey, 1978).

Trapping methods (see Chapter 2, *this volume*) generally provide an indirect measurement of insect populations (Byrne and Bishop, 1979) but can be valuable for use with highly mobile or nocturnal pests. Widely used designs employ trapping media ranging from water in Möericke pan traps (Möericke, 1951) to adhesives such as Tack-Trap® (Animal Repellents Inc., Griffin, GA, U.S.A.) (Taylor and Palmer, 1972). Trap efficiency can be greatly improved in many cases by use of semiochemicals (Eckenrode and Arn, 1972), ultraviolet radiation (Roach, 1975), and attractive colors (Owens and Prokopy, 1978).

Differences between individual observers, sampling equipment operators, and sampling equipment can often contribute significantly to experimental error. Ideally, all treatments and replications (blocks) of a given experiment should be sampled by the same individual or piece of equipment. This idealized scheme is seldom possible, particularly in large field trials, but the confounding action of these types of sampling error on treatment comparisons can be minimized by use of several procedures.

First, observers can be randomly assigned treatments through a replicated trial, thus reducing the probability that one individual will sample the same treatments in each replication. A second, superior tactic is to assign one individual to all treatments of the first replicate, another person to all treatments of the second replicate, and so on. This technique coupled with proper experimental design will allow partitioning of replicate-observer error from treatment effects in analysis of variance.

Finally, several points regarding storage of living insect samples deserve mention. The accuracy of counts obtained from such collections can be affected seriously by the presence of general predators such as coccinellids and spiders. The target species may also be subject to cannibalism. In cases where cooling is ineffective in immobilization, living arthropods should be killed by exposure to freezing temperatures or toxic substances. A 5-gallon, widemouth polyethylene carboy with a tight-fitting screwcap (Nalge Company, Rochester, NY, U.S.A.) is useful for killing samples confined in vacuum collection nets. Porous material such as paper towelling moistened with 5 ml of ethyl acetate is placed in the carboy along with insect samples and most species will be killed within 60 min. Killing vessels employing potassium cyanide or dichlorvos (Shell No-Pest Strip®) are usually superior to those containing ethyl acetate vapors for killing resistant arthropods such as spiders and some Coleoptera.

B. Developmental Biology and Population Dynamics

Plant resistance to insects is often expressed in terms of its deleterious impact on key elements of insect developmental and population biology. Although it may be possible to screen effectively for resistance by measurement of only one criterion, e.g., total mortality from egg (or stadium 1) to the adult stage, more exhaustive collection of data is advisable for studies focused on the plant characteristics conferring resistance and their collective, cumulative impact on insect performance. For comprehensive examination of these phenomena, the collection of data appropriate for life table or similar exhaustive analysis may be justified.

Techniques appropriate for such studies are common to many aspects of entomological and ecological research (Hughes, 1972; Ruesink and Kogan, 1975; Southwood, 1978). The types of information useful in such studies can include age-, time-, or life stage-specific data on survival and mortality, duration of immature and adult stages, and weight of life stages. Data on adult life span, fecundity, duration of the reproductive period, sex and life-stage ratios, and the frequency or phenology of diapause can be valuable in the measurement of resistance. In the analysis of resistance conditioned by nutritional factors, use of nutritional indices (Beck and Reese, 1976) and measurement of number, size, or volume of feces (Kastings and McGinnis, 1962) or honeydew (Auclair, 1958; Pagnia et al., 1980) may be useful. Detailed information on specific techniques and applications of these procedures is presented in Chapter 6, *this volume*.

C. Behavioral Responses

The analysis of behavior in response to plant stimuli can be an important element in defining the nature of resistance. Many techniques have been developed for use in the study of insect behavior and the selection of an appropriate methodology will depend on the specific insect, host plant, and behavior in question. In general, such bioassays are directed toward analysis of plant-directed shifts in host recognition, selection, and acceptance. Techniques for evaluation of insect behavior can employ plants, plant organs, fractions of plant tissues, or plant chemicals of suspected behavioral activity. Typical responses for evaluation include directed movement toward or away from the host; restlessness and irritability; acceptance or rejection of the host for feeding and oviposition; magnitude and duration of feeding and oviposition; and diurnal or phenological shifts in behavior patterns. In measurement of locomotor responses to plant stimuli, the choice of applicable techniques and apparatus design range from flight mills (Chambers, et al., 1976), wind tunnels (Wilson, 1968), actographs (motion detectors) (Edwards, 1964), and olfactometers (Nation, 1975), to traps of varying complexity. At more fundamental levels of investigation, electrophysiological techniques (see Chapter 10, *this volume*) have proven useful in analysis of insect sensory behavior, particularly in regard to vision (Agee, 1977), taste (Cook, 1976), and olfaction (Jacobson, 1972). Finally, electronic technology has been successfully employed for evaluation of probing, salivation, and ingestion behavior in insects with piercing-sucking feeding habits such as aphids and leafhoppers (McLean, 1977). Many additional techniques and applications for evaluation of insect behavior are described in Chapters 1–3, *this volume*.

6.2. Plant Growth and Damage Assessment

A. Direct Feeding Damage

The ultimate criteria in analysis of plant resistance are crop yield and quality in response to insect attack. These parameters are best measured through use of standardized yield trials, replicated over time and location. Experimental designs should be appropriate for analysis of variance. Split-plot designs utilizing insect infestation levels as the main effects and plant genotypes as subtreatments are extremely valuable in comparing plant cultivars, lines, or clones for resistance when their intrinsic yield or quality potential differ. With this method, comparison of yield or quality differences between insect-free and insect-infested treatments of a plant genotype will indicate the value of resistance in protection of yield, and allow comparison of several genotypes having different intrinsic yield or quality potential, exclusive of insect resistance.

In cases where susceptibility to insect damage is associated with delay in crop maturity, use of multiple harvest dates can provide useful information. First harvest yield of resistant genotypes would be expected to constitute a larger fraction of total yield than that of more susceptible types.

Yield and quality performance relative to resistance are usually assessed in a sequential fashion, beginning with small plot research station trials and culminating in larger pilot type experiments in conjunction with co-operating growers. Needless to say, standard susceptible cultivars should be liberally employed in all trials.

Yield and quality assessment is time- and labor-intensive and best suited for advanced stages of germplasm development. At earlier stages of evaluation, measurements of insect damage are preferable to yield analysis. These can include evaluation of plant mortality, defoliation, tissue removal, abscission of fruiting forms, tissue necrosis, discoloration, and premature senescence. Analysis of tunnel length and number in addition to plant lodging and breakage are useful in evaluation of resistance to stalk boring and leaf mining species. Obviously, damage criteria selected for examination should be closely associated with ultimate loss in crop yield and quality.

Researchers may wish to utilize rating scales in measurement of criteria difficult to quantify otherwise. Rating scales can be valuable in processing large quantities of germplasm, especially in preliminary stages of mass evaluation. All scaling methods contain subjective elements that may vary between observers. Researchers should use experimental and statistical designs that allow removal of observer variation from that of treatments.

Several types of commercially available instrumentation can be useful in measuring gross or subtle feeding injury. Area sensing devices are valuable in measuring tissue removal and defoliation of planar surfaces such as leaves (Li-Cor Inc., Lincoln, NE, U.S.A.). Feeding injury expressed by discoloration of plant tissue (necrosis, premature senescence, delayed maturity) is amenable to measurement by use of reflectance sensors and infrared or fluorescence photography (Hart and Myers, 1968; Chiang et al., 1973; Kogan, 1972).

B. Altered Morphology, Development, and Physiology

In the case of insects that do not produce immediate or obvious feeding damage in the form of plant mortality, defoliation, and other types of tissue removal, injury can be difficult to evaluate prior to crop maturity. This problem may be especially serious for various hemipterous and homopterous pests whose damage potential is associated with removal of plant sap, injection of disruptive salivary secretions, or mechanical blockage of the vascular system. For example, plant bugs (Miridae) that feed selectively on rapidly growing meristematic tissues of fruiting organs, foliage, or roots often produce insidious and delayed types of damage (Tingey and Pillemer, 1977). Although gross damage symptoms may not be obvious in these cases until late in plant development, subtle effects on growth and development can be evident soon after injury and measurement of these responses can effectively complement or in some cases supplant analysis of losses in yield and quality. Techniques for plant growth analysis are useful in these situations and can be drawn from the discipline of plant physiology (see Evans, 1975). Plant growth criteria useful for measurement of response to insect attack include the following: (1) plant weight, volume, area, and growth habit; (2) number, size, weight, and location of vegetative and fruiting organs; (3) growth, elongation, and maturity of plant organs, (4) ability for recovery or replacement of damaged organs. Sequential evaluation of these parameters can be valuable in the case of insect injury that results in premature senescence or the opposite reaction—delayed maturity. In addition, sequential evaluation of these phenomena is critical in establishing rates of change in growth processes and in the development of damage symptoms.

Measurement of plant recovery or replacement of damaged organs may be extremely useful in identifying tolerance modalities of resistance. A novel method for measuring compensatory root growth of maize in response to feeding by the western corn rootworm, *Diabrotica virgifera* LeConte, was reported by Zuber et al. (1971). Whereas tolerance components of resistance are usually measured in the continuous presence of the pest, analysis of compensatory growth and recovery following removal of the pest (Capinera and Roltsch, 1980) may be useful in cases of moderate tolerance.

Insect feeding injury can seriously alter fundamental processes of plant growth and metabolism. Analysis of insect-induced changes in absolute levels or rates of intrinsic physiological processes such as photosynthesis (Hall and Ferre, 1975), transpiration (de Angelis et al., 1982), and respiration (Sances et al., 1979) may be useful in the study of resistance modalities, particularly the phenomenon of tolerance, when coupled with information on transport and localization of the biochemical products of these reactions (Sanders et al., 1977).

C. Simulated Feeding Injury

In an effort to streamline screening procedures or reduce the variability associated with use of assays employing insects, methods intended to duplicate insect feeding injury have been developed. These have focused largely on simulating defoliation or abscission of fruiting organs. In the latter case, use of plant growth regulators that accelerate abscission may be helpful. Carbaryl and naphthaleneacetic acid (NAA) (Luckwill, 1953) are two well-known examples of abscission-promoting chemicals. Attempts to simulate defoliation have centered on mechanical excision and analysis of hail-induced removal of plant organs or parts of organs (Poston et al., 1976; Howell, 1978; Capinera, 1979; Detling et al., 1979). These techniques may be helpful in evaluating resistance to pests whose sole impact is restricted to mechanical removal of plant tissue. Simulation of injury by introduction of toxic salivary secretions into plants (Miles, 1972) has received little attention and certainly merits further investigation. Toxins derived from fungal pathogens have been successfully used in screening germplasm for disease resistance (Luke and Wheeler, 1964; Schertz and Tai, 1969).

D. Plant Defensive Traits

If the qualitative and quantitative effects of plant defensive factors on insect performance are well defined, it may be possible to screen or evaluate germplasm for resistance in the absence of a target pest bioassay. In the case of resistance conditioned by a physical trait such as leaf pubescence, measurement of trichome density, length, shape, and angle of insertion may suffice for preliminary screening purposes (Parnell et al., 1949; Webster et al., 1975; Tingey and Laubengayer, 1981). Measurements of tissue strength, stiffness, density (Tanton, 1962; Hewitt, 1968), and ease of penetration (Beckwith and Helmers, 1976) can be appropriate when these plant characteristics are closely associated with the expression of resistance (Djamin and Pathak, 1967; Agarwal, 1969; Martin et al., 1975). In the case of allelochemical defensive factors (releasing stimuli, physiological inhibitors, and nutrients), measurement of presence versus absence, localization, and concentration relative to plant age (Klun and Robinson, 1969), diurnal periodicity (Loper and Lapioli, 1972), and seasonality (Thompson et al., 1971) may be useful in evaluation of resistance.

VII. Experimental Design and Statistical Procedures

7.1. General Considerations

The selection of experimental and statistical designs for use in studies of plant resistance to insects should be made after consideration of time, labor, and facility resources, variability of the parameters to be measured, and especially the degree of precision necessary to achieve the desired objectives.

In mass screening, techniques adaptable for rapid processing of large populations are essential. In unselected populations, the frequency distribution is usually heavily skewed toward susceptibility and the plant or insect criteria assessed may vary over a wide range. The researcher is most interested in selecting those few genotypes at the extreme end of the resistance continuum. The majority of the susceptible germplasm can be effectively discarded if the researcher adheres to rigorous cut-off levels. This approach will minimize selection of genotypes bearing low levels of resistance, or those that falsely appear resistant because of temporal escape, spatial evasion, or other factors.

In mass screening, elaborate experimental designs are unnecessary and usually wasteful of resources. Replication of genotypes may also be unnecessary, and in some instances impossible because of limited availability of botanical or vegetative propagules. Repetitive evaluation of resistance criteria, e.g., sequential sampling throughout the seasonal period of host plant–insect interaction, can serve as an effective alternative to genotype replication. Repetitive sampling is also useful in identifying those plants that falsely appear resistant because of spatial or brief temporal phenomena.

Although location effects can usually be ignored in mass screening, the random placement of genotypes in field plots, glasshouse trials, and controlled environmental chambers may be worthwhile, particularly if the plant population contains groups of several closely related families, lines, or clones. Simple descriptive statistics such as means and estimates of variability (standard deviation, standard error) usually provide adequate precision in mass screening.

Most other types of research involving plant resistance to insects require greater statistical precision and, correspondingly, more sophisticated experimental designs than those used in mass screening. Analysis of variance procedures have proven highly adaptable in research on plant and insect relationships and are widely used for their flexibility and precision. In addition, analysis of variance provides information critical in selection of appropriate methods for comparison of treatment means, e.g., multiple-range, Least Significant Difference (LSD), and F tests. Experimental designs appropriate for use in analysis of variance range in sophistication from the simple, completely randomized design, to complex factorial arrangements. The major difference in these designs is the manner in which experimental units are arranged or grouped. In the case of a completely randomized design, the number of experimental units is equal to the number of treatments multiplied by the number of replications. Treatments are assigned position at random within the set of experimental units. This design is extremely flexible with regard to physical arrangement of the experimental units but lacks precision if variation other than that associated with treatments is present.

In practice, the most useful arrangement particularly for field plot trials is the randomized complete block design. Treatments are randomly assigned to each experimental unit or "block." The principal advantage of this design is that the variation of blocks can be separated from that of treatments. Thus, the randomized complete block design becomes increasingly more precise as variability among blocks increases and is the design of choice if major sources of variability in addition to those associated with treatments are expected. Typical sources of such variation include those associated with location (soil type and fertility, drainage patterns, direction of air movement, proximity to reservoir hosts), sampling equipment, and operator error. Whenever possible, blocks should be arranged across the gradient of principal variation while treatment plots within blocks should be laid parallel to the gradient. The value of split-plot designs in yield trials was discussed previously. For information on applicability of increasingly complex blocked designs such as latin square, split-split plot, and split-block designs, the reader is referred to Little and Hills (1972) and Snedecor and Cochran (1967).

A wide variety of other statistical procedures can be effectively used in studies of plant resistance to insects. For example, Student's t and chi-square procedures are often useful for behavioral and developmental bioassays in which the effects of paired experimental treatments are compared. Correlation and regression techniques may be helpful in evaluating relationships between two or more variables. Covariance analysis can be useful in removing major sources of variation among experimental units.

The precision of analysis of variance and many other statistical procedures is dependent on several important assumptions, i.e. (1) random, independent, and normal distribution of error terms, (2) homogeneous variance of different samples, and (3) independence of means and their variances. If these assumptions are not met, data transformation or use of nonparametric procedures may be advisable. It should be noted that insect population data from field plots seldom satisfy all of these assumptions. Specific information on these and other considerations in the selection of appropriate experimental designs and statistical procedures is provided by Murdie (1972), Sokal and Rohlf (1969), and Snedecor and Cochran (1967).

7.2. Free-Choice versus Isolation Assays

In the evaluation of multiple or paired plant genotypes using free-choice methods, the target insect is free to discriminate in its selection of hosts for feeding and oviposition. Free-choice experimental designs are widely used in mass screening for resistance and are extremely useful in preliminary evaluation of resistance modalities affecting insect behavior. However, insect behavioral responses sometimes differ between free-choice and no-choice assays. A plant genotype classified as "resistant" when the test insect is given a choice between other genotypes may not enjoy this advantage when isolated (Wiseman et al., 1961; Overman and MacCarter, 1972). This is an important consideration in the design of screening methods because the economics of crop production often dictate uniformity in crop maturity, growth habit, quality, and other horticultural properties. For individual fields, growers, and sometimes regions, this frequently means use of a single cultivar—a situation in which the insect's opportunity to exercise choice in short-range selection of hosts is limited. To maximize the identification and measurement of resistance modalities effective in these situations, the use of isolation procedures is recommended to supplement free-choice screening methods. Caged confinement techniques or spatially isolated field plots of single genotypes are useful in minimizing free-choice bias.

VIII. Research Objectives and Approach

A comprehensive approach to the study of plant resistance to insects requires inputs from many disciplines including but not restricted to those of entomological science (insect ecology, behavior, physiology, and pest management); plant breeding, genetics, horticulture, and physiology; and chemistry. Well-balanced research programs generate information relevant to the development and release of resistant cultivars as well as knowledge about the fundamental nature of plant-insect interactions. A generalized listing of research objectives appropriate for a broadly based approach to the study of plant resistance to insects is provided below:

1. Review pest problems of the crop with an interdisciplinary group composed of entomologists, plant breeders, plant pathologists, and other crop production specialists.
2. Determine the economic need for management of insect pests and define the nature of plant damage. Review the biology and ecology of the target pests and the crop host.
3. Collect adapted and wild germplasm for evaluation of resistance.
4. Determine methodology appropriate in mass screening for resistance and initiate the screening program.
5. Intermate parental germplasm and select resistant progeny. Recycle germplasm for recombination as long as reasonable progress is made each generation. Initiate selection for desired horticultural properties.
6. Initiate studies of plant traits conferring resistance, their influence on pest performance, mechanisms of reducing damage, and inheritance.

7. Determine the permanence and stability of resistance by evaluation of advanced genotypes in diverse cultural and environmental conditions.
8. Evaluate the influence of resistance traits on key predators and parasitoids of the cropping system.
9. Determine the vulnerability of advanced resistant germplasm to non-target pests.
10. Determine the economic value of resistance in elite germplasm using infested versus noninfested comparisons.
11. Multiply seed of elite resistant germplasm and release to public and commercial organizations.
12. Monitor post-release performance of resistant cultivars to determine the appearance of host-specific biotypes and varietal reaction to other commercial production hazards and practices.
13. Publicize the economic and environmental benefits of insect-resistant cultivars to grower groups using demonstration plots, press releases, and other extension information devices.
14. Continue breeding and selection programs for intensified levels and alternate modalities of resistance.
15. Publish research findings to familiarize the scientific community with progress and to establish a formal reference base for justification of continued financial support.

References

Abernathy CO, Thurston R (1969) Plant age in relation to the resistance of *Nicotiana* to the green peach aphid. J Econ Entomol **62**:1356–1369.

Adams JB, van Emden HF (1972) The biological properties of aphids and their host plant relationships. In: Aphid Technology. van Emden HF (ed), Academic Press, New York. pp. 47–104.

Agarwal RA (1969) Morphological characteristics of sugarcane and insect resistance. Entomol Exp Appl **12**:767–776.

Agee HR (1977) Instrumentation and techniques for measuring the quality of insect vision with the electroretinogram. USDA-ARS Bull S-162.

Auclair JL (1958) Honeydew excretion in the pea aphid, *Acyrthosiphon pisum* (Harr.) (Homoptera: Aphididae). J. Insect Physiol **2**:330–337.

Beck SD, Reese JC (1976) Insect-plant interactions: nutrition and metabolism. In: Biochemical Interaction Between Plants and Insects. Wallace JW, Mansell RL (eds), Recent Adv Phytochem Vol. 10. Plenum Press, New York. pp. 41–92.

Beckwith RC, Helmers AE (1976) A penetrometer to quantify leaf toughness in studies of defoliators. Environ Entomol **5**:291–294.

Byrne DN, Bishop GW (1979) Comparison of water trap pans and leaf counts as sampling techniques for green peach aphids on potatoes. Am Potato J **56**:237–241.

Capinera JL (1979) Effects of simulated insect herbivory on sugarbeet yield in Colorado. J Kans Entomol Soc **52**:712–718.

Capinera JL, Roltsch WJ (1980) Response of wheat seedlings to actual and simulated migratory grasshopper damage. J Econ Entomol 73:258–261.

Cartier JJ, Isaak I, Painter RH, Sorensen EL (1965) Biotypes of pea aphid *Acyrthosiphon pisum* (Harris) in relation to alfalfa clones. Can Entomol 97:754–760.

Chada HL (1962) Toxicity of cellulose acetate and vinyl plastic cages. J Econ Entomol 55:970–972. ·

Chambers DL, Sharp JL, Ashley TR (1976) Tethered insect flight: a system for automated data processing of behavioral events. Behav Res Meth Instr 8:352–356.

Chesnokov PG (1962) Methods of investigating plant resistance to pests. Natl Tech Inform Serv, US Dept Commerce.

Chiang HC, Latham R, Meyer MP (1973) Aerial photography: use in detecting simulated insect defoliation in corn. J. Econ Entomol 66:779–784.

Clark RA, Casagrande RA, Wallace DB (1982) Influence of pesticides on *Beauveria bassiana*, a pathogen of the Colorado potato beetle. Environ Entomol 11:67–70.

Cook AG (1976) A critical review of the methodology and interpretations of experiments designed to assay the phagostimulatory activity of chemicals to phytophagous insects. In: The Host-Plant in Relation to Insect Behavior and Reproduction. Jermy T (ed), Plenum Press, New York. pp. 47–54.

Creech JL, Reitz LP (1971) Plant germ plasm now and for tomorrow. In: Advances in Agronomy, Vol 23. Brady NC (ed), Academic Press, New York, pp 1–49.

Dahms RG (1972) Techniques in the evaluation and development of host-plant resistance. J. Environ Qual 1:254–259.

Davis FM, Williams WP (1980) Southwestern corn borer: comparison of techniques for infesting corn for plant resistance studies. J Econ Entomol 73:704–706.

de Angelis JD, Larson KC, Berry RE, Krantz GW (1982) Effects of spider mite injury on transpiration and leaf water status in peppermint. Environ Entomol 11:975–978.

Detling JK, Dyer MI, Winn DT (1979) Effect of simulated grasshopper grazing on CO_2 exchange rates of western wheatgrass leaves. J Econ Entomol 72:403–406.

Dietrick EJ (1961) An improved backpack motor fan for suction sampling of insect populations. J Econ Entomol 54:394–395.

Djamin A, Pathak MD (1967) The role of silica in resistance to Asiatic rice borer, *Chilo suppressalis* (Walker) in rice varieties. J Econ Entomol 60:347–351.

Duke WB, Hagin RD, Hunt JF, Linscott DL (1975) Metal halide lamps for supplemental lighting in greenhouses: crop response and spectral distribution. Agron J 67:49–53.

Eckenrode CJ, Arn H (1972) Trapping cabbage maggots with plant bait and allyl isothiocyanate. J Econ Entomol 65:1343–1345.

Edwards DK (1964) Activity rhythms of lepidopterous defoliators. I. Techniques for recording activity, eclosion, and emergence. Can J Zool 42:923–937.

Edwards LJ, Patton RL (1965) Effects of carbon dioxide anesthesia on the house cricket, *Acheta domesticus* (Orthoptera:Gryllidae). Ann Entomol Soc Am 58:828–832.

Evans LT (ed) (1975) Crop physiology. Cambridge University Press, England.

Eveleens KG, van den Bosch R, Ehler LE (1973) Secondary outbreak induction of beet armyworm by experimental insecticide applications in cotton in California. Environ Entomol 2:497–503.

Fery RL, Cuthbert FP Jr, Perkins WD (1979) Artificial infestation of the tomato with eggs of the tomato fruitworm. J Econ Entomol 72:392–394.

Fye RE, Bouham CC, Leggett JE (1969) Modification of temperature by four types of insect cages. J Econ Entomol 62:1019–1023.

Gibson RW (1976) Glandular hairs on *Solanum polyadenium* lessen damage by the Colorado beetle. Ann Appl Biol 82:147–150.

Guthrie WD, Rathore YS, Cox DF, Reed GL (1974) European corn borer: virulence on corn plants of larvae reared for different generations on a meridic diet. J Econ Entomol 67:605–606.

Hall FR, Ferre DC (1975) Influence of two-spotted spider mite populations on photosynthesis of apple leaves. J Econ Entomol 68:517–520.

Hammond RB, Pedigo LP, Poston FL (1979) Green cloverworm leaf consumption on greenhouse and field soybean leaves and development of a leaf-consumption model. J Econ Entomol 72:714–717.

Harlan JR, Starks KJ (1980) Germplasm resources and needs. In: Breeding Plants Resistant to Insects. Maxwell FG, Jennings PR (eds), John Wiley, New York. pp. 253–273.

Hart WG, Myers VI (1968) Infrared aerial color photography for detection of populations of brown soft scale in citrus groves. J Econ Entomol 61:617–624.

Hartman HT, Kester DE (1975) Plant Propagation: Principles and Practices. Prentice-Hall, New Jersey.

Heathcote GD (1972) Evaluating aphid populations on plants. In: Aphid Technology. van Emden HF (ed), Academic Press, London. pp. 105–145.

Hewitt GB (1968) An instrument for measuring the resistance of the leaf and culm of forage and crop plants to cutting and chewing insects. J Econ Entomol 61:1114–1115.

Hollingsworth CS, Berry RE (1982) Twospotted spider mite (Acari: Tetranychidae) in peppermint: population dynamics and influence of cultural practices. Environ Entomol 11:1280–1284.

Horton DL, Carner GR, Turnipseed SG (1980) Pesticide inhibition of the entomogenous fungus *Nomuraea rileyi* in soybeans. Environ Entomol 9:304–308.

Howell JF (1978) Spotted cutworm: simulated damage to apples. J Econ Entomol 71:437–439.

Hower AA Jr, Ferrer FR (1970) An artificial oviposition technique for the alfalfa weevil. J Econ Entomol 63:761–764.

Hughes PR, Hunter RE, Leigh TF (1966) A light-weight leaf cage for small arthropods. J Econ Entomol 59:1024–1025.

Hughes RD (1972) Population dynamics. In: Aphid Technology. van Emden HF (ed), Academic Press, London. pp. 275–293.

Hutt RB, White LD (1972) Polystyrene toxic to codling moths. J Econ Entomol 65:615.

Irwin ME, Yeargan KV, Marston NL (1979) Spatial and seasonal patterns of phytophagous thrips in soybean fields with comments on sampling techniques. Environ Entomol 8:131–140.

Jacobson M (1972) Insect Sex Pheromones. Academic Press, New York.

Jermy T, Hanson FE, Dethier VG (1968) Induction of specific food preference in lepidopterous larvae. Entomol Exp Appl 11:211–230.

Kaloostian GH (1955) A magnetically suspended insect cage. J Econ Entomol 48:756–757.

Kastings R, McGinnis AJ (1962) Quantitative relationship between consumption and excretion of dry matter by larvae of the pale western cutworm, *Agrotis orthogonia* Morr. (Lepidoptera: Noctuidae). Can Entomol 94:441–443.

Kennedy GG, Yamamoto RT (1979) A toxic factor causing resistance in a wild tomato to the tobacco hornworm and some other insects. Entomol Exp Appl 26:121–126.

Kennedy GG, Yamamoto RT, Dimock MB, Williams WG, Bordner J (1981) Effect of daylength and light intensity on 2-tridecanone levels and resistance in *Lycopersicon hirsutum* f. *glabratum* to *Manduca sexta*. J Chem Ecol 7:707–716.

Kindler SD, Staples R (1969) Behavior of the spotted alfalfa aphid on resistant and susceptible alfalfas. J Econ Entomol 62:474–479.

Kindler SD, Staples R (1970) Nutrients and the reaction of two alfalfa clones to the spotted alfalfa aphid. J Econ Entomol 63:938–940.

Kinzer HG, Ridgill BJ, Reeves JM (1972) Response of walking *Conophthorus ponderosae* to volatile attractants. J Econ Entomol 65:726–729.

Klun JA, Robinson JF (1969) Concentration of two 1,4-benzoxazinones in dent corn at various stages of development of the plant and its relation to resistance of the host plant to the European corn borer. J Econ Entomol 62:214–220.

Koehler PG, Pimentel D (1973) Economic injury levels of the alfalfa weevil (Coleoptera: Curculionidae). Can Entomol 105:61–74.

Kogan M (1972) Fluorescence photography in the quantitative evaluation of feeding by phytophagous insects. Ann Entomol Soc Am 65:277–278.

Kring JB (1970) Dialyzing membrane for aphid cages. J Econ Entomol 63:1032–1033.

Laster ML, Meredith WR Jr (1974) Evaluating the response of cotton cultivars to tarnished plant bug injury. J Econ Entomol 67:686–688.

Leigh TF, Gonzalez D, van den Bosch R (1970) A sampling device for estimating absolute insect populations on cotton. J Econ Entomol 63:1704–1706.

Leppla NC (1976) Circadian rhythms of locomotion and reproductive behavior in adult velvetbean caterpillars. Ann Entomol Soc Am 69:45–48.

Little TM, Hills FJ (1972) Statistical methods in agricultural research. University of California Ext Bull.

Loper GM, Lapioli AM (1972) Photoperiod effects on the emanation of volatiles from alfalfa, *Medicago sativa* L. florets. Plant Physiol 49:729–732.

Lorenz OA, Maynard DN (1980) Knott's Handbook for Vegetable Growers. John Wiley & Sons, New York.

Luckwill LC (1953) Studies of fruit development in relation to plant hormones. II. The effect of naphthaleneacetic acid on fruit set and fruit development in apples. J Horticult Sci 28:25–40.

Luke HH, Wheeler H (1964) An intermediate reaction to victorin. Phytopath 54:1492–1493.

Mack TP, Smilowitz Z (1981) The vertical distribution of green peach aphids and its effect on a model quantifying the relationship between green peach aphids and a predator. Am Potato J **58**:345–353.

Mackinnon JP (1961) Preference of aphids for excised leaves to whole plants. Can J Zool **39**:445–447.

Marston NL, Hennessey MK (1978) Extracting arthropods from plant debris with xylene. J Kan Entomol Soc **51**:239–244.

Marston NL, Morgan CE, Thomas GO, Ignoffo CM (1976) Evaluation of four techniques for sampling soybean insects. J Kan Entomol Soc **49**:389–400.

Martin FA, Richard CA, Hensley SD (1975) Host resistance to *Diatraea saccharalis* (F.): relationship of sugarcane internode hardness to larval damage. Environ Entomol **4**:687–688.

Maxwell FG, Jennings PR (eds) (1980) Breeding Plants Resistant to Insects. John Wiley & Sons, New York

Mayse MA, Kogan M, Price PW (1978) Sampling abundances of soybean arthropods: comparison of methods. J Econ Entomol **71**:135–141.

McLean DL (1977) An electrical measurement system for studying aphid probing behavior. In: Aphids as Virus Vectors. Harris KF, Maramorosch, K (eds), Academic Press, New York. pp. 277–290.

McMillian WW, Wiseman BR (1972) Separating egg masses of the fall armyworm. J Econ Entomol **65**:900–902.

Mihm JA, Peairs FB, Ortega A (1978) New procedures for efficient mass production and artificial infestation with lepidopterous pests of maize. CIMMYT Rev, 138 pp.

Miles PW (1972) The saliva of Hemiptera. Adv Insect Physiol **9**:183–255.

Mitchell ER (ed) (1981) Management of Insect Pests with Semiochemicals. Plenum Press, New York.

Möericke V (1951) Eine Farbfalle zur Kontrolle des Fluges von Blättlausen, insbesondere der Pfirsichblattus, *Myzodes persicae* (Sulz.) Nachrich Deutsch Pflanzenschutz **3**:23–24.

Murdie G (1972) Problems of data analysis. In: Aphid Technology. van Emden HF (ed), Academic Press, New York. pp 295–318.

Nanne HW, Radcliffe EB (1971) Green peach aphid population on potatoes enhanced by fungicides. J Econ Entomol **64**:1569–1570.

Nation JL (1975) The sex pheromone blend of Caribbean fruit fly males: isolation, biological activity, and partial chemical characterization. Environ Entomol **4**:27–30.

Noble MD (1958) A simplified clip cage for aphid investigation. Can Entomol **90**:760.

Overman JL, MacCarter LE (1972) Evaluating seedlings of cantaloupe for varietal nonpreference-type resistance to *Diabrotica* spp. J Econ Entomol **65**:1140–1144.

Owens ED, Prokopy RJ (1978) Visual monitoring trap for European apple sawfly. J Econ Entomol **71**:576–578.

Pagnia P, Pathak MD, Heinrichs EA (1980) Honeydew excretion measurement techniques for determining differential feeding activity of biotypes of *Nilaparvata lugens* on rice varieties. J Econ Entomol **73**:35–40.

Palmer DF, Windels MB, Chiang HC (1977) Artificial infestation of corn with western corn rootworm eggs in agar-water. J Econ Entomol 20:277–278.

Parnell FR, King HE, Ruston DF (1949) Jassid resistance and hairiness of the cotton plant. Bull Entomol Res 39:539–575.

Pathak MD, Saxena RC (1980) Breeding approaches in rice. In: Breeding Plants Resistant to Insects. Maxwell FG, Jennings PR (eds), John Wiley, New York. pp 421–455.

Pedigo LP, Leutz GL, Stone JD, Cox DF (1972) Green cloverworm populations in Iowa soybean with special reference to sampling procedure. J Econ Entomol 65:414–421.

Peterson AG (1963) Increases of the green peach aphid following the use of some insecticides on potatoes. Am Potato J 40:121–129.

Pielou DP (1961) A volumetric method for the determination of numbers of apple aphids, *Aphis pomi* DeGeer, on samples of apple foliage. Can J Plant Sci 41:442–443.

Pillemer EA, Tingey WM (1978) Hooked trichomes and resistance of *Phaseolus vulgaris* L. to *Empoasca fabae* (Harris). Entomol Exp Appl 24:83–94.

Poston FL, Pedigo LP, Pearce RB, Hammond RB (1976) Effects of artificial and insect defoliation on soybean net photosynthesis. J Econ Entomol 69:109–112.

Radcliffe EB, Lauer FI (1966) A survey of aphid resistance in the tuber-bearing *Solanum* (Tourn.) L. species. Minn Agr Exp Stn Tech Bull 253.

Raman KV, Tingey WM, Gregory P (1979) Potato glycoalkaloids: effect on survival and feeding behavior of the potato leafhopper. J Econ Entomol 72:337–341.

Roach SH (1975) *Heliothis zea* and *H. virescens:* moth activity as measured by blacklight and pheromone traps. J Econ Entomol 68:17–21.

Roberts DWA, Tyrrell C (1961) Sawfly resistance in wheat. IV. Some effects of light intensity on resistance. Can J Plant Sci 41:457–465.

Ruesink WG, Kogan M (1975) The quantitative basis of pest management: sampling and measuring. In: Introduction to Pest Management. Metcalf RL, Luckmann WH (eds), John Wiley & Sons, New York. pp 309–351.

Sams DW, Lauer FI, Radcliffe EB (1975) An excised leaflet test for evaluating resistance to green peach aphid in tuber-bearing *Solanum* germplasm. J Econ Entomol 68:607–609.

Sances FV, Wyman JA, Ting IP (1979) Physiological responses to spider mite infestation on strawberries. Environ Entomol 8:711–714.

Sanders TH, Ashley DA, Brown RH (1977) Effects of partial defoliation on petiole phloem area, photosynthesis, and C^{14} translocation in developing soybean leaves. Crop Sci 17:548–550.

Saxena KN (1967) Some factors governing olfactory and gustatory responses in insects. In: Olfaction and Taste II. Hayashi T (ed), Pergamon Press, Oxford. pp 799–820.

Schalk JM, Stoner AK (1976) A bioassay differentiates resistance to the Colorado potato beetle and tomatoes. J Am Soc Horticult Sci 101:74–76.

Schertz KF, Tai YP (1969) Inheritance of reaction of *Sorghum bicolor* L. Moench. to toxin produced by *Periconia circinata*. Crop Sci 9:621–624.

Schoonhoven LM (1967) Loss of host plant specificity by *Manduca sexta* after rearing on an artificial diet. Entomol Exp Appl 10:270–272.

Searls EM, Harris HH (1936) Handy insect cages made from cellophane. J Econ Entomol **29**:1158–1160.

Shands WA, Gordon CC, Simpson GW (1972a) Insect predators for controlling aphids on potatoes. 6. Development of a spray technique for applying eggs in the field. J Econ Entomol **65**:1099–1103.

Shands WA, Simpson GW, Murphy HJ (1972b) Effects of cultural methods for controlling aphids on potatoes in northeastern Maine. Maine Agr Exp Stn Tech Bull 57.

Shepard M, Carner GR, Turnipseed SG (1974) A comparison of three sampling methods for arthropods in soybeans. Environ Entomol **3**:227–232.

Shepard M, Carner GR, Turnipseed SG (1977) Colonization and resurgence of insect pests of soybean in response to insecticides and field isolation. Environ Entomol **6**:501–506.

Shorey HH, McKelvey JJ (eds) (1977) Chemical Control of Insect Behavior: Theory and Application. John Wiley & Sons, New York.

Sinden SL, Schalk JM, Stoner AK (1978) Effects of daylength and maturity of tomato plants on tomatine content and resistance to the Colorado potato beetle. J Am Soc Horticult **103**:595–600.

Singh P (1977) Artificial diets for insects, mites, and spiders. IFI-Plenum Data Co., New York.

Smith CM (1978) Factors for consideration in designing short-term insect-host plant bioassays. Bull Entomol Soc Am **24**:393–395.

Smith CM, Frazier JL, Knight WE (1976) Attraction of clover head weevil *Hypera meles,* to flower bud volatiles of several species of *Trifolium.* J Insect Physiol **22**:1517–1521.

Snedecor GW, Cochran WG (1967) Statistical methods. Iowa State University Press, Ames, IA.

Sokal RR, Rohlf FJ (1969) Biometry. W. H. Freeman, San Francisco.

Soper RS, Holbrook FR, Gordon CC (1974) Comparative pesticide effects on *Entomophthora* and the phytopathogen *Alternaria solani.* Environ Entomol **3**:560–562.

Sorensen EL, Wilson MC, Manglitz GR (1972) Breeding for insect resistance. In: Alfalfa Science and Technology. Hanson CH (ed), Am Soc Agron, Madison, WI. pp 371–390.

Sosa O Jr, Foster JE (1976) Temperature and the expression of resistance in wheat to the Hessian fly. Environ Entomol **5**:333–336.

Southwood TRE (1978) Ecological Methods with Particular Reference to the Study of Insect Populations. John Wiley & Sons, New York.

Sparks MR (1970) A surrogate leaf for oviposition by the tobacco hornworm. J Econ Entomol **63**:537–540.

Starks KJ, Burton RL (1977) Greenbugs: determining biotypes, culturing and screening for plant resistance. US Dept Agric ARS Tech Bull 1556.

Stern VM (1969) Interplanting alfalfa in cotton to control lygus bugs and other insect pests. Tall Timbers Conf Ecol Animal Control Habitat Manage **1**:55–69.

Stern VM, Dietrick EJ, Mueller A (1965) Improvements on self-propelled equipment for collecting, separating, and tagging mass numbers of insects in the field. J Econ Entomol **58**:949–953.

Stride GO (1969) Investigations into the use of a trap crop to protect cotton from attack by *Lygus vosseleri* (Heteroptera: Miridae). J Entomol Soc S Afr **32**:469–477.

Tanton MT (1962) The effect of leaf "toughness" on the feeding of larvae of the mustard beetle *Phaedon cochleariae* Fab. Entomol Exp Appl **5**:74–78.

Taylor LR, Palmer JMP (1972) Aerial sampling. In: Aphid Technology. van Emden HF (ed), Academic Press, New York. pp 189–234.

Thewka SE, Puttler B (1970) Aerosol application of lepidopterous eggs and their susceptibility to parasitism by *Trichogramma*. J Econ Entomol **63**:1033–1034.

Thomas JG, Sorensen EL, Painter RH (1966) Attached vs. excised trifoliates for evaluation of resistance in alfalfa to the spotted alfalfa aphid. J Econ Entomol **59**:444–448.

Thompson AC, Baker DN, Gueldner RC, Hedin PA (1971) Identification and quantitative analysis of the volatile substances emitted by maturing cotton in the field. Plant Physiol **48**:50–52.

Tingey WM, Laubengayer JE (1981) Defense against the green peach aphid and potato leafhopper by glandular trichomes of *Solanum berthaultii*. J Econ Entomol **74**:721–725.

Tingey WM, Leigh TF, Hyer AH (1973) Three methods of screening cotton for ovipositional nonpreference by lygus bugs. J Econ Entomol **66**:1312–1314.

Tingey WM, Leigh TF, Hyer AH (1975) *Lygus hesperus:* growth, survival, and egg laying resistance of cotton genotypes. J Econ Entomol **68**:28–30.

Tingey WM, Pillemer EA (1977) Lygus bugs: crop resistance and physiological nature of feeding injury. Bull Entomol Soc Am **23**:277–287.

Tingey WM, Singh SR (1980) Environmental factors influencing the magnitude and expression of resistance. In: Breeding Plants Resistant to Insects. Maxwell FG, Jennings PR (eds), John Wiley & Sons, New York. pp. 87–113.

Tingey WM, Plaisted RL, Laubengayer JE, Mehlenbacher SA (1982) Green peach aphid resistance by glandular trichomes in *Solanum tuberosum* x *S. berthaultii* hybrids. Am Potato J **59**:241–251.

Tingey WM, van de Klashorst G (1976) Green peach aphid: magnification of field populations on potatoes. J Econ Entomol **69**:363–364.

van Emden HF (1969) Plant resistance to aphids induced by chemicals. J Sci Food Agric **20**:385–387.

van Emden HF (ed) (1972) Aphid Technology. Academic Press, New York.

van Emden HF, Bashford MA (1976) The effect of leaf excision on the performance of *Myzus persicae* and *Brevicoryne brassicae* in relation to the nutrient treatment of the plants. Physiol Entomol **1**:67–71.

van Steenwyk RA, Stern VM (1975) Air temperature modification in Lumite field cages in alfalfa hay. J Econ Entomol **68**:795–796.

Villacorta A, Bell RA, Callenbach JA (1971) An artificial plant stem as an oviposition site for the wheat stem sawfly. J Econ Entomol **64**:752–753.

Webster JA, Smith DH, Rathke E, Cress CE (1975) Resistance to cereal leaf beetle in wheat: density and length of leaf-surface pubescence in four wheat lines. Crop Sci **15**:199–202.

Widstrom NW, Burton RL (1970) Artificial infestation of corn with suspensions of corn earworm eggs. J Econ Entomol **63**:443–446.

Wilson DM (1968) The flight-control system of the locust. Sci Am **218**:83–90.

Wiseman BR, Hall CV, Painter RH (1961) Interactions among cucurbit varieties and feeding responses of the striped and spotted cucumber beetles. Proc Am Soc Horticult Sci **68**:379–384.

Wiseman BR, Widstrom NW (1980) Comparison of methods of infesting whorl-stage corn with fall armyworm larvae. J Econ Entomol **73**:440–442.

Zuber MS, Musick GJ, Fairchild ML (1971) A method of evaluating corn strains for tolerance to the western corn rootworm. J Econ Entomol **64**:1514–1518.

Chapter 10

Electrophysiological Recording and Analysis of Insect Chemosensory Responses

James L. Frazier[1] and Frank E. Hanson[2]

I. Introduction

The elegant behavioral experiments of Dethier (1955) on the chemosensory control of feeding in the blow fly set the stage for the first recordings of individual cell responses by Hodgson et al. (1955). The animals clearly accepted some substances and rejected others, leading Dethier to infer that one chemosensory cell coded for acceptance and another for rejection. When electrophysiological techniques became available, the resulting data indicated that more than one sensory cell mediates each of these behaviors. Further electrophysiological experimentation led to the elucidation of how salt-, sugar-, and water-sensitive cells can interact to regulate feeding behavior (Dethier, 1976). Since then many investigators have incorporated electrophysiology as an invaluable tool in conjunction with behavioral experiments for elucidating the chemosensory basis of behavior.

The central role of chemosensory cells in the behavior of phytophagous insects is well documented (Visser and Minks, 1983). Orientation to host plants, the selection of oviposition sites, and the choice of food plants and feeding sites, all involve chemosensory cues. Yet critical details are currently lacking in our understanding of how chemosensory cells detect

[1]Agricultural Products Department, EI DuPont de Nemours, Wilmington, Delaware 19898, U.S.A.
[2]Department of Biological Sciences, University of Maryland Baltimore County, Catonsville, Maryland 21228, U.S.A.

chemicals, and the type of information that is actually furnished to the central nervous system (CNS). Our knowledge of what constitutes a behaviorally meaningful message is as yet incomplete; at best, only "candidate codes" exist. The goal of further neurobiological research on phytophagous insects is to understand a specific behavior in terms of chemosensory inputs and associated neural circuit activity.

This chapter will focus on current techniques used to record and analyze the responses of individual sensory cells associated with various types of sensilla. Three basic recording techniques are used for recording from a wide variety of sensilla. Although recording techniques have changed little over the years, amplifiers and data analysis techniques have seen recent improvements. For example, earlier recordings omitted the first few hundred milliseconds of spikes due to amplifier blockage (Hodgson et al., 1955). This problem has been overcome by the use of a clamping preamplifier that reduces the loss of data to a few milliseconds.

Perhaps the most welcomed advances have been made in the area of data analysis. Equipment of any desired level of complexity can be assembled. Simple analog circuits are used individually for threshold or window discrimination and counting, or coupled with microcomputers. More complex analyses of insect chemosensory data are accomplished with large-scale laboratory computer systems. This ability to process large amounts of data helps overcome the inherent variability in insect chemosensory data that has long impeded our understanding of the function of chemosensory cells.

II. Structure/Function Dictates the Type of Recording

Obtaining information about individual sensory cells has proved to be very difficult. The small size of chemosensory cells, together with their clustering in variable numbers below a rigid cuticular process, has precluded intracellular recording. All recordings of chemosensory cell activity are thus extracellular, with all the attendant drawbacks. Since the electrode may vary in position from one preparation to another, caution must be exercised in interpreting the resulting recordings. Resolving individual unit activity from a mixed recording is difficult, if not impossible, when the number of cells is greater than four. This number is exceeded in many insect sensilla; thus, current capabilities for analysis are exceeded. In addition, sensilla are often multimodal, so that recorded pulse trains may include activity from cells other than chemoreceptors. These problems underscore the importance of obtaining ultrastructural details of the sensillum of interest before electrical recordings can be interpreted properly.

2.1. Contact Chemosensory Cells

A. *Morphology*

The ultrastructure of chemosensory and accessory cells and their associated cuticular processes have been reviewed recently (Altner and Prillinger, 1980; Zacharuk, 1984; van der Wolk et al., 1984). The basic elements of contact chemosensory sensilla are shown in Figure 1A. The sense cells are contained within a cuticular process of variable shape with a single terminal pore. The dendrites of these cells are enclosed by a dendritic sheath that separates the extracellular space surrounding them from the remainder of the hair lumen; both are filled with receptor lymph fluid. Several accessory cells envelop the sense cells and have septate junctions with them. One or two inner sheath cells (thecogen or trichogen) have gap junctions between them and with an outer sheath cell (tormogen). The adjoining epidermal cells are similarly connected, forming a high-resistance ionic barrier between the hemolymph and the receptor lymph cavity.

B. *Electrical Analog*

In the usual tip recording technique for sensilla with contact chemosensory cells, the recording electrode makes contact through the pore at the tip of the cuticular process, and the reference electrode is placed in the hemolymph at some distal point in the body. Figure 1B depicts the sensillum together with a simplified equivalent circuit between the recording electrodes. On contact, the tip electrode measures a large standing potential difference of -30 to -70 mV depending on the sensillum under study (Thurm and Wessel, 1979; Thurm and Kuppers, 1980). This potential has two main components, the transepithelial potential (E_t) and the dynamic sensory dendritic potential (E_s). The gap junctions among the accessory cells form an epithelial layer of high resistance separating the lumen of the sensillum from the hemolymph. The resistances of the receptor lymph (R_l) and the sensory cell cytoplasm (R_s) are low compared to those of the sensory dendrite membrane (R_d) and the accessory cell membranes (R_t). There are thus two parallel pathways for current flow: one from the outer sheath cell through the receptor lymph, and one through the dendrite cytoplasm (Erler and Thurm, 1981).

In the classical theory, the receptor potential is generated by the interaction of stimulus molecules with receptors on the distal dendritic membrane. This induces an increase in cation conductance (decrease in R_d), allowing cations to enter and depolarize the dendrite (Broyles and Hanson, 1976). This depolarization conducts passively to the proximal dendritic or axonal regions, where the spike initiation zone most likely

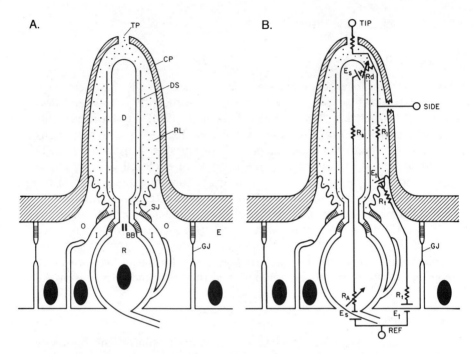

Figure 1. A generalized insect contact chemosensillum. (A) Structural features include the cuticular process (CP), terminal pore (TP), receptor lymph (RL), receptor cell body (R), dendrite (D), dendritic sheath (DS), basal body, (BB), inner sheath cell (I), outer sheath cell (O), septate junction (SJ), epidermal cell (E), gap junction (GJ). (B) Equivalent circuits for recording by the tip and sidewall techniques. Major sources of resistance and Emf are indicated with explanations in the text. For simplification, membrane capacitances have been omitted.

exists (Morita and Yamashita, 1959). This model explains the behavior of chemosensory cells with high specificity (e.g., sugar); however, those cells with broader specificity (e.g., salt) may require other explanations (Rees, 1970; den Otter, 1972; Kurihara et al., 1986

The initial receptor potential is recorded as a negative baseline shift (Morita and Takeda, 1959). This depolarization is conducted passively to the proximal dendritic region where the resulting action potential travels in two directions: orthodromically down the axon, and antidromically up the dendrite. The depolarization at the proximal dendrite appears at the tip recording electrode as the initial positive phase of the recorded action potential because the resistance of the intradendritic pathway $(R_d + R_s)$ is less than that of the extradendritic one $(R_1 + R_t)$. As the action potential travels antidromically, it passes the large R_t and reduces the R_d, thus providing a low-impedance pathway between the negative interior of the dendrite and the tip recording electrode. This results in the negative half of the biphasic action potential that is normally recorded. The back-conducting potential involves ion channels distinct from those participating in receptor potential generation (Wolbarsht, 1965; Wolbarsht and Hanson, 1965). The negative phase is abolished by local anesthetics such as procaine and the Na^+ channel blocker tetrodotoxin (TTX), while the excitation of the cell and the positive phase of the action potential remains unchanged. The back-conducting potential terminates the positive phase, so the recorded biphasic action potential has a broader negative-phase (Figs. 11b and 15). Recording from labellar hairs of decreasing length in the blow fly results in action potentials of larger negative-phase amplitudes, indicating an inverse relationship between sensillum length and the size of the negative phase (Wolbarsht and Hanson, 1965). Other interpretations of these events have been put forward (Maes, 1977a; de Kramer et al., 1984).

Several components of the sensillum may vary and result in modulation of the recorded action potentials. Variation in dendritic back-conduction results in changes in spike shape (Fujishiro et al., 1984). The unknown substance filling the pore at the tip of the sensillum of the locust appears to be produced after feeding and interferes with reception (Bernays and Chapman, 1972). Similar variations in responses of fly sensilla have also been reported and ascribed to this phenomenon (Sturckow, 1967; Omand and Zabara, 1981; de Kramer and van der Molen, 1983). The close apposition of the dendrites in sensilla with multiple cells may lead to electrical interactions. Variation in the transepithelial potential, as shown in mechanosensory cells by Erler and Thurm (1981), may modulate individual cell activity in ways that are not currently understood. Thus, the reliability of recording action potentials with extracellular techniques is limited by the dynamics of the components of the sensillum. Since the sensillum is an integral part of the recording circuit, any cellular changes contribute to uncontrolled variability in the recorded signals.

2.2. Olfactory Cells

A. Morphology

Sensilla containing olfactory cells have a similar basic arrangement, except the cuticular process has multiple pores in the wall (Fig. 2A). The distal dendrites are multiply branched and communicate with the pores in specific arrangements, the ultrastructural details of which are difficult to interpret (Altner and Prillinger, 1980). As in contact sensilla, the pores contain unknown material that may offer resistance to the entry of stimulus molecules. Many multiporous sensilla contain large numbers (10–30) of sensory cells (Zacharuk, 1984). This morphological arrangement seriously constrains investigation with electrophysiological techniques, since separation of spike trains resulting from more than four cells cannot currently be resolved into component responses. Other multiporous sensilla contain fewer cells, particularly those responsive to pheromones.

B. Electrical Analog

Figure 2B depicts a sensillum containing cells with an olfactory function and simplified equivalent circuits for recording electrodes at two sites. Action potentials may be obtained from an electrode inserted near the base of a sensillum. Placement of the electrode tip is uncertain, but presumably is located in the receptor lymph cavity as shown in Figure 2B. Another technique is to cut off the tip of the sensillum and place a capillary electrode over the cut end (Kaissling, 1974; van der Pers and den Otter, 1978). The electrical recording circuits for either electrode position differ only by the small R_1 and therefore are essentially equivalent. A possible

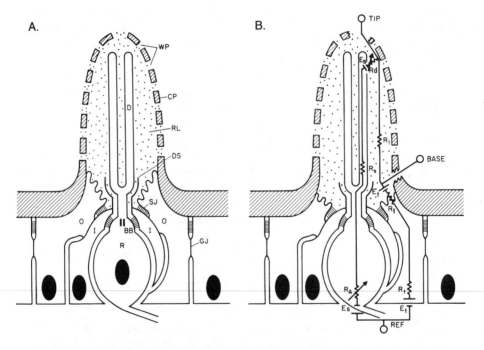

Figure 2. A generalized insect olfactory chemosensitive sensillum. (A) Structural features include the cuticular process (CP), numerous wall pores (WP), receptor lymph (RL), receptor cell body (R), branched dendrites (D), short dendritic sheath (DS), basal body (BB), inner sheath cell (I), outer sheath cell (O), septate junctions (SJ), gap junctions (GJ). (B) Equivalent circuits for recording by the tip and base techniques. Major sources of resistance and Emf are indicated with explanations in the text. For simplification, membrane capacitances have been omitted.

mechanism to account for the shape of the spike in olfactory sensilla is discussed in de Kramer and Kaissling (1984).

III. Equipment for Recording Chemosensory Cell Potentials

Equipment typically used for recording chemosensory cell potentials involves variations of standard electrophysiological equipment described by Miller (1979). A compound microscope with long working-distance objectives (100–500 × working range) coupled with fiber optics and a high-intensity light source (quartz-halogen) is most often used. We have found that the Leitz Diavert inverted microscope without the stage, but with focusable optics cluster, gives a high-resolution system that is adaptable to a variety of recording conditions. A pair of high-resolution micromanipulators, such as Leitz or Zeiss-Jena, is commonly used. The insect preparation is often mounted directly on the reference electrode and held in one manipulator so that fine adjustments in the orientation of the sensilla can be made. The other manipulator is used to position the recording electrode. Some applications require additional lower resolution micromanipulators and/or flexible arm stands to support fiber optics and other devices.

The single component most crucial to high-quality recording is the preamplifier. In addition to high input impedance, it must have a high-frequency response, and must not be blocked by large voltage changes when the tip electrode is applied to a sensillum that may have a large static (epithelial) potential. A nonblocking filter amplifier (e.g., Maes, 1977b) solves this problem, but distorts the low-frequency components of the waveform. This can be solved with an amplifier having two filters and automatic selection between them. We use the Johnson amplifier (George Johnson Electronics, 506 Woodside Road, Baltimore, MD 21119) shown in Figure 3, which is similar to the one described by de Kramer and van der Molen (1980). This unit provides all of the desirable features to permit recording from chemosensory cells of any type, and particular features specifically designed for the tip recording method. It provides a high input impedance of 4×10^9 ohms for use with microelectrodes, and a driven shield for reducing noise pickup by the active input lead. Distortion of the shape of action potentials by capacitance loading is minimized by a compensation circuit that can be adjusted for maximal risetime for a given preparation. A calibration pulse of selectable amplitude can be delivered through the indifferent electrode so that variation in resistance among sensilla or preparations can be monitored. A bucking potential is available to counteract any standing potentials such as epithelial and diffusion potentials. The large offset potential that occurs when the recording

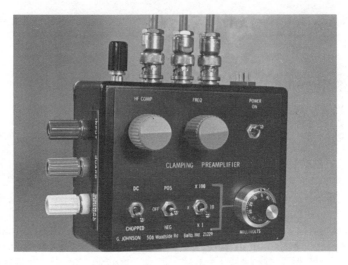

Figure 3. The clamping preamplifier used for recording from insect chemosensory sensilla.

electrode makes contact with the tip of a sensillum is counteracted by a circuit that rapidly returns the system to baseline, thereby allowing the monitoring of spikes within a few milliseconds after contact. The pulse that activates this "clamping" circuit is available as a separate output and can be recorded on a separate channel of the tape recorder. On playback it may be used as a synchronizing pulse to trigger data sampling by various devices. Spikes are monitored through a $50 \times$ AC output having a long time constant (20 msec) for minimum distortion of the spike shape. This signal is normally fed into a bandpass amplifier (100–2000 Hz) before being recorded on magnetic tape. A DC $1 \times$ output is also available from the preamp, permitting simultaneous recording of receptor potentials or visual monitoring of offset potentials.

A separate timing circuit (George Johnson Electronics, Baltimore, MD), couples with the above preamplifier and allows the automatic delivery of calibration pulses at the preselected time after contact with the sensillum. These calibration pulses are used by automated procesing programs to scale the data from different experiments. This compensates for differences in recorded spike size resulting from variability in resistance among sensilla, different stimulus solutions, or different amplifier settings. Another circuit permits the quality of the electrical contact with the preparation to be monitored visually via a red-green diode placed in the visual field of the experimenter.

To complete the set-up, an FM tape recorder is used for minimal distortion of spike shapes, with voice commentary conveniently logged with a boom microphone attached to earphones or by a miniature microphone placed between the oculars of the microscope. The hands are thus free for focusing and micromanipulation during the recording trials. A block diagram of the recording equipment is shown in Figure 4.

IV. Techniques of Recording Chemosensory Potentials

4.1. Contact Chemosensilla

The classical method of recording from insect chemosensilla is the tip recording method first used by Hodgson et al. (1955). Standard glass capillary microelectrodes are filled with stimulus plus an electrolyte. The low-resistance reference electrode is made by breaking off the tip of the glass micropipette produced by the puller. The active electrode requires an internal diameter just large enough to permit placement over the tip of the sensillum. To achieve the correct diameter, the pipette may have the tip broken or ground to a flat tip (Ogden, 1978). If the tip opening is too large relative to the sensillum, or if the surface of the insect is hydrophylic, solution will flow freely on contact. This shunts the signal and prevents recording. This problem can be obviated by using dilute agar in the pipette to increase viscosity of the solution (Dethier and Hanson, 1965). A tip-recorded spike train from the maxillary styloconic sensillum of the caterpillar *Manduca sexta* is shown in Figure 5A.

A second technique is to record through the side wall. This was first used by Morita and co-workers on sensilla of the flesh fly and the butterfly *Vanessa* (Morita and Takeda, 1959). This technique is more difficult than tip recording, but offers the advantages of allowing continuous monitoring of cell activity in the absence of stimulation, and of permitting stimulation with an electrolyte-free stimulus. The major disadvantage of this technique

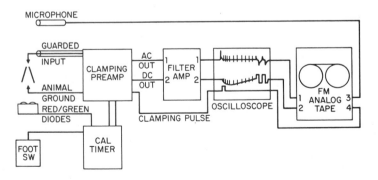

Figure 4. Block diagram of the equipment used to record from insect chemosensory sensilla.

is that the lifetime of individual preparations is comparatively short, ranging from a few minutes to an hour at best. Side-wall recording requires a supporting pipette placed opposite the recording microelectrode. One such system uses a dual microelectrode holder fitted with a tungsten needle and a recording glass capillary. The supporting pipette is brought against the side of the sensillum, and the tungsten needle is used to puncture the wall. The tungsten needle can be used as the recording electrode or a glass recording electrode may be placed in contact with the puncture. Stimulation is done with another micropipette applied to the tip. The technique is simplified if longer sensilla are used, although it has been applied successfully to the styloconic sensilla of the caterpillar, *Manduca* (Fig. 5B).

Side-wall recording allows detection of both spontaneous activity and receptor potentials of cells in the absence of an interfering electrolyte. An artifact, apparent spontaneous activity from otherwise silent cells, may be generated by excessive pressure from the recording electrode on the sensory dendrites. There are no good controls for this phenomenon. In spite of the precise information that can be obtained by this method, its mechanical difficulty, often high spontaneous activity, and short lifetime preparations have limited its use (Morita and Yamashita, 1959; Broyles and Hanson, 1976; Fujishiro et al., 1984).

A.

B.

Figure 5. Action potentials recorded from the styloconic sensilla of *Manduca sexta* larva. (A) Tip recorded response from medial styloconicum to 0.3 *M* glucose in 0.5 *M* NaCl with calibration pulses injected near the end of the stimulus period. Time from onset (arrow) to onset of calibration (arrow) is one second (Frazier, unpublished.) (B) Sidewall recorded response from lateral styloconicum to wheat germ medium. Arrow indicates onset of stimulation. Time mark is 1 sec (Hanson, unpublished.)

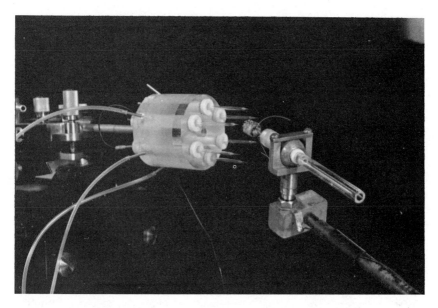

Figure 6. Stimulus turret used for recording from the styloconic sensilla on a *Manduca sexta* isolated larval head preparation.

Application of the stimulus solution is normally done using a micro-pipette in contact with a small reservoir of stimulus solution. For experiments requiring many stimuli in a specified order, such as determining dose–response relationships, a multiple elctrode holder that can be rotated in the manipulator is used (Fig. 6). It is constructed from a 2-cm plexiglass rod, with six chambers machined on each end to accept Microcap® silicone rubber septa. Gold pin connectors on chlorided silver wires can be sealed into one rubber septum to constitute the electrode surface. A recording pipette can be easily replaced through the septum at the other end. Stainless steel tubes are connected with polyvinyl tubing to a syringe that allows air pressure to force solution through the pipette tip just prior to stimulation. Excess solution can be easily blotted with a Q-tip touched to the tip of the pipette. Evaporation of stimulus solution from the pipette tips and crystallization of electrolyte are minimized by producing high humidity in the immediate area by a constantly flowing moist air stream, or by lining the Faraday cage with large sheets of moistened chromatography paper.

If highly volatile compounds are used, the solution should flow continuously during stimulation. Maintained air pressure in the polyvinyl tubing will produce a continuous flow out the electrode and back along the taper only if the electrodes have been thoroughly cleaned prior to pulling. One satisfactory method is to soak them in alcohol.

Many techniques of producing a suitable insect preparation for chemosensory recording are used. The most satisfactory results for long-term experiments are obtained with restrained whole insects. The difficulties of this method are preventing unwanted movements of body areas from which recording is intended, and obtaining the proper orientation of the sensilla for application of the stimulus. As an example, recording from fly tarsi is accomplished by mounting the fly directly on the reference capillary. The pipette is inserted into the dorsal neck membrane so that the tip is in the thoracic cavity. The head and prothorax are affixed with beeswax to the pipette. The prothoracic legs can be brought forward with the femoro tibial joint waxed to the sides of the pipette, leaving the tarsal chemosensory sensilla perpendicular to the pipette. Movement of the tarsi can be prevented by careful waxing of the basal tarsal segment to the tibia. The other legs can be waxed to the body to prevent movement. This preparation leaves the mouth parts free so the insect can be fed. The same individual can be used for several days, far exceeding the duration of excised leg preparations used in earlier studies.

Recording from the chemosensilla associated with mouth appendages is often hampered by excessive movements. These are reduced by using isolated heads, and if necessary by destroying the suboesophageal ganglion. Successful preparations of caterpillar heads for recording from styloconic sensilla are obtained by ligaturing the neck and cutting off the head, inserting the reference electrode into the membrane at the base of the labium, and forcing the tip into the lobe of the hypopharynx. The labium is waxed to the electrode shank at the point of penetration. The head is tilted dorsally, and the occipital region is waxed to the pipette, thus bringing the lobes of the maxilla away from the mandibles leaving both sets of sensilla pointing toward the front of the head. This preparation allows recording from both sets of styloconic sensilla, but only for an hour or two. Whole caterpillar preparations require special immobilization procedures, e.g., see Devitt (1983).

Electrophysiological studies of the contact chemosensory cells of phytophagous insects provide some insight into their sensory capabilities (see review by Stadler, 1984). Several species of caterpillars have been studied, leading to the generalization that they possess cells responsive to sugars, sugar alcohols, salts, alkaloids, steroids, glucosinolates, and other compounds (Ishakawa, 1966; Schoonhoven and Dethier, 1966; Dethier and Kuch, 1971; Ma, 1972; Dethier, 1973; Stadler and Hanson, 1975; Wieczorek, 1976; de Boer et al., 1977; van Drongelen, 1978; Dethier and Crnjar, 1982; Schoonhoven, 1981). Three species of phytophagous beetles have been studied, including both adults and larvae, revealing specificities for amino acids, sugars, salts, and glucosinolates (Rees, 1969; Mitchell and Schoonhoven, 1974; Mitchell, 1978; Mitchell and Gregory, 1979; Mitchell and Harrison, 1984; Mitchell and Sutcliff, 1984). Taste cells of the locust are sensitive to sugars, salts, and nicotine (Winstanley and Blaney, 1978). One species of fly has been found to respond to sucrose, salts, and glucosinolates, (Stadler, 1978), and three species to oviposition deterring pheromone (Behan and Schoonhoven, 1978; Crnjar and Prokopy, 1982; Bowdan, 1984; Stadler, personal communication). Tarsal receptors of lepidopterous adults have been shown to respond to salts and pheromones albeit other plant compounds are likely stimuli (Ma and Schoonhoven, 1973; Mitchell and Seabrook, 1974; Behan and Schoonhoven, 1978; Calvert and Hanson, 1983; Walaade, 1983; Renou, 1983).

4.2. Olfactory Chemosensilla

A. Single-Cell Recording

Recording from insect olfactory cells has been accomplished by two methods. The first method utilizes sharpened, uninsulated tungsten electrodes inserted at the base of a selected sensillum (Schneider, 1957; Schoonhoven and Dethier, 1966; Boeckh, 1967). Some workers have used platinized tungsten electrodes (O'Connell, 1975), or platinum iridium (Seelinger, 1983). The technique requires a rigid preparation and ultra-micromanipulators together with high-resolution optics. The reference electrode is placed close to the sensillum under study, usually in the same antennal segment. The active electrode is gently touched to the base of the sensillum and allowed time to settle into position. In some preparations nothing more is necessary; for others, a sharp gentle tap on the end of the manipulator is needed for penetration. Successful contact allows spikes to be monitored continuously in the absence of stimulation, since many cells are spontaneously active (Fig. 7). The orientation of the sensillum relative to the active electrode is critical since the electrode tip must settle close to the cell bodies. The major disadvantage of this type of recording is the extreme mechanical sensitivity of the preparation. Any drift of the manipulators or vibration through the table may result in loss of contact.

A more recently developed technique of tip recording requires the removal of the tip of the sensillum prior to insertion into a saline-filled capillary (Kaissling, 1974; van der Pers and den Otter, 1978). Such a preparation is more stable to mechanical disturbance, permitting reliable recording for longer times (Fig. 7A). A device for holding the preparation and a micro-knife assembly for cutting the tips of sensilla has been developed (Murphy Developments, P. O. Box 1547, NL-1200BM, Hilversum, The Netherlands).

There are two basic designs for odor delivery systems. One injects odor-laden air into a constantly flowing stream. This method dilutes each stimulus, thereby adding another variable in quantitation. The second method

A.

B.

Olfactory inhibition

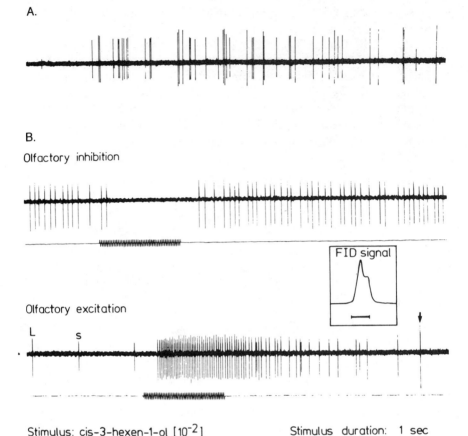

FID signal

Olfactory excitation

L   s

Stimulus: cis-3-hexen-1-ol [10^{-2}]   Stimulus duration: 1 sec

Figure 7. (A) Record of action potentials from olfactory sensilla on the antenna of *Agrotis segetum* male using the tip recording technique (van der Pers J C and Lofstedt 1983.) (B) Record of action potentials from olfactory sensilla on the antenna of a Colorado potato beetle adult using the base recording technique (Ma and Visser 1978).

involves switching from a clean air stream to an odor stream with as brief an interruption as possible. Both systems use highly purified air passed over activated charcoal and/or through molecular sieve, then rehumidified to 50% in gas washing bottles over saturated salt solutions. The air stream passes over the preparation and is exhausted to the outside.

Valves used for switching gas streams are mounted at some distance from the preparation so that considerable dead time exists before an odor pulse arrives at the sensillum. This arrival can be monitored by a thermistor

mounted directly behind the preparation, thereby providing a more accurate estimate of the onset of stimulation.

A wide variety of stimulus control devices has been used in insect studies
(Fig. 8). The simplest system uses filter paper in glass cartridges that can
be replaced relatively easily between trials and stored in the freezer between uses (Albert et al., 1974). Miniature switching valves may be arranged to give multiple stimuli (Davis, personal communication). A glass
tube calibrated by gas-liquid chromatography (GLC) and used for both
behavioral and sensory experiments offers the advantage of insuring the
same stimuli for comparative studies (Mayer et al., 1984). Visualization
of the actual odor plume (e.g., with $TiCl_4$) is required for each design to
ensure that the stimulus actually arrives at the desired part of the preparation.

A more complicated system uses a gas chromatograph to quantify each
stimulus (Frazier and Heitz, 1975). A sample of saturated headspace of
pure odorant is injected into rapidly flowing carrier gas with an open column. The outlet is mixed into the air flowing over the preparation. The
GLC may also be used for separation of unknown biological material and
a determination of sensory activity may be made for each peak as it elutes
(Arn et al., 1975; van der Pers and Lofstedt, 1983).

Whole insects are used whenever possible for preparations. The ap-
pendage bearing the sensilla for study is restrained. Moth antennae bear
scales on the side opposite the sensilla, which can be removed with a
brush before affixing to double-stick tape, 5-min epoxy, or beeswax (O'-
Connell, 1975; Rumbo, 1981).

The olfactory cells from a few phytophagous insects have been characterized and grouped into odor ''generalists'' or ''specialists'' depending
on the variety of compounds that stimulate them (Ma and Visser, 1978;
Selzer, 1981; Bromley and Anderson, 1982). Many volatile C_6 alcohols,
aldehydes, and derivative acetates vary in composition to produce different
green leaf odors and stimulate olfactory cells of several species of phytophagous insects (Visser, 1983; Dickens et al., 1984). Complex plant odors
stimulate caterpillar olfactory cells in ways that have made identification
of ''candidate codes'' difficult (Dethier, 1980a; Dethier and Crnjar, 1982).
Some olfactory cells respond to both plant odors and pheromones in adult
insects (den Otter et al., 1978; Dickens et al., 1984).

B. EAG Recording

The electroantennogram (EAG) was first used by Schneider (1957) in
studies of pheromone reception by the silkworm moth, *Bombyx*. Theoretically it results from the summation of the receptor potentials of all the
excited chemosensory cells on the antenna between the recording elec-

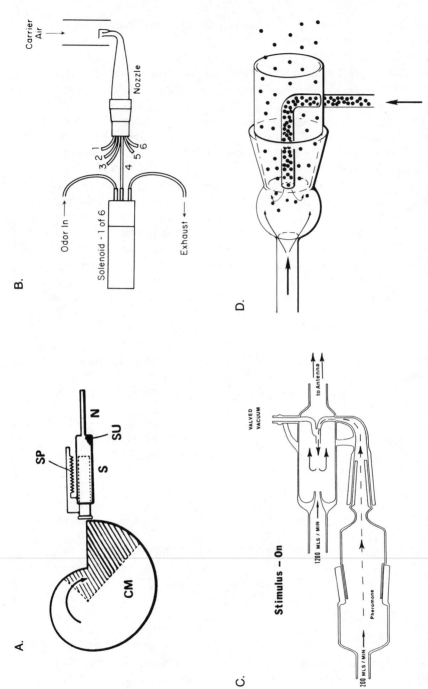

Figure 8. Various types of stimulus delivery systems used in studies of insect olfactory sensilla. (A) Diluted odorant in glass cartridge (Lacher, 1971). (B) Multiple valves for delivering mixed odors (Davis, 1984). (C) Glass tube used for pheromone delivery in both sensory and behavior studies (Mankin and Grant, unpublished). (D) Olfactory jet used on Gypsy moth sensilla (Hanson, unpublished).

trodes (Mayer et al., 1984). Recording the EAG requires an amplifier with a low band pass (e.g., 0.01–10 Hz). With electrically stable preparations, direct coupled amplification can be used. Glass electrodes with Ag-AgCl junctions are filled with insect saline. Stimulus systems are the same as those described above. A typical example of an EAG from the antenna of the Colorado potato beetle is shown in Figure 9. After a brief delay, a change in DC voltage is seen during stimulation, the amplitude of which increases with odor intensity. Since there are several conditions that may result in artifactual voltage fluctuations, one can verify that the recorded potential is biological in origin by showing that it can be abolished by HCN. The geometry of the stimulus delivery system relative to the antenna can affect the size and shape of the EAG and should be held constant. Preparation techniques are simpler than those described for single-cell studies; the distal end of the antenna is clipped off and a saline-filled electrode is placed over the tip.

The EAG has been widely used in studies of pheromone reception (Hummel and Miller, 1984), but relatively few studies have been published on plant volatile perception by phytophogous insects. The responses of several species of bark beetles to both pheromones and host tree volatiles have been comparatively well studied (Payne, 1974; Dickens et al., 1983). Visser (1979) found the Colorado potato beetle to respond to 2-hexene-1-ol with a low threshold and to other green leaf volatiles that are common to many plants. A comparative study of eight moth species to 57 plant volatiles and six host-plant extracts revealed similar patterns among closely related species, and no differences between males and females (van der

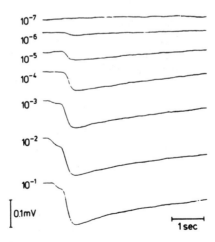

Figure 9. Electroantennograms recorded from the antenna of the Colorado potato beetle in response to increasing concentrations of cis-3-hexen-1-ol (Visser 1979).

Pers, 1981). In another study, both sexes of boll weevils were found to have green leaf odor receptors as well as pheromone receptors (Dickens, 1984). The carrot fly responds to unique components of its host plant (Guerin et al., 1983). The effects of age, time of day, and behavioral inhibitors that affect EAG responses to pheromones have yet to be determined for plant volatile perception (Roelofs and Comeau, 1971; Payne, 1974; Preisner et al., 1977; Fu-shun and Visser, 1982).

4.3. Chemosensory Nerves

Recording from insect chemosensory nerves has much potential, yet has been utilized little to date. In preparations with extremely dense fields of chemosensilla, or those with extremely short sensilla, conventional methods of recording may not be possible. Perhaps the use of a suction electrode on the afferent nerve may allow some chemosensory information to be recorded as has been done in Crustacea (Thompson and Ache, 1980). Although it may not be possible to record single units, recording at the level of the afferent nerve permits one to begin building a picture of multiple sensory cell inputs. Information arriving from separate sensilla simultaneously, or from those separated spatially and temporally, could be investigated this way. Extracellular recording with metal electrodes placed near a nerve may also be used with the inherent drawback that most insect sensory nerves are mixed, so that interference from mechanical stimuli is likely (Ave et al., 1978; Seelinger, 1983).

4.4. Ganglia

As a further extension of studies of chemosensory cell input from individual cells, the arrival of messages at ganglionic integrative centers provides a rich area for study. Such an approach has begun for olfactory cell inputs in the American cockroach (Boeckh et al., 1984) and for pheromone sensitive cell inputs in *Bombyx* (Olberg, 1983a; Light, 1986) *Manduca* (Matsumoto and Hildebrand, 1981), and Crustacea (Hamilton and Ache, 1983). Such studies are just beginning on phytophagous insects (Visser, personal communication).

Convergence of both modality and numbers of cells in the cockroach suggest that many important features of chemosensory processing can be unraveled at this level. These studies involve selective stimulation of one or more individual cells, while recording from the same ganglionic site. Standard intracellular recording and staining techniques are employed, coupled with thorough electron and light microscopic studies (Strausfeld and Miller 1980). The size of the ganglionic cell bodies may pose some limitations on the cells that can be investigated, although recordings from cells as small as 5 μm have been achieved (Matsumoto and Hildebrand, 1981).

V. Data Reduction and Analysis

5.1. Introduction

An advantage of electrophysiological techniques is that, once the apparatus is assembled, a large amount of raw data can be obtained rapidly. This appears to be an ideal experimental situation, yet this initial impression is only a silver lining upon a dark cloud. Much data reduction is required before the electrophysiological information can be interpreted.

Historically, the method of recording has been to film the data with an oscilloscope camera and manually count the action potentials as a function of time. This is preferred if small amounts of data are to be obtained, e.g., a project having a duration of less than a year (some researchers say 3). The investigator will soon discover, however, that this method gives new meaning to the term "drudgery." If a project of greater magnitude is envisioned, a wide range of automated and semiautomated techniques are available for action potential collection, sorting, and analysis. These have a number of advantages: large quantities of data can be manipulated rapidly, unbiased criteria are used in sorting action potentials, and the same data can be repeatedly sorted with different criteria or subjected to many types of analysis procedures in a minimum amount of time.

5.2. Semiautomated Data Analysis

A basic assumption in the analysis of electrophysiological data is that an action potential is an unitary event in a neuron, and is the equivalent of a bit of data in a channel. Action potentials are recorded on a baseline of varying amounts of electrical noise, and therefore the most essential elemental building block for data analysis is the threshold device to separate the data from the background. One such device is a Schmitt trigger (monostable pulse generator), which emits a standardized square pulse when its threshold has been exceeded by an action potential (Fig. 10A). The fidelity of this transformation is monitored by displaying the pulse on the second channel of the oscilloscope directly beneath the analog data (Fig. 10B). An electronic counter provides raw counts of action potentials during the time period in which the counter is enabled. This simple system, however, does not preserve the temporal characteristics of the response. This task is accomplished by an integrator, or counter with a cumulative count mode, coupled to a chart recorder. The cumulative sum of the activity is represented by the height of the chart trace (Fig. 10C).

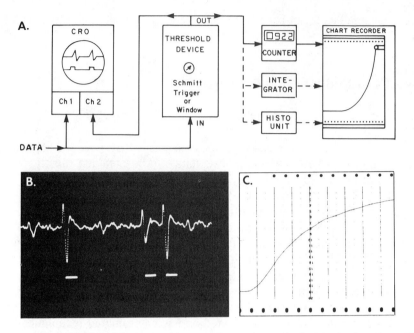

Figure 10. A semiautomated data acquisition system. (A) Block diagram. (B) Typical oscilloscope display. (C) Chart recorder output of the integrator/counter. The cumulative number of spikes is displayed with time. (Hanson, unpublished).

Additional data reduction may be accomplished by adding other hardware such as rate meters and units that directly produce histograms of either pulse height or pulse interval. The beauty of these systems is their simplicity; even a novice can have them up and running in a few hours. The units can be obtained for a few hundred dollars from many companies that produce modular instruments for neurophysiology (see references in Brown et al. 1973, 1982). Alternatively, since the main components are now standard chips, the units can be fabricated for a minimum of expense. Commercial sources for electronic parts and modules are found in Science's annual *Guide to Scientific Instruments,* the Journal of Electrophysiological Techniques, and in references in Brown et al (1973, 1982). Several technical publications give details of circuit design and applications (Vanderkooi, 1973; Jung, 1974; Lancaster, 1975).

The above systems work well with "clean" data comprised of action potentials that consistently rise above baseline or other undesired neural activity. If two spikes are only slightly different in height, some researchers cautiously suggest that nonlinear amplifiers may help in exaggerating those differences and may simplify setting thresholds for separation (caution: noise that rides on top of the action potential will also be magnified disproportionately, thus contributing more to the variability). If one wishes to count a medium-sized spike rather than the largest spike, or if two action potentials differ in shape but not in size, then more sophisticated devices must be employed. For example, a window discriminator used in place of the Schmitt trigger produces an output pulse only for spikes having peak voltages above one threshold but below another, i.e., they must fall within a "window" prescribed by the operator. More than one spike size can be counted using multiple discriminators having two or three windows. Also available are modules with "time windows" that separate spikes of different durations. Combining height and time windows may provide even more discrimination.

5.3. Automated Data Analysis

The above are effective systems for counting action potentials in simple recording situations; however, the resulting partially reduced data must still be handled manually for averaging, analysis, and graphing. Elimination of this manual intervention is accomplished by directly connecting the threshold device to the laboratory computer.

A word of caution is in order at this point. Automated data collection and processing usually require a significant investment of time for most biologists. Unless programs are already available for experiments of the type the investigator is contemplating, it may not be efficient to use computers for small projects. The uninitiated may require months of learning about computers, writing and debugging programs, etc., before the first yields are obtained. Nevertheless, the serious electrophysiologist must consider the possibility of using computer analysis, since it clearly adds significantly to the power of techniques in this area of research.

A. Hybrid Analog and Digital Systems

Laboratory computers may augment the semiautomated devices already discussed (Fig. 11). At the onset of the recording interval (e.g., beginning of a stimulus) a gating pulse starts the computer internal clock. When a pulse representing an action potential arrives, the computer stores the time at which the event occurred. If a multichannel window discriminator is used, the computer also saves the channel number in which the event occurred. When recording is finished for that trial, the data are stored on mass storage for later recall. The computer is then readied for another trial. When the entire session is terminated and the data are stored, the analysis programs are loaded into the computer. These programs might, for example, ask the operator which data are to be selected from within the trials (e.g., those spikes from 0.1 to 3.0 sec into the trial) and proceed with counting, averaging, plotting, etc.

An analog discrimination system coupled with a digital data accumulation/storage system is superior to the semiautomated system. The primary advantage is that data are stored in a form accessible to the computer. This permits immediate sorting and display which helps one determine whether the data are worth saving or whether another trial is needed. Data manipulations and analyses can be programmed, thereby freeing the investigator for more experiments. The equipment is relatively simple. An analog threshold device is coupled to a microcomputer with either an analog-to-digital (A-D) input or pulse/level interface, printer, and floppy disks. A plotter is also desirable. The time investment required to develop software can be considerable, although some appropriate software is available commercially, or through user groups.

The limitations of this hybrid system are primarily those of the threshold devices. Action potentials having similar heights are not easily separated, nor are those that vary in height or rise from a fluctuating or noisy baseline (a situation often encountered in insect chemosensory recording). If more than two spikes are to be separated under such circumstances, it is difficult to maintain a proper setting of the window thresholds in real time, since data arrive more rapidly than the operator can react. One solution is to utilize a tape recorder for playback at reduced speed; however, this not only makes the voice commentary useless, but also greatly extends the time required for analysis. Thus, for analyzing complex data in a reasonable time period the computer is used for spike recognition and separation as well as storage and analysis. Several systems have been developed for these tasks, some of which are described below.

B. Computer Systems

(a) Data collection. In contrast to hybrid systems which reduce the action potential to a pulse, the computer system digitizes the full action potential and stores it, thus facilitating subsequent reconstruction and scrutiny. This requires an A-D converter that can obtain 10–30 voltage samples over the duration of the action potential. The number of voltage samples required depends on the analyses desired. Although 10 samples are sufficient to portray the overall shape of the spike, they are not enough to resolve the peak reliably. Some investigators have used over 100 samples which give excellent resolution but require extra memory. We find that 23 samples/spike provides a good compromise. Thus, a 2.3-msec action potential is sampled at intervals of 100 μsec for an effective conversion rate of 10 KHz.

If each action potential is sampled and stored individually (see below), this 100-μsec interval must include the software time for storing the sample and re-initiating the next sample. Although most A-D converters can maintain a 10-KHz rate, software "overhead" requirements often reduce overall throughput to 1 KHz or less, especially when using interpreted high-level languages (e.g., BASIC). Consequently, special machine language routines or direct memory access hardware must be used for digitizing and storing if data are to be collected at real-time speeds. Compiled languages (e.g., FORTRAN and CBASIC) are not fast enough to keep up with high digitization rates, but as faster machines become available this may be possible. For the investigator with software/hardware limitations for real time acquisition the tape recorder can be played back at lower speeds if the loss of voice information is not critical.

Given the availability of hardware/software systems having adequate speeds, the next decision is how to digitize the data. Two approaches are currently used, both of which are amenable to on-line or off-line data capture: (1) digitize only the action potentials; (2) digitize the entire record. A discussion of the merits of each is outlined below.

(1) Digitization of single action potentials. This method is the most efficient for small machines with limited memory, but requires some analog equipment and operator control. The configuration in Figure 11A has been effectively used by Hanson et al. (1986). The data output from the amplifier or tape recorder follows two pathways: in one, an action potential triggers the A-D converter to begin digitizing; in the other, the action potential itself is digitized. In the first pathway, the nerve spikes are differentiated to provide a sharp, clean pulse (Fig. 11B, middle trace). This is fed into a threshold device having a Schmitt trigger to provide a pulse when a suprathreshold event occurs, just as in the analog system discussed earlier. This pulse triggers the A-D converter to initiate the sampling for a preset number of samples. In the second pathway, the amplifier or tape recorder output is connected directly to the input of the A-D converter to permit digitization of action potentials. In many cases, some analog filtration prior to digitization (as shown in Fig. 11A) may be desirable.

Incorporating an analog delay device between the tape playback and the A-D converter permits saving the early portion of the action potential as well as the post-triggered portion. The action potential destined for the A-D converter is delayed until sampling is initiated (Fig. 11B, bottom trace). Thus, any portion of the action potential can be used to initiate digitization (top pathway of Fig. 11A) and still capture the entire action potential (Fig. 11C). For example, we found the negative slope (the peak-to-trough transition) has the best signal-to-noise ratio. Passing the action potential through an analog differentiating circuit or a narrow band filter ahead of the threshold device enhances the negative pulse (Fig. 11B, middle trace). By carefully adjusting the differentiation constants or bandwidth limits of the filter, surprisingly small action potentials are "rescued" from noisy baselines.

(2) Digitization of the entire record. If large computer memories are available or if individual segments of data are short, the entire record

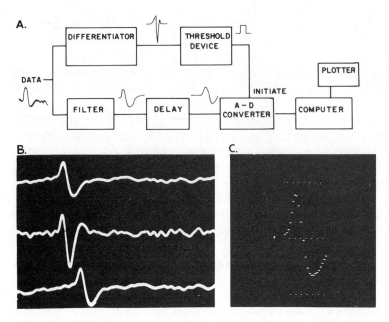

Figure 11. An automated data acquisition and analysis system. (A) Block diagram with an indication of how the signals are modified by each unit. (B) Analog signal input (top), a differentiated signal (middle), and a delayed signal (bottom). (C) Computer display of digitized trace during data acquisition. (Hanson, unpublished.)

consisting of spikes and baseline can be digitzied and stored (Fig. 12, top record). Most of the above analog hardware is eliminated: the amplifier or tape recorder is connected (through a filter, if desired) directly to the A-D converter. The hardware requirement is that the A-D converter have its own memory buffer or direct access to the computer's memory (DMA). Sampling is initiated at the onset of stimulation and each sample is stored as it is obtained.

Digitization continues until the memory buffers are full or a preset time has elapsed. This can be accomplished with programs written in high level languages, thus simplifying programming. Software threshold and peak-finding programs then search for spikes resembling action potentials (Fig. 12, bottom trace). This operation shrinks the data by a factor of 5–50, depending on action potential frequency. Only the processed data are permanently saved, thereby reducing mass storage requirements.

The major advantage of this approach is that data may be visually inspected in their digitized form and repeatedly scanned with different thresholds or peak-finding criteria to give maximal signal recovery from noise. This is superior to the first method, in which an improper hardware threshold setting loses data that cannot be recovered. Additionally, this

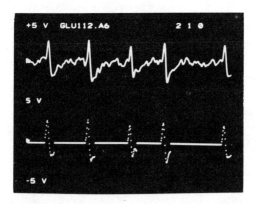

Figure 12. Entire digitized record of both spikes and baseline noise (top) and the results of a filtering program that selects spikes by threshold and peak-finding criteria (bottom). (Frazier, unpublished.)

method permits one to apply the same criteria to data from different digitized files. Normalization is done by using calibration voltage pulses that were inserted through the preparation at a preset time during the original recording. During analysis, calibration pulses are automatically separated from spikes by their unique waveform. They are then averaged and used to scale the neural data to correct for any differences in amplification or resistances among preparations.

(b) Parameters for sorting action potentials. (1) Height of action potentials. The above procedures store signals only some of which are action potentials. Another routine is required to reject the noise spikes and to separate action potentials according to an appropriate criterion.

The most commonly used separation parameter is the height of the action potential. A peak-finding algorithm can pass through the data and retrieve the voltage values of each peak. Any slow fluctuations in the baseline, however, will be reflected as variations in these peak values. An improvement is to use a routine that retrieves the total height of the action potential and is thus independent of the slow baseline fluctuations. Such peak-to-trough heights are then plotted as frequency histograms (Fig. 13B). Spikes of similar heights fall into clusters. Separation of these clusters is accomplished manually by inserting limits for each spike class or automatically by an algorithim that separates classes by determining the point of overlap of the clusters (O'Connell, 1975).

These approaches of separating spikes have two main disadvantages. First, they require operator attention after each trial. This is highly repetitive work which is reduced or eliminated by using more automated routines. Second, a simple parameter of the spike such as height is a crude discriminator because it does not utilize all the information available in

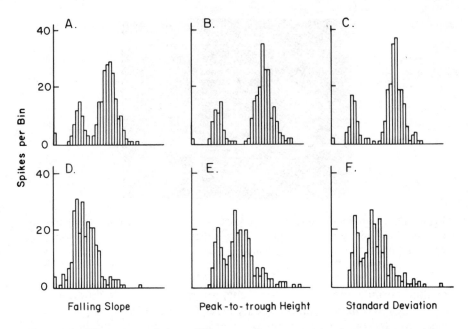

Figure 13. Histograms of three different spike parameters that may be used for spike separation. (Top row) Two cells of differing heights are easily separated by any of the three parameters; (bottom row) three cells not readily separable by these parameters. Data from the medial maxillary styloconicum of *Manduca sexta* in response to leaf stimuli: *Canna generalis,* a non-host plant (A, B, C) and tomato, a host plant (D, E, F). (Hanson, unpublished.)

the spike. Sometimes spikes originating from different cells are shaped differently, but have nearly the same height and are not separable by height alone (Fig. 14A, C left). Adding to this problem, height distributions of spikes of similar size overlap when baseline noise is added. If more than two spike types are present in high frequency, the algebraic addition of imperfectly coinciding action potentials results in spikes of almost any size, which confounds separation attempts based on height alone.

(2) Slope and standard deviation. The most noise-free portion of the spike is the falling slope. Accordingly, we hoped this would be a good parameter for spike separation, but this was not always the case. Slope is not as good a discriminator as peak height (compare Fig. 13D vs. 13E) and neither is as good as standard deviation (Fig. 13F), probably because it is more indicative of spike shape. O'Connell (1975) used standard deviation to separate two spikes of obviously different sizes from a pheromone-sensitive sensillum. If spikes are considerably different in size, any of the above will serve as suitable parameters for separation. In our experience, more complex approaches are needed to separate three or more spikes, as well as two spikes of similar size.

(3) Shape. Several algorithms for discrimination make use of the shape information of a spike. For separating spikes of two pheromone-sensitive cells in *T. ni,* Mankin (personal communication) obtains two shape-sensitive indices from each spike. One is the total area under the positive and negative peaks of the spikes. The other index is the area between the full spike outline and a five-point skeleton of the spike. A scatterplot of these two indices separates the action potentials into four clusters: spikes from cell A, spikes from cell B, double spikes, and noise pulses.

Even more shape-sensitive are template-matching algorithms. The templates may be selected manually (Gerstein and Clark, 1964) or automatically (Milechia and McIntyre, 1978; Hanson et al., 1986). These routines determine the least-squares difference between the voltage samples of the template and those of a spike. If this difference is less than an operator designated threshold, the spike is included in the class defined by that template. Each remaining spike in the data set is compared with the template in the same manner. Similar interations are made with each remaining template until all spikes are classified or rejected as unclassifiable.

An example of the sensitivity of this method is seen in Fig. 14. In this experiment a house fly tarsis is stimulated with NaCl and sucrose. Two action potential classes are seen in the response, a "salt spike" and a "sugar spike." Close inspection of Figure 14A shows that the two are nearly the same height, but have slightly different shapes: the predominant sugar spike has a longer-lasting negative phase. The template algorithm successfully resolves these two into separate classes (Fig. 14B). For comparison, Figure 14C, left, shows the histogram of spike amplitudes of all the spikes before classification. A continuous distribution is seen. Separation by spike amplitude alone would be impossible. Following classification by the template algorithm, two distinct classes differing only slightly in height are seen (Fig. 14C, middle and right). This illustrates the advantages of using shape information as a discriminator, rather than height alone.

(4) Multiple parameters. Another approach (Piesch and Wieczorek, 1982) uses four or more spike parameters. First, a histogram is made of a single parameter, such as spike height or area. The operator designates only the midpoints of the clusters of the histogram. The computer then finds the individual spikes that represent those midpoints ("safe spikes") and uses them as models for each class. The computer attempts to find the best match for each of the remaining spikes with a "trainable pattern classifiers" algorithm (Nilsson, 1965) using as training data four features derived from the raw spike. These are projected into multidimensional space, and the computer looks for hyperplanes that separate the clouds of points.

(c) Monitoring the classification process. The results of separation routines must be carefully monitored to ensure that classification is accomplished in accordance with the operator's overall experience and

A.

0.1 SUCROSE/0.5 NACL

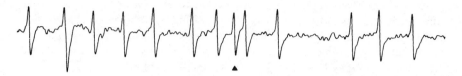

B.

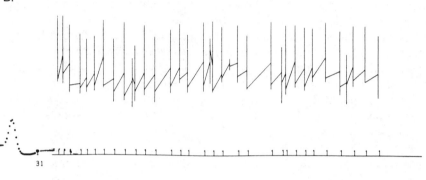

Figure 14. Automated sorting of a mixed spike train into the component neuronal responses using spike shape criteria. (A) Spikes from a tarsal D hair of the house fly in response to 0.1 M sucrose mixed with 0.5 M NaCl. Salt cell spike indicated by arrow. (B) Output from the CLASIF program showing separation of the two types of spikes into separate classes.

knowledge of chemosensory cell properties. Allowing the computer to make selections without continuous monitoring is inviting large scale generation of artifacts. Feedback at each step is not only essential, but should be done with many types of displays, so the investigator can apply independent criteria in determining that the classification process was successful. If the operator is not satisfied with the separation, the classification program must be run again with different parameters, or the experiment must be repeated to improve the signal-to-noise ratio. The problem may be too many different kinds of spikes at high frequencies, resulting in many partial coincidences. One solution to this problem is to repeat the experiment with more dilute stimuli giving lower response frequencies.

The utility of displays for collecting and processing data is illustrated

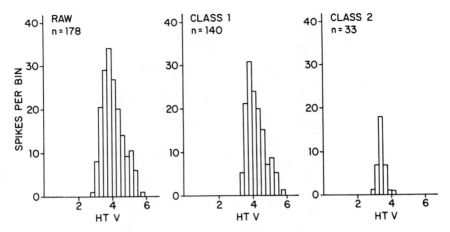

Figure 14.(C) Histograms of the data from which 14A, B were taken before separation by shape criteria (raw), and after separation into sugar cell (class 1), and salt cell, (class 2). Abscissa is spike height in volts after amplification of ca. 4000. (Frazier, unpublished.)

by the following examples. During the digitizing step, the analog signal is monitored on an oscilloscope. If the single spike collection option is used, the program returns a pulse when a digitization occurs (Fig. 10B, bottom trace). If the entire record collection option is used, the software collection system returns a display of all the digital data (Fig. 12, top) as well as selected spikes (Fig. 12, bottom) on completion of digitization. These displays allow the investigator to optimize parameters to reject noise and retain spikes. When all the spikes are digitized and stored, they may be easily retrieved and displayed to assure that the A-D conversion was satisfactory (Fig. 15A).

Displays also return information about the automatic construction of the templates in the classification procedure. The templates are displayed for each spike class, and if desired, the spikes that comprise each template. This provides the operator with the opportunity to control makeup of the template. Erroneous signals that may have been included in the template construction process can be deleted selectively to improve the template in the subsequent reclassification.

Another display provides information on the efficacy of spike separation. Rapid visual comparison of all the action potentials in a class indicates the degree of homogeneity (Fig. 15B,C). For an overview, plotting one shape parameter against another as a scattergram displays class homogeneity in a more compact form (Fig. 16A). Perhaps the most useful and intuitively satisfying display is a scatter plot of the spike height versus time (Fig. 16B). This display further confirms the quality of the classification. Second, it provides the operator with information on any temporal

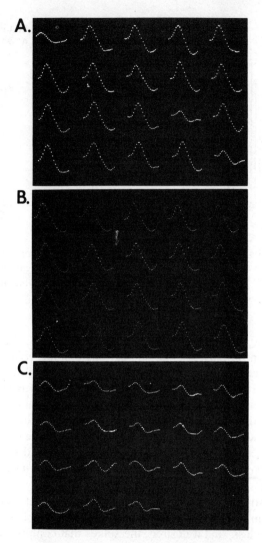

Figure 15. Computer displays verifying spike separation. (A) All spikes in a train are displayed sequentially, 20 at a time on the computer screen. (B, C) Display of class 1 and class 2 spikes, respectively, after separation by the computer. Data are from the file histogrammed in Figure 13 A, B, C.

pecularities of the spike train, such as the occurrence of especially large or small spikes, or gradual changes in spike sizes (e.g., larger spike Fig. 16B). Knowledge of the latter is especially important if the spike gradually changes from one class into another, resulting in erroneous counts in each class.

   (d) Data analysis. Once the spikes are successfully sorted into classes,

each representing responses of single cells, the stored data are analyzed. For example, a program that retrieves a class of spikes from mass storage and plots instantaneous frequency (1/interspike interval) illustrates adaptation of an individual cell (Fig. 17A). The resulting plot shows the real variability in the cell's firing rate. A smoother curve is obtained by averaging repetitions from the same cell or from homologous cells of different sensilla. One method of accomplishing this is to average interspike intervals. A more robust method is to average the frequency bins of post-stimulus time histograms. Such a curve based on data from multiple repetitions on many sensilla is shown in Figure 17B (linear time axis), and in Figure 17C (logarithmic axis). This shows the phasic and tonic portions

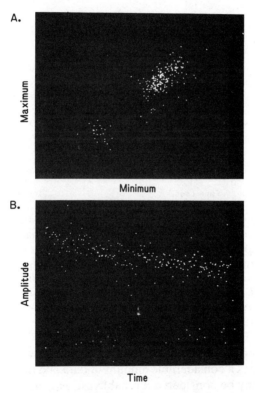

Figure 16. Computer display of scattergrams of spike parameters to verify spike separation. Each class can be displayed separately or together and examined for overlap with other classes. Data are the same as in Figure 13 A, B, C. (A) Magnitudes of positive peaks (maximum) plotted against magnitudes of negative peaks (minimum). (B) Total spike amplitude (maximum-minimum) plotted against elapsed time after onset of stimulus. Total time = 4 sec. Note decrease in amplitude of the larger spike with time, which is a characteristic of rapidly firing insect chemosensory cells. (Hanson, unpublished.)

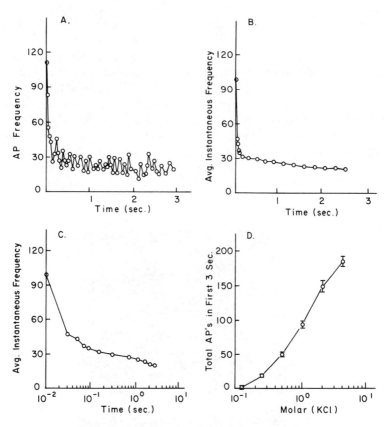

Figure 17. Computer plots of chemosensory cell data illustrating the use of averaging. (A) Instantaneous frequency of a single response trial of a single cell (i.e., one class of spike) from a blow fly tarsal sensillum to 0.5 M RbCl. (B) Same, but average of two trials from each of 10 cells using the average post stimulus time histogram method. (C) Same, plotted on logarithmic time axis. (D) Average dose–response curve. Each point represents the average response of two or three repetitions on eight to ten cells of eight animals (total trials = 170). (Hanson, unpublished.)

of the adaptation curve as well as the transition between them. Such a curve summarizes a considerable amount of data and therefore the shapes of the curves may be considered a reliable estimate of the response. This represents the temporal component of a cell's response to one concentration of a stimulus compound.

Another characteristic parameter of a cell is its response to different concentrations of a stimulus. This is most reliably summarized by the averaged dose–response curve. A single measurement, such as the total number of action potentials in a selected time interval during the tonic phase, is obtained from each response of each cell. These measures are then averaged across each repetition of every trial, sensillum, animal, and concentration. Plotting these averages results in a dose–response curve, such as that shown in Figure 17D.

In addition to coding the concentration of a stimulus, chemosensory cells must also provide information that allows the animal to distinguish one type of stimulus compound from another. The coding for quality of a stimulus may be found in any aspect of the cell's response. More information probably exists in a spike train than simple frequencies of the responding sensory cells. Consequently, averaging interspike intervals over a long time period (e.g., 1 sec) may obscure stimulus-specific patterns. Some stimuli may trigger irregular intervals while others elicit more regular firing. Such interval coding is one of several "candidate codes" for stimulus quality. This is demonstrated by comparing the distributions of interspike intervals of *Manduca* taste receptors resulting from stimulation by different host plants (Fig. 18). The distributions of interspike intervals are different for each class of spike in response to each stimulus. This is evidence that there is complex temporal information in spike trains of this chemosensory organ. This supports similar conclusions of Dethier and Crnjar (1982).

The above are only some examples of the methods of analyses that are used. In addition, multivariate analyses such as cluster analysis (van der Molen et al., 1978), factorial analysis, and multidimensional scaling (Schiffman et al., 1981) may also be useful. Once the data are in a computer file, they are amenable to rapid treatment by many standard statistical and display programs.

VI. Characterization of Chemosensory Cells

One of the goals of chemosensory physiology is to characterize completely the response capabilities of each sensory organ of each important insect. This involves many descriptors. The purpose of this section is to indicate which electrophysiological parameters will lead toward such a characterization.

Attempts to compare data from chemoreceptors of many insect species are often futile because each investigator measures different parameters or does so in different ways (Mitchell and Gregory, 1979). A more systematic approach is necessary to categorize basic parameters in a manner useful for comparative physiology. The following are some key parameters that should be described for each chemosensory cell.

6.1. Number of Sensory Cells per Sensillum

The number of sensory cells in the organ to be studied can be determined from electron micrographs. The following electrophysiological studies then have a predictable number of classes or shapes of action potentials.

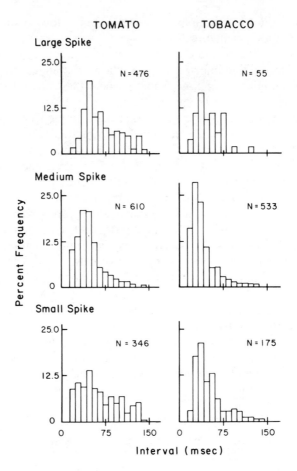

Figure 18. Computer plots of interspike intervals of three classes of action potentials from the medial maxillary styloconica of *Manduca sexta* larva in response to two highly acceptable host plants. Data are accumulated from two trials on each of three animals. Note that intervals from the large spike (and possibly also the small spike) have different distributions, indicating that information from this sensillum is different for these two plants. (Hanson, unpublished.)

6.2. Adaptation and Disadaptation

The first important response parameter to determine is the adaptation rate of the cell to each stimulus used. For example in Figures 17B and 17C, the cell adapts phasically for about 100 msec, then the tonic portion of the curve shows a slower adaptation. Different exponential decay rates are seen for each portion of the curve, suggesting that each is governed by a different process (Hanson, unpublished). The time of transition between the phasic and tonic portions of the curve is useful information, particularly if the measure of response magnitude is to be consistently taken from the tonic phase uncontaminated by the more variable phasic phase.

The disadaptation rate is also important to determine for a proper design of the experimental protocol. This test should be conducted with the highest concentration of the most active compound. Stimuli of constant duration are repeated ca. 10 times at each of several intervals, e.g., 1, 2, 5, and 10 min apart. An intertrial interval should be adopted that permits complete disadaptation between stimuli, i.e., recovery to the level of the initial trial. If this criterion is met, repeated responses can then be compared for different concentrations of each compound.

6.3. Normalization

Comparison of responses among sensory cells from many animals requires some type of normalization. The usual method involves a standard stimulus repeated several times at specific points in the protocol for each cell. Thus, all responses are normalized to an average of the responses to this standard. Periodic inclusion of a standard stimulus also serves the important function of monitoring the constancy of response to the standard over the duration of a single experiment. If deviations are small, the data are corrected for the trend; if serious deterioration of response to the standard occurs, data from that preparation are eliminated.

6.4. Dose–Response Parameters

Once the foregoing basic characteristics are determined, repetitive experiments can be done with confidence that the responses will remain constant over many trials. The average dose response curve, for example, requires many repetitions of several concentrations of solutions on many sensory cells of many animals (Fig. 17D). The shape of the curve is thus representative of the "species response." From this several key parameters can be obtained that are useful in cross-species comparisons.

One such parameter is the threshold. Although this is difficult to determine exactly (Mankin and Mayer, 1983), an approximation is still useful. Most workers select a point at which the extrapolated dose–response curve crosses the abscissa, even though the actual mean frequency is not zero (e.g., this point is approximately $0.1\ M$ in Fig. 17D).

Other useful parameters are the maximum response, the effective concentration range, the Hill coefficient, and the K_b (the concentration that elicits a half-maximal response). In classical enzyme-substrate models of receptor function, the K_b represents the concentration where half the receptor sites are filled. This may be the most useful single comparative parameter, since it probably best represents the effective sensitivity of the receptor to a compound (Hansen and Wieczorek, 1981).

6.5. Cell Specificity (Stimulus-Best Type)

One of the major tasks in characterizing a chemosensory cell is determining its cell type. Electrophysiological research on chemosensory systems in vertebrates, for example, has searched for the specific cell types responsible for sweet, salt, sour, and bitter tastes, while being watchful for others as well (Smith and Travers, 1979). Such stimulus specificity has not been found at the receptor cell level in mammalian systems; instead, each sensory cell responds to nearly all stimuli, and thus terms like "sugar-best," "salt-best," etc. more properly describe a cell's response by indicating its highest relative sensitivity.

Broadly tuned cells are also found in insects. Because of the difficulty in challenging such a cell with a wide variety of stimuli, very few have been well characterized. Insects are perhaps better known for their narrowly tuned cells, such as the gustatory "sugar cell" and certain pheromone-sensitive cells. Many of these have been studied extensively, although very few have been well characterized. For example, in a comparative review of sugar-sensitive cells in insects, Mitchell and Gregory (1979) list responses of each cell to all of the known stimuli as well as other response parameters discussed above. This review concludes that not enough information is available to permit a complete characterization of the cells in question.

Perhaps the most difficult aspect of a cell to characterize is its breadth of tuning because this is dependent on the number of potential stimuli with which the cell is challenged. The usual approach is to begin with the behaviorally relevant stimuli grouped in functional mixtures. For example, a group of potential stimulants such as sugars and amino acids, or a group of potential deterrents such as alkaloids or terpenes, may be used as stimuli. Alternatively, one could group chemicals by structures, such as mixtures of only salts or amino acids. Once a mixture is found to stimulate a particular cell, the individual components can be assayed individually. First, qualitative information concerning cell responsiveness is obtained using each component stimulus at a single dose; subsequently, quantitative characterizations are made using a range of concentrations for each stimulus type.

If the effective stimuli are of similar structures, such as hexoses, a cell of high specificity is indicated. In the most extreme case, this specificity may be conferred by only a single receptive site, such as certain olfactory cells exquisitely sensitive to an insect pheromone (Mustaparta et al., 1984). Most gustatory cells appear to be less specific, which can be the result of a cell having one type of receptive site that is relatively unspecific or multiple receptive sites of different specificities. For narrowly tuned cells, a name is selected that describes the effective structurally similar chemical group(s) (e.g., "sugar cell") whereas for broadly tuned cells a name is usually selected that describes the behaviorally relevant parameter (e.g., "deterrent cell").

Individual receptive sites can be further characterized. Verification that two compounds of similar structure interact with the same receptor site results from applying their mixtures in reciprocally varying concentrations (Gillary, 1966). This is a laborious process, and as a result the specificities of receptor sites of most insect chemosensory cells are poorly documented.

6.6. Pharmacological Properties

The susceptibiliy of insect chemosensory cells to chemical modification by various agents is potentially a very useful approach to their characterization. Many enzymes, toxins, group-specific reagents, and detergents are now available for determining biochemical differences among membrane components. These are becoming increasingly important in many areas of neurochemistry, yet only a handful of studies on insect chemosensory cells have taken advantage of them (Shimada et al., 1974; Frazier and Heitz, 1975). The first verification that separate receptor sites for pyranose and furanose sugars reside in the same cell resulted from the finding of their differential sensitivity to p-chloromercuribenzoate (PMB) (Shimada et al., 1974). We anticipate that many other important properties of insect chemosensory cells will be determined by studies with such chemical agents. These properties can then be used as additional parameters to characterize the cells.

VII. Applications of Electrophysiology to Insect–Plant Interactions

7.1. The Role of Chemosensory Cells in Regulating Behavior

Understanding the chemosensory regulation of specific behaviors, such as feeding or host-plant finding, is of considerably greater magnitude than understanding a single sensory modality. Many simultaneous sensory inputs and sequential segments of behavior must be taken into account. This complex problem can only be analyzed by subdividing it into components. Accordingly, one must begin by describing a small portion of the total behavior, such as chemically stimulated feeding during a single meal in experimental conditions that are rigidly defined and reproducible.

A simple model for chemosensory regulation of feeding in the caterpillar, *Pieris brassicae*, was proposed by Ma (1972) and further clarified by Blom (1978). Electrophysiological records from gustatory organs, the two maxillary styloconica and the labral sensilla, show that the responses to sucrose by the individual organs do not correlate well with feeding behavior. Only when the responses from all of these sensilla are algebraically added does a linear input/output relationship emerge. Although this is but a simple example using a single stimulus chemical, it illustrates that combined input from multiple organs must be considered in understanding chemosensory regulation of behavior.

Knowledge of the sensory code can be helpful in understanding the physiological basis of a particular behavior. For example, deterrence of feeding by a food or chemical could be coded by one or more of the following: (1) stimulation of a "deterrent" cell; (2) inhibition of an "acceptance" cell; or (3) some "complex code" involving many cells, such as rapid or irregular volleying of many or all sensory cells in the sensillum.

Evidence for each of these has been obtained using electrophysiological techniques. For an example of the first type of candidate code, Ma (1972) reports that strychnine stimulates a single cell in the sensillum styloconicum of *Pieris brassicae* at concentrations found in nature. This compound was also shown to decrease feeding in behavioral assays, suggesting that the cell responsive to strychnine is a deterrent cell.

The second method of coding for feeding deterrence, inhibition of an "acceptance cell," has been demonstrated by Mitchell and Sutcliff (1984) in the red turnip beetle. Three plant alkaloids (sparteine, nicotine, and quinine) all inhibit activity of the sucrose-sensitive cell. Behavioral assays also show sparteine to be an effective feeding deterrent.

The third candidate code, a complex pattern of spike activity, has been seen in many preparations. For example, Mitchell and Harrison (1984) showed that tomatin, which is rejected by the Colorado potato beetle in behavioral assays, causes rapid volleying of all cells in the chemosensilla of this animal. Also Dethier (1980) showed that all unacceptable saps elicite multicellular responses in *Danaus plexippus*, suggesting that this may be a common deterrence code.

The above are examples of sensory input to the CNS from single organs that are correlated with a particular type of feeding behavior. Accordingly, such input is described as a candidate code for this behavior. Acceptance behavior of single substances, for example, is often correlated with activity in a single cell, which suggests "labeled line" coding. A more complex code involving many cells is, "across fiber patterning," which appears to be employed for complex stimuli such as leaves (Dethier, 1976).

7.2. Problems in Understanding Chemosensory Codes

Deciphering sensory coding properly requires both an accurate measure of the sensory input and a correlated behavioral response that is unambiguous. The variabilities associated with assessing these two factors lead to significant problems (Smith et al., 1983). For example, even though two animals may have the same sensory input, their behavior may be different depending on physiological state, satiety, nutritional state, developmental state, etc.

Prior experience also affects subsequent insect behavior (Jermy et al., 1969; Prokopy et al., 1982; de Boer and Hanson, 1984). Plasticity in feeding behavior has been shown to correlate with modified sensory input for the tobacco hornworm (Stadler and Hanson, 1976; Schoonhoven, 1969) and the armyworms *Spodoptera exempta* and *littoralis* (Simmonds and Blaney, 1983; Blaney and Simmonds, 1985) but probably not for the promethia moth (Stadler and Hanson, unpublished observations). Thus, the mechanisms of these gross behavioral changes likely involve peripheral and central modifications, both of which complicate the understanding of sensory coding.

One of the underlying assumptions in all of these studies is that the activity of each sense cell can be determined with certainty. For many preparations, this is not yet possible. Despite improving techniques for recording and analysis, many uncertainties remain in identifying which recorded spikes are associated with which sensory cells. One problem is that spike shapes may vary with time during a trial, particularly if the cell is responding at high frequencies (e.g., Fig. 16B). Second, spike shapes of the same cell may vary with different stimuli, with different concentrations of the same stimuli, or with interactions of many stimuli (Fujishiro et al., 1984; Bowdan, 1984; Frazier, unpublished observations). All of these conditions, in addition to the endemic problems of poor signal-to-noise ratio in these high-resistance preparations, adversely affect our ability to identify the cellular origin of the recorded action potentials.

Because of these difficulties, additional ways of identifying the activity of individual cells in the electrophysiological recordings are needed. As a supplement to characterizing the cell, as discussed above, one can catalog the sensory cell's responses to various stimuli and reagents, much as the pharmacologists have done for the reagent sensitivities of neurons and synapses. Such a list might include (1) response spectra to stimuli and inhibitors; (2) reagent specificities, e.g., the effects of protein modifying agents, etc; and (3) special characteristics of responses to mixed stimuli, e.g., the effects of different salt concentrations on the shape of the "sugar spike." Other investigators attempting to identify a cell by electrophysiological techniques could compare their results to the library of such tests for verification of their proposed identification.

The physiological dynamics of the insect and the uncontrolled experimental variables enumerated above dictate that multiple approaches to understanding coding must be employed. The sheer magnitude of the data required for such an analysis will likely make insect sensory biology even more dependent on computer assistance in the future. But as computers evolve from special installations to standard laboratory instruments, many of the data manipulations will be simplified by hardware and software innovations. The investigator's time previously needed for data analysis may then be spent on experimentation. The result will be a more accurate and complete picture of the capabilities and response repertoire of the insect sensory system. This in conjunction with more precise behavioral observations will help clarify input/output relationships and improve our understanding of the role played by the chemical senses in controlling behavior.

References

Albert PJ, Seabrook WD, Paim U (1974) Isolation of a sex pheromone receptor in males of the Eastern Spruce Budworm, *Choristoneura fumiferana*. J Comp Physiol **91**:79–89.

Altner H, Prillinger L (1980) Ultrastructure of invertebrate chemo-, thermo-, and hygroreceptors and its functional significance. Int Rev Cytol **67**:69–139.

Arn H, Stadler E, Rauscher S (1975) The electroantennographic detector—a selective and sensitive tool in the gas chromatographic analysis of insect pheromones. Z Naturforsch **30**:722–725.

Ave D, Frazier JL, Hatfield LD (1978) Contact chemoreception in the tarnished plant bug, *Lygus lineolaris*. Entomol Exp Appl **24**:17–27.

Behan M, Schoonhoven LM (1978) Chemoreception of an oviposition deterrent associated with eggs in *Pieris brassicae*. Entomol Exp Appl **24**:163–179.

Bernays EA, Chapman RF (1972) The control of changes in peripheral sensilla associated with feeding in *Locusta migratoria*. J Exp Biol **57**:755–765.

Blaney WM, Simmonds MSJ (1985) Experience of chemicals alters the tast sensitivity of lepidopterous larvae. Chem Senses **8**:245.

Blom F (1978) Sensory activity and food intake: a study of input-output relationships in two phytophagous insects. Neder J Zool **28**:277–340.

Boeckh J (1967) Reaction thresholds and specificity of an odor receptor on the antenna of *Locusta*. Z Vergl Physiol **55**:378–406.

Boeckh J, Ernst KD, Sass H, Waldow U (1984) Anatomical and physiological characteristics of individual neurones in the central antennal pathway of insects. J Insect Physiol **30**:15–26.

Bowdan E (1984) Electrophysiological responses of tarsal contact chemoreceptors of the apple maggot fly *Rhagoletis pomenella* to salt, sucrose and oviposition-deterrent pheromones. J Comp Physiol **154**:143–152.

Bromley AK, Anderson M (1982) An electrophysiological study of olfaction in the aphid *Nasonovia ribis-nigri*. Entomol Exp Appl **32**:101–110.

Brown PB, Franz GN, Moraff H (1982) Electronics for the Modern Scientist. Elsevier, New York.

Brown PB, Marfield BW, Moraff H, (1973) Electronics for Neurobiologists. MIT Press, Cambridge.

Broyles JL, Hanson FE (1976) Ion dependence of the tarsal sugar receptor of the blowfly, *Phormia regina*. J Insect Physiol **22**:1587–1600.

Calvert WH, Hanson FE (1983) The role of sensory structures and preoviposition behavior in oviposition by the patch butterfly, *Chlosyne lacinia*. Entomol Exp Appl **33**:179–187.

Crnjar RM, Prokopy RJ (1982) Morphological and electrophysiological mapping of tarsal chemoreceptors of oviposition-deterring pheromone in *Rhagolettis pomonella* flies. J Insect Physiol **28**:393–400.

Davis EE (1984) Regulation of sensitivity in the peripheral chemoreceptor systems for host-seeking behavior by a hemolymph-borne factor in *Aedes aegypti*. J Insect Physiol **30**:179–183.

de Boer G, Dethier VG, Schoonhoven LM (1977) Chemoreceptors in the preoral cavity of the tobacco hornworm, *Manduca sexta,* and their possible function in feeding behavior. Entomol Exp Appl **22**:287–298.

de Boer G, Hanson FE (1984) Foodplant selection and induction of feeding preference among host and non-host plants in larvae of the tobacco hornworm, *Manduca sexta*. Entomol Exp Appl **35**:177–193.

de Kramer JJ, Kaissling KE (1984) Passive electrical properties of insect olfactory sensilla may produce biphasic shape of spikes. Chem Senses **8**:289–295.

de Kramer JJ, van der Molen JN (1980) Special purpose amplifier to record spike trains of insect taste cells. Med Biol Eng Comput **18**:371–374.

de Kramer JJ, van der Molen JN (1983) The pore mechanism of the contact chemoreceptors of the blowfly, *Calliphora vicina*. In: Olfaction and Taste VII: Pasveer FJ (ed), IRL Press, London. pp 61–64.

den Otter CJ (1972) Interactions between ions and receptor membrane in insect taste cells. J Insect Physiol **18**:389–402.

den Otter CJ, Schuil HA, Sander-van Oosten A (1978) Reception of host plant odors and female sex pheromone in *Adoxophyes orana* (Lepidoptera: Tortricidae): electrophysiology and morphology. Entomol Exp Appl **24**:370–378.

Dethier VG (1955) The physiology and histology of the contact chemoreceptors of the blowfly. Rev Biol **30**:348–371.

Dethier VG (1973) Electrophysiological studies of gustation in lepidopterous larvae II. Taste spectra in relation to food plant discrimination. J Comp Physiol **82**:103–134.

Dethier VG (1976) The Hungry Fly. Harvard University Press, Cambridge

Dethier VG (1980a) Responses of some olfactory receptors of the Eastern Tent Caterpillar *(Malacosoma americanum)* to leaves. J Chem Ecol **6**:213–220.

Dethier VG (1980b) Evaluation of receptor sensitivity to secondary plant substances with special reference to deterrents. Am Nat **115**:45–66.

Dethier VG, Crnjar RM (1982) Candidate codes in the gustatory system of caterpillars J Gen Physiol **79**:549–569.

Dethier VG, Hanson FE (1965) Taste papillae of the blowfly. J Cell Comp Physiol **65**:93–100.

Dethier VG, Kuch JH (1971) Electrophysiological Studies of gustation in lepidopterous larvae. Z Vergl Physiol **72**:343–363.

Devitt B (1983) The contact chemosensory system of the dark-sided cutworm *Euxo messoria* (Harris) (Lepidoptera: Noctuiidae), and its function in feeding behavior: an ultrastructural, electrophysiological and behavioral study. Ph.D Thesis, University of Toronto.

Dickens JC (1984) Olfaction in the boll weevil: *Anthonomus grandis* Boh. (Coleoptera: Curculionidae): electroantennogram studies. J Chem Ecol **10**:1759–1785.

Dickens JC, Gutman A, Payne TL, Ryker LC, Rudinski JA (1983) Antennal ol-factory responsiveness of Douglas fir beetle, *Dendroctonus pseudotsugae* Hopkins (Coleoptera: Scolytidae) to pheromones and host odors. J Chem Ecol **9**:1383–1395.

Dickens JC, Payne TL, Ryker LC, Rudinsky JA (1984) Single cell responses of the Douglas-fir beetle *Dendroctonus pseudotsugae* Hopkins to pheromones and host odors. J Chem Ecol **10**:583–600.

Erler G, Thurm U (1981) Dendritic impulse initiation in an epithelial sensory neuron. J Comp Physiol **142**:237–249.

Frazier JL, Heitz JR (1975) Inhibition of olfaction in the moth, *Heliothis virescens* by the sulfhydryl reagent fluorescein mercuric acetate. Chem Senses Flavor **1**:271–281

Fujishiro N, Hiromasa K, Morita H (1984) Impulse frequency and action potential amplitude in labellar chemosensory neurones of *Drosophila melanogaster*. J Insect Physiol **30**:317–325.

Fu-shun Y, Visser JH (1982) Electroantennogram responses of the cereal aphid *Sitobion avenae* to plant volatile components. In: Proc 5th Int Symp Insect–Plant Relationships, Pudoc, Wageningen. Visser JH, Minks AK (ed), pp 387–388.

Gerstein GL, Clark WA (1964) Simultaneous studies of firing patterns in several neurons. Science **143**:1325–1327.

Gillary HL (1966) Stimulation of the salt receptors of the blowfly. J Gen Physiol **50**:337–350.

Guerin PM, Stadler E, Buser HR (1983) Identification of host plant attractants for the carrot fly. J Chem Ecol **9**:843–861.

Hamilton KA, Ache BW (1983) Olfactory excitation of interneurons in the brain of the Spiny lobster *Panulirus argus*. J Comp Physiol **150**:129–140.

Hansen K, Wieczorek H (1981) Biochemical aspects of sugar reception in insects. In: Biochem of Taste and Olfaction. Cagan R, Kare M (eds), Academic Press, New York. pp 139–162.

Hanson FE, Kogge S, Cearley C (1986) Computer analysis of chemosensory signals. In: Mechanisms in Insect Olfaction. Payne T, Birch M and Kennedy C (eds), Oxford University Press, Oxford (in press).

Hodgson ES, Lettvin JY, Roeder KD (1955) Physiology of a primary chemoreceptor unit. Science **122**:417–418.

Hummel HE, Miller TA (eds) (1984) Techniques in Pheromone Research. Springer Series in Experimental Entomology Springer-Verlag, New York.

Ishakawa S (1966) Electrical response and function of a bitter substance receptor associated with the maxillary sensilla of the larva of the silkworm, *Bombyx mori*. J Cell Physiol **67**:1–12.

Jermy T, Hanson FE, Dethier VG (1969) Induction of specific food preference in lepidopterous larvae. Entomol Exp Appl **11**:211–230.

Jung WG (1974) IC Op-amp Cookbook. Sams, Indianapolis.

Kaissling KE (1974) Sensory transduction in insect olfactory receptors. In: Biochemistry of Sensory Function. Jaenicke L (ed), Springer-Verlag, New York. pp 244–273.

Kurihara K, Yoshii K, Kashiwayanagi (1986) Transduction mechanisms in chemoreception. J Comp Physiol Biochem (in press).

Lacher V (1971) Electrophysiological equipment for measuring the activity of single olfactory nerve cells on the antenna of mosquitoes. J Econ Entomol **64**:3313–3314.

Lancaster D (1975) Active-filter Cookbook. Sams, Indianapolis.

Light D (1986) Central Interpretation of Sensory Signals: An exploration of processing of pheromonal and multimodal information in lepidopteran brains. Mechanisms in Insect olfaction Oxford Univ Press Payne TL, Birch M, and Kennedy C (Eds) (In press)

Ma WC (1972) Dynamics of feeding responses in *Pieris brassicae* L as a function of chemosensory input: a behavioral, ultrastructural and electrophysiological study. Med Landbouwhogesch Wageningen **72-11**:1–162.

Ma WC, Schoonhoven LM (1973) Tarsal contact chemosensory hairs of the large white butterfly *Pieris brassicae* and their possible role in oviposition behaviour. Entomol Exp Appl **16**:343–357.

Ma WC, Visser JH (1978) Single unit analysis of odor quality coding by the olfactory antennal receptor system of the Colorado beetle. Entomol Exp Appl **24**:320–333.

Maes FW (1977a) Simultaneous chemical and electrical stimulation of labellar taste hairs of the blowfly, *Calliphora vicinia* J. Insect. Physiol **23**:453–460.

Maes FW (1977b) Nonblocking AC preamplifier for tip recording from insect hairs. Med Biol Eng Comput **15**:470–471.

Mankin RW, Mayer MS (1983) Stimulus-response relationships of insect olfaction: correlations among neurophysiological and behavioral measure of reponse. J Theor Biol **100**:613–630.

Mayer MS, Mankin RW, Lemine CF (1984) Quantitation of the insect electroantennogram: measurement of sensillar contributions, elimination of background potentials, and relationship to olfactory sensation. J Insect Physiol **30**:757–763.

Matsumoto SG, Hildebrand JG (1981) Olfactory mechanisms in the moth *Manduca sexta:* response characteristics and morphology of central neurons in the antennal lobes. Proc R Soc Lon **B213**:249–277.

Milechia R, McIntyre T (1978) Automatic nerve impulse identification and separation. Comput Biomed Res **11**:459–468.

Miller TA (1979) Insect Neurophysiological Techniques. Springer Series in Experimental Entomology. Springer-Verlag, New York.

Mitchell BK (1978) Some aspects of gustation in the larval red turnip beetle *Entomocelis americana* related to feeding and host plant selection. Entomol Exp Appl **24**:340–349.

Mitchell BK, Schoonhoven LM (1974) Taste receptors in Colorado potato beetle larvae. J Insect Physiol **20**:1787–1793.

Mitchell BK, Gregory P (1979) Physiology of the maxillary sugar sensitive cell in the red turnip beetle, *Entomocelis americana*. J Comp Physiol **132**:167–178.

Mitchell BK, Harrison GD (1984) Characterization of galeal chemosensilla in the adult Colorado beetle *Leptinotarsa decemlineata*. Physiol Entomol **9**:49–56.

Mitchell BK, Seabrook WD (1974) Electrophysiological investigations on tarsal chemoreceptors of the Spruce Budworm *Choristoneura fumiferana*. J Insect Physiol **20**:1209–1218.

Mitchell BK, Sutcliffe JF (1984) Sensory inhibition as a mechanism of feeding deterrence: effects of three alkaloids on leaf beetle feeding. Physiol Entomol 9:57–64.

Morita H, Takeda K (1959) Initiation of spike potentials in contact chemosensory hairs of insects II. The effect of electric current on tarsal chemosensory hairs of *Vanessa*. J Cell Comp Physiol 54:177–187.

Morita H, Yamashita S (1959) The backfiring of impulses in a labellar chemosensory hair of the fly. Mem Fac Sci Kyushu Univ Ser E (Biol) 3:81–87.

Mustaparta H, Tommeras BA, Baeckstrom P, Bakke JM, Ohloff G (1984) Ipsdienol-specific receptor cells in bark beetles: structure-activity relationships of varous analogues and of deuterium-labelled ipsdienol. J Comp Physiol 154:591–595.

Nillson NJ (1965) Learning Machines. McGraw Hill, New York.

O'Connell RJ (1975) Olfactory receptor responses to sex pheromone components in the redbanded leafroller moth. J Gen Physiol 65:179–205.

Ogden TE (1978) The jet stream microbeveler: an inexpensive way to bevel ultrafine glass micropipettes. Science 201:469–470.

Olberg RM (1983a) Interneurons sensitive to female pheromone in the deutocerebrum of the male silkworm moth, *Bombyx mori*. Physiol Entomol 8:419–428.

Olberg RM (1983b) Pheromone-triggered flip-flopping interneurons in the ventral nerve cord of the silkworm moth, *Bombyx mori*. J Comp Physiol 152:297–307.

Omand E, Zabara J (1981) Response reduction in dipteran chemoreceptors after sustained feeding or darkness. Comp Biochem Physiol 70A:469–478.

Payne TL (1974) Pheromone perception. In: Pheromones. Birch M (ed), American Elsivier, New York. pp 35–61.

Piesch D, Wieczorek H (1982) Computer analysis of multi-unit spike trains with a low signal to noise ratio. Verh Dtsch Zool Ges 1982:325.

Preisner E, Bestmann HJ, Vostrowsky O, Rosel P (1977) Sensory efficacy of alkyl-branched pheromone analogs in Noctuid and Tortricid Lepidoptera. Z Naturforsch 32:979–991.

Prokopy PR, Averill AL, Cooley SS, Roitberg CA, Kallet C (1982) Variation in host acceptance pattern in apple maggot flies. In: Proc 5th Int Symp Insect–Plant Relationships, Pudoc, Wageningen, Netherlands. Visser JH, Minks AK (eds). pp 123–129.

Rees CJC (1969) Chemoreceptor specificity associated with choice of feeding site by the beetle *Chrysolina brunsvicensis* on its foodplant *Hypericum hirsuitum*. Entomol Exp Appl 12:565–583.

Rees CJC (1970) The primary process of reception in the type 3 (water) receptor cell of the fly *Phormia terranovae*. Proc R Soc Lond B 174:469–490.

Renou M (1983) Gustatory chemoreceptors of the anterior tarsus of the female *Heliconus charitonius*. Ann Soc Entomol Fin 19:101–106.

Roelofs W, Comeau A (1971) Sex pheromone perception: synergists and inhibitors for the red-banded leafroller attractant. J Insect Physiol 17:435–448.

Rumbo ER (1981) Study of single sensillum responses to pheromone in the light-brown apple moth, *Epiphyas postvittana*, using an averaging technique. Physiol Entomol 6:87–98.

Sass H (1983) Production release and effectiveness of two female sex pheromone components of *Periplaneta americana*. J Comp Physiol 152:309–317.

Schiffman SS, Reynolds ML, Young, FW (1981) Introduction to multidimensional scaling: theory, methods, and applications. Academic Press, New York.

Schneider D (1957) Electrophysiologische Untersuchungen von chemo und mechanorezeptoren der antenne des Seidenspinners, *Bombyx mori* L. Z Vergl Physiol 40:8–14.

Schoonhoven LM (1969) Sensitivity changes in some insect chemoreceptors and their effect on food selection behavior. Proc Kon Ned Series C 72:491–498.

Schoonhoven LM (1981) Chemical mediators between plants and phytophagous insects. In: Semiochemicals: their Role in Pest Control. John Wiley & Sons, New York, Nordlund DA (ed) pp 31–50.

Schoonhoven LM, Dethier VG (1966) Sensory aspects of host plant discrimination by lepidopterous larvae. Arch Neerl Zool 16:497–530.

Seelinger G (1983) Response characteristics and specificity of chemoreceptors in *Hemilepistus reaumuri* (Crustacea, Isopoda). J Comp Physiol 152:219–229.

Selzer R (1981) The processing of a complex food odor by antennal olfactory receptors of *Periplaneta americana*. J Comp Physiol 144:509–519.

Shimada S, Shiraishi A, Kijima H, Morita H (1974) Separation of two receptor sites in a single labellar sugar receptor of the flesh fly by treatment with p-chloromercuribenzoate. J Insect Physiol 20:605–615.

Simmonds MSJ, Blaney W (1983) Some effects of Azadirachtin on Lepidopterous larvae. Proc 2nd Neem Conf Rauischholzhausen. 163–180.

Smith DV, Bowdan E, Dethier VG (1983) Information transmission in tarsal sugar receptors of the blow fly, *Phormia regina*. Chem Senses 8:81–102.

Smith DV, Travers J (1979) A metric for the breadth of tuning in gustatory neurons. Chem Senses 4:215–229.

Stadler E (1978) Chemoreception of host plant chemicals by ovipositing females of *Delia* (Hylemya) *brassicae*. Entomol Exp Appl 24:511–520.

Stadler E (1984) Contact chemoreceptors In: Chemical Ecology of Insects. Bell WJ, Carde RT (eds), Chapman and Hall, London. pp 3–35.

Stadler E, Hanson FE (1975) Olfactory capabilities of the "gustatory" chemoreceptors of the tobacco hornworm larvae. J Comp Physiol 104:97–102.

Stadler E, Hanson FE (1976) Influence of induction of host preference on chemoreception of *Manduca sexta:* behavioral and electrophysiological studies. Symp Biol Hung 16:267–273.

Strausfeld N J (1980) Neuroanatomical Techniques. Springer Series in Experimental Entomology. Springer-Verlag, New York.

Sturckow B (1967) Occurance of a viscous substances at the tip of a labellar taste hair of the blowfly. In: Olfaction and Taste II. Hayashi T (ed), Pergamon Press, Oxford. pp 707–720.

Thompson H, Ache BW (1980) Threshold determination for olfactory receptors of the spiny lobster *Panulirus argus*. Mar Behav Physiol 7:249–260.

Thurm U, Kuppers J (1980) Epithelial physiology of insect sensilla. Insect Biology in the Future. Locke M, Smith DS (eds), Academic Press, New York. pp 735–764

Thurm U, Wessel G (1979) Metabolism-dependent transepithelial potential differences at epidermal receptors of Arthropods. J Comp Physiol 134:119–130.

Tichy H, Loftus R (1983) Relative excitability of antennal olfactory receptors in the stick insect *Carasius morosus* L: in search of a simple concentration-independent odor-coding parameter. J Comp Physiol 152:459–473.

Vanderkooi MR (1973) Linear Applications Handbook. National Semiconductor Corp., Santa Clara, California.

van der Molen N, Meulen JW, deKramer JJ, Pasveer FJ (1978) Computerized classification of taste cell responses. J Comp Physiol 128:1–11.

van der Pers JNC (1981) Comparison of electroantennogram response spectra to plant volatiles in seven species of *Yponomeuta* and in the tortricid *Adoxophyes orana*. Entomol Exp Appl 30:181–192.

van der Pers JNC, den Otter K (1978) Single cell responses from olfactory receptors of small ermine moths to sex attractants. J Insect Physiol 24:337–343.

van der Pers JNC, Lofstedt C (1983) Continuous single sensillum recording as a detection method for moth pheromone components in the effluent of a gas chromatograph. Physiol Entomol 8:203–211.

van der Wolk FM, Menco B, van der Starre H (1984) Freeze-fracture characteristics of insect gustatory and olfactory sensilla II. Cuticular features. J Morph 179:305–321.

van Drongelen W (1978) The significance of contact chemoreceptor sensitivity in the larval stage of different *Yponomeuta* species. Entomol Exp Appl 24:143–147.

Visser JH (1979) Electroantennogram responses of the Colorado beetle *Leptinotarsa decemlineata* to plant volatiles. Entomol Exp Appl 25:86–97.

Visser JH (1983) Differential sensory perceptions of plant compounds by insects. In: Plant Resistance to Insects ACS Symp. 208. Hedin P (ed), ACS Press. Washington. pp 215–230.

Visser JH, Minks AK (eds) (1983) Proceedings of the Fifth International Symposium Insect–Plant Relationships. Pudoc, Wageningen, 464 pp.

Walaade SM (1983) Chemoreception of adult stem-borers tarsal and ovipositor sensilla in *Chilo partellus* and *Eldana saccharina*. Insect Sci Appl 4:159–165.

Wieczorek H (1976) The glycoside receptor of the larva of *Mamestra brassicae* L. J Comp Physiol 106:153–176.

Winstanley C, Blaney WM (1978) Chemosensory mechanisms of locusts in relation to feeding. Entomol Exp Appl 24:550–558.

Wolbarsht ML (1965) Receptor sites in insect chemoreceptors. Cold Spring Harbor Symp Quant Biol 30:281–288.

Wolbarsht ML, Hanson FE (1965) Electrical activity in the chemoreceptors of the blowfly III. Dendritic action potentials. J Gen Physiol 48:673–683.

Zacharuk RY (1984) Antennae and sensilla. In: Comprehensive Insect Physiology, Biochemistry, and Pharmacology. Vol 6. Kerkut GA, Gilbert LI (eds), Pergamon Press, Oxford. pp 1–69.

Index